Women in Early British and Irish Astronomy

Mary Brück

Women in Early British and Irish Astronomy

Stars and Satellites

Springer

Mary Brück

ISBN 978-90-481-2472-5 e-ISBN 978-90-481-2473-2
DOI 10.1007/978-90-481-2473-2
Springer Dordrecht Heidelberg London New York

Library of Congress Control Number: 2009931554

Printed on acid-free paper

Springer is part of Springer Science+Business Media (www.springer.com)

For my sisters (nées Conway)
Eithne, Meadhbh and Aoife

Foreword

The role of women in the growth of science has become an important area of modern historical scholarly research. And as far as the study of the role of women in astronomy is concerned, Dr. Mary Brück is an established and illustrious pioneer, with an international reputation and acclaimed books and articles already to her credit. Her present book, *Stars and Satellites*, brings together many figures who worked over a 300-year time scale, and whose relationship with astronomy ranged from informed assistants to independent researchers to major writers and interpreters of contemporary astronomy to, eventually, paid professionals. But what is more, Mary Brück is the undoubted pioneer in the study of Irish women scientists, several of whom appear in the present volume, for Ireland was one of the most astronomically-active regions of the British Isles in the nineteenth century. And as she is a professional astronomer herself, with a University College Dublin training, combined with a love of history and an Irishwoman's genius for narrative and the gift of making people come to life, her latest book is both a mine of information and a joy to read.

In the present-day world, where it is accepted that capable girls will have full access to secondary and then higher education, and will proceed to the professions with a first-rate training behind them, we tend to have a distorted view of earlier scientific women. It is true that most girls and women in the past were not expected to become involved in science and scholarship, and as the elderly Mary Somerville recorded in her *Personal Recollections* (as mentioned in Chapter 6), access to learning in the days of her youth, around 1800, was not easy. But I would argue that the opportunity for women to make significant contributions to science in general, and to astronomy in particular, was perhaps greater in the British Isles (and perhaps the USA) than it would have been in continental Europe. And this derived from the social organisation of British science, especially after the French Revolutionary and Napoleonic wars which began in the 1790s.

During and after these devastating wars, several European countries remodelled their ancient universities, while the French Revolutionary École Polytechnique (1794) and the new universities of Berlin (1809), Bonn (1818), and the re-founding of Munich brought fresh powerful research institutions into being. Indeed, many of the countries of post-Napoleonic continental Europe came to re-invent key parts of their national identities by means of high culture: music, literature, critical scholarship, scientific research, and higher education. These universities, *conservatoires, instituts, académies*, and similar bodies generally received state funds, usually from taxation, to be spent in the cause of advancing national cultural prestige. They were hierarchic institutions, often under the patronage of ministers of state and operated via officially-appointed career directors and professors who were there to draw up and supervise research plans. And yes, they paid intellectual dividends in terms of student numbers and discoveries, with major research schools in chemistry, physiology, pure mathematics, and other sciences by 1840. Their most successful graduates emerged with the new, prestigious Ph.D. degree, and then hoped in their turn to ascend their respective academic ladders to senior chairs and directorships. In many ways, the internal organisation of these academic institutions had been influenced by the lurch towards political absolutism, starting with Bonaparte and continuing with Europe's Kings, Emperors, and Czars who, regaining their old authority after 1815, were determined to prevent popular libertarian ideas from breaking out again. Yet the education system that arose from Europe's revolutionary and Napoleonic maelstrom was very much of a man's world, to which women had no access.

In Great Britain, however, things were very different. Emerging as the victor from the Napoleonic wars, it managed to channel its home-grown popular unrest peacefully through a stream of reform legislation, and there was no need for a powerful Parliament and a now largely constitutional monarchy to re-establish national identity through centralised institutions of high culture. Britain had an already firmly-rooted and vigorous intellectual and cultural life, but it was expressed through private, self-governing institutions. Seven ancient independent universities existed across England, Scotland, and Ireland before 1828, together with great national learned societies, such as The Royal Society (1660), The Society of Antiquaries (1717), The Royal Academy (1768), The Royal Astronomical Society (1820), and The British Association for the Advancement of Science (1831). And then by 1830 there were scores of literary and philosophical societies across the kingdom, as well as a burgeoning number of mechanics' institutions. Yet not a single penny of government money went to any of them, nor could any minister of state exert any control, for the universities and even those societies with royal titles were still constitutionally independent, paid their own way, and elected their own members without any official interference.

Consequently, Britain's cultural and scientific organisation worked on a self-electing rather than on a hierarchical basis, and while the above-mentioned institutions were undoubtedly male-dominated, the system itself contained an inherent flexibility. I have styled this system 'Grand Amateur', for the scientists and other cultured people operating within it were self-confessed 'Amateurs' in so far as they pursued their subjects for love and not for gain, and 'Grand' because they supplied from their own means the financial and technical resources with which to address great projects, be they the advancement of British art or the discovery of nebulae in deep space.

And curiously enough, this was a world into which intelligent, original women astronomers could enter, and where they could make a serious contribution, for the holding of an M.A. or Ph.D. degree or of an academic post was not necessary. In the early days this entry would usually be effected by means of a husband, brother, or father, who was already active in astronomy. But then, because of the flexibility of the British social system, more women started to come through in their own right, such as the mathematician, astronomer, and writer Mary Somerville. Some of these women had particularly enlightened parents who gave them a first-class private education, for example the Irish 'Scholarly Sisters' Agnes and Mary Ellen Clerke, and Lady Margaret Huggins (née Lindsay), whom we will meet in Chapters 11 and 12. Others picked up their astronomy from working with male relatives, such as Caroline Herschel, Elizabeth Brown, and Caroline Lassell. By the 1880s, moreover, women were beginning to enter astronomy on the strength of a university training in science, such as Annie Maunder and Alice Everett, though their subsequent scientific careers were still very often in the private research tradition. But by the early twentieth century, as Mary Brück makes clear, both astronomy and astronomers were changing. The escalating cost and technological complexity of the 'new astronomy' were moving the science, especially in the USA, towards an academic professionalism, leading to the British-born Cecilia Payne-Goposhkin becoming the first woman Harvard Ph.D. astronomer and full Harvard professor in astronomy.

With the exception of special membership granted to both Caroline Herschel and Mary Somerville in 1835, the Royal Astronomical Society was to elect no woman to its Fellowship until a change was made to its statutes in 1916. Long before that, however – since the early 1880s, in fact – the new, large 'metropolitan' amateur astronomical societies of Liverpool, then Leeds, Manchester, Belfast, and Cardiff, had been electing women on equal terms to men, and even on to their governing councils, as also had the professional Royal Meteorological Society. Indeed, these high-level amateur societies were the first scientific bodies to give a voice to women who enjoyed serious astronomy but were not perhaps in the full 'Grand Amateur' league, but worked as teachers or were independent ladies of means. And it was Elizabeth Brown of Cirencester – who with her sister Jemima was co-heiress to a family vintner's business – who was the driving force behind the founding of The British Astronomical Association in 1891.

It is clear that Mary Brück's chief interest lies in the incidents and circumstances that form individual human lives. For she comes from a narrative tradition of historical scholarship, and is not concerned with issues in feminist theory. Her interest is individual people and, in this particular study, how a couple of dozen women, born across three centuries, all come to engage with astronomy on a serious, ground-breaking level. And it is due to Mary Brück's eye for personality, motivation, originality, and persistence in the face of obstacles that the women whose lives and careers she traces and analyses take on the fascination that they do.

This is also a very positive, uplifting, and beautifully-written book, and while fully acknowledging the difficulties with which many of these women astronomers had to grapple, it nonetheless gives us a fresh vision of what determined individuals could, and still can, achieve.

Allan Chapman

Acknowledgements

Many friends old and new have provided encouragement as well as enlightenment to me while putting together this book. I would like to thank in particular Barbara Becker, Mary Creese and Ian Elliott, helpful correspondents for many years, and Allan Chapman who has also very kindly contributed the Foreword. I owe a special debt of gratitude to Vanda Morton, who generously allowed me to read and use the unpublished memoirs of Thereza Story-Maskelyne, and to Michael Hoskin, who kindly advised me on my chapter on Caroline Herschel. Peter Hingley, the immensely erudite librarian and archivist at the Royal Astronomical Society has generously found me illustrations from the Society's archives. Others who helped in various ways and whom I wish to thank are Richard Baum, John Birt, Tom Bogden, Andrew Brück, Mary Croarken, Anthony Cross, Clive Davenhall, Rickard Deasy, David Gavine, Gerard Gilligan, Ian Glass, Mark Hurn, Kevin Kilburn, Anthony Kinder, Sir Patrick Moore, Richard Morris, Willie Soon, Terry Moseley, Stephen Wainright, George Wilkins. I would also like to remember the late Sheelagh Grew of Armagh Observatory, a dear friend and a valuable contributor to the history of Irish astronomy.

I also thank the archivists who kindly provided material on this and on earlier occasions especially Adam Perkins, Royal Greenwich Observatory archives, Cambridge University Library; Anthony Kinder, British Astronomical Association; Kate Perry, Girton College Cambridge; the archivists of Lick Observatory, California; Wellesley College, USA, Vassar College, USA. Most of all, I am deeply grateful to Karen Moran, librarian at the Royal Observatory Edinburgh, for making her knowledge and expertise so generously and cheerfully available, and to Jason Cowan who copied most of the illustrations, the sources of which are credited individually.

Andrew Lawrence, Regius Professor of Astronomy at the University of Edinburgh, is warmly supportive of historical research and of the value of archive conservation. It is a pleasure to thank him and my former colleagues at the Royal Observatory Edinburgh for their friendship. Finally, I gratefully acknowledge the labours of the editors, Sue Bowler and Simon Mitton, of the Royal Astronomical Society, who brought my offering to fruition.

Penicuik
May 2008

Contents

Introduction

Astronomy today is Big Science. Thousands of scientists, men and women, in observatories, in laboratories and even in outer space investigate the universe with the aim of discovering not only its extension and its contents, but its history back to its very origin.

A hundred years ago the world of professional astronomy had much more limited dimensions, and in Britain was virtually closed to women until well into the twentieth century. Long before this, however, women were participating in various ways. A few made independent names for themselves: Caroline Herschel, amanuensis to her brother William, achieved fame and admiration in her own right as a discoverer of several comets. Others, such as Mary Somerville, mathematician and social celebrity, found a vocation as educators and interpreters of science. Most, though, served unostentatiously as assistants to their menfolk. In the eighteenth and nineteenth century much of astronomy in Britain was carried out by independent amateurs in their own homes and at their own expense. Family members were naturally drawn in as helpers and collaborators, especially wives, sisters and unmarried daughters who had no option but to live at home and few outlets for their talents and interests.

The tradition of the gentleman-amateur scientist in these westerly islands of Europe, as is described by Allan Chapman in his opus on amateur astronomy in Victorian Britain,[i] contrasted with the official state-supported organisation of science in France and Germany. Astronomy, a combination of leisure pursuit and learning, flourished to a remarkable degree in the British amateur system. The Royal Observatory at Greenwich, the country's only publicly-maintained astronomical institution, had the utilitarian duty of supplying data for navigation. Astronomy in the universities was chiefly concerned with its traditional mathematical aspect. Excursions into new practical fields were left to the independent researcher, working without interference from any quarter, a state of affairs supported by Sir George Airy, the great nineteenth century Astronomer Royal himself. Many of these amateur astronomers were skilled observers, who in spite of the unfavourable climate were very successful in terms of discoveries and innovations.

The period from the mid-eighteenth century to the end of the nineteenth was one of fast and exciting growth of knowledge of the physical universe. It was the age in which the foundations of modern astrophysics and cosmology were laid. Within the solar system it saw the discovery of previously unknown planets; of several planetary satellites, and of a new genre of object, the asteroids. Comet Halley returned as predicted, and several new comets came to light. The distances of the nearest stars were measured for the first time. Larger and better telescopes revealed multitudes of faint stars apparently without limit, and resolved many nebulous blobs into stars. As the nineteenth century progressed, the human eye was powerfully aided by the new art of photography which could record celestial objects in permanent form, and in large numbers. This leap forward was followed by the development of spectroscopy, a technique by which the chemical constitution of the Sun and stars could be inferred by examining their light, knowledge that had previously been believed beyond the reach of Man. In all these activities, British astronomers, including its talented independent amateurs, were leading participants. Successful astronomers of all hues, professional and amateur, associated freely and exchanged experiences. They belonged to various learned societies and published in the scientific journals. Leading amateurs were awarded honorary degrees and knighthoods, while, conversely, it was not unknown for holders of academic posts never to have attended university.

Women, however, were not admitted to membership of learned societies, and could not – except by rare special arrangement – publish in their official journals. They were thus without an entree into the public world of science. Yet, since astronomy was a labour intensive pursuit, every observation worth reporting required assistance behind the scenes from more humble mortals. If a complete picture of its development is to be put on record, the efforts of these helpers, including the women among them, ought not to be overlooked.

Traditional histories of science – to quote the historian Patricia Fara in her study of women and science in the Enlightenment[ii] – "have been written like schoolboy adventure novels". Progress was depicted as a succession of discoveries and inventions by individual heroic figures. Women were absent from these accounts. Dr. Fara challenges this view of the scientific past. She places the science of that era in a wider human context, in the home and in everyday society, and contends that the role of women – educated, intelligent, supportive, active – is an essential part of our understanding of the way in which science flourished and developed. The women we describe in these chapters, who were far from few, are a contribution to that understanding.

During the nineteenth century, women still remained conspicuously absent from official accounts of astronomy, as indeed was inevitable if the sources of information were the publications of learned societies and elite institutions from which women were excluded. Agnes Clerke, chief and most trusted authority on the astronomy of that period to which she herself belonged, endeavoured in her famous *History of Astronomy in the Nineteenth Century* to "give each individual discoverer, strictly and impartially, his due". But though she was a friend and encourager of other women in astronomy, her *History* is practically devoid of female names. Of a 500-long chronological list of significant events in astronomy between 1775 and the end of the nineteenth century, it is only in the last decade that female names (chiefly American) make their appearance.

The women whom we meet in the earlier part of our account were probably happy enough to remain in the background and to conform with the social customs of their day. In fact, to some extent they colluded in their role, dreading being labelled bluestockings. Later, however, women would strive to gain entrance to university and to secure parity with men on a professional level. They were almost universally supported in their ambitions by their male fellow-astronomers, but came up against officialdom's obstacles to careers for women in science and entrenched attitudes in the male-dominated academic institutions. As will be told, progress was slow. It took 30 years, for example, from the first proposal for women's admission to the Royal Astronomical Society for this to be realised. It took even longer for them to rise above the level of humble computers to full professional level in employment. It must be said that women elsewhere in Europe fared no better.

The women in our pantheon had diverse personalities, motivations and achievements. They had few opportunities for advancing knowledge of the universe and were disadvantaged by social conventions. However, they contributed more than they have received credit for, and their contributions are now being more justly appreciated. All their stories are interesting from the human point of view. They also illustrate the slow progress of women towards equal status with men in the world of science.

It is fitting that they should emerge from the shadows.

Notes

[i] Allan Chapman. *The Victorian Amateur Astronomer. Independent astronomical research in Britain 1820–1920*. Chichester: Praxis Publishing (and John Wiley & Sons) 1998.

[ii] Patricia Fara. *Pandora's Breeches, Women Science and Power in the Enlightenment*. London: Pimlico 2004.

Chapter 1
A Clever and Determined Wife

Astronomy is one of the oldest of the sciences. From earliest times it has been useful to humankind for timekeeping, calendar planning, navigation and geography. These activities required observations of the heavens, construction of star maps and catalogues, and a considerable measure of organisation. Beginning in the sixteenth century – the century that saw the first use of the telescope on the sky by Galileo – national observatories were founded which could provide important practical knowledge for such useful purposes, as well as bringing lustre and prestige on vying nations. One of the first of these was the Royal Observatory at Greenwich near London, instituted by King Charles II in 1675 and erected on the spot which since 1881 is the zero point of longitude for the whole world.[i] Only Paris Observatory, founded in 1667, was older. Paris and Greenwich Observatories were followed by Berlin in 1701 and St Petersburg in 1725.[ii]

The Royal Observatory was designed by the great astronomer and architect Sir Christopher Wren in consultation with "England's Leonardo", the ingenious experimental scientist Robert Hooke.[iii] This beautiful building, known as Flamsteed House after its first occupant John Flamsteed (1646–1719), is today still part of the old Royal Observatory and the associated nearby National Maritime Museum (Fig. 1.1). In this historic place lived for 27 years Flamsteed's wife Margaret (née Cooke) (1670–1739), the first woman on record to be associated with astronomy in Britain.

The very specifically designated function of the Royal Observatory was to deal with the vitally important matter of finding a method of determining a ship's location at sea. A fundamental necessity to solve this problem was to establish the exact positions on the sphere of the sky of celestial bodies that mariners would use as navigational beacons when out of sight of land. The astronomer charged by the king in 1675 to carry out this task "with the most exact care and diligence" was the young and largely self-taught astronomer John Flamsteed,[iv] a contemporary of Isaac Newton and of other great scientists of that "heroic age of British science" such as Robert Boyle, Robert Hooke and Edmond Halley.[v] John Flamsteed, designated Astronomical Observator, became known later as the Astronomer Royal, a title that remained associated with the Royal Observatory until the late twentieth century. Flamsteed moved into his observatory one year later, in 1676, to work and to live there for the rest of his life.

M. Brück, *Women in Early British and Irish Astronomy*,
DOI: 10.1007/978-90-481-2473-2_1, © Springer Science+Business Media B.V. 2009

Fig. 1.1 Flamsteed House, Greenwich. 1680. The beautiful house still stands in the midst of later observatory buildings. (National Maritime Museum)

Flamsteed set about producing a catalogue of stars, and over a period of 13 years made single-handedly some 20,000 observations. He achieved this in spite of a meagre salary and insufficient funds that forced him to take pupils in mathematics to subsidise the enterprise. He also had to supply most of his own instruments, and pay his assistants out of his own pocket. The great Catalogue, when finally published in 1725, after his death, established him as "the greatest systematic observer since the invention of the telescope."[vi] The eventual appearance of this important catalogue of 3,000 stars owed much, indeed everything, to the efforts of Margaret whom he married in 1692 when he was 46.[vii]

Margaret, who was only 22 or 23 at the time of her marriage, was the daughter of a London lawyer and granddaughter of the Reverend Dr. Ralph Cooke, former Rector of Burstow, a country parish in Surrey about 25 miles from Greenwich. Flamsteed, who had been ordained a clergyman in the Church of England with a view to a possible career in the Church before his great opportunity in astronomy came, succeeded Margaret's grandfather as Rector in 1684, eight years after his Greenwich appointment. The post provided a useful supplement to his income without occupying too much of his time. A curate carried out the normal parish duties, but Flamsteed nevertheless took his clerical responsibilities seriously, and visited his parish especially at harvest-time and Christmas. The Rectory where he (with Margaret) lived on these occasions, and the beautiful old Church of St Bartholomew at Burstow which he served, still stand and are still in use.[viii]

Margaret was well educated and, though not wealthy, was reasonably well provided for. The discrepancy in the couple's ages was not unusual among the middle class in that era, when men tended to postpone marriage until they were established in life and could afford it.[ix] In Flamsteed's case, he had come into an inheritance from his father's estate a few years previously; this money had also paid for an important new instrument (a mural arch) for the observatory.[x] The Flamsteeds were to enjoy 27 years of what is likely to have been a happy and compatible marriage in their unique observatory home until the husband's death. Margaret was interested and competent in mathematics and astronomy. Notebooks in her handwriting in the Royal Observatory archives show that she was studying these subjects within months of her marriage. They contain theorems and diagrams in geometry; notes on gravitation, on the geometry of the solar system, and on optics. More advanced topics are also there, such as a demonstration of the method of finding the distance of the Sun through observations of the planet Mars, and some problems in pure mathematics, including worked examples of Newton's method of fluxions (i.e. calculus, as invented by Newton and using his notation), hardly a common skill among young ladies of the day. A page of impressive calculus from among her notes is reproduced in Rob Iliffe and Frances Willmott's study of Margaret's work.[xi]

Flamsteed's programme at this time involved not only the preparation of his catalogue, but also observations of the Sun, Moon and planets (objects that move with respect to the fixed stars) and the drawing up from them of tables of their positions in the sky. He made use of assistants and calculators, and in the later part of his career hired extra hands to expedite the work. Margaret was capable of performing the desk-bound element of the assistants' work, but it is not clear how much part she took in actual observations. In one entry, Flamsteed notes an observation of the Sun, done "solus cum sponsa" (alone with my wife): observations as well as scientific writings were still all recorded in Latin. In another instance, an observation of the moon is recorded (by an assistant) as having been made by the "Clarissima Domina M. Flamsteed", and another, of Jupiter, made with the assistance of the same noble lady. Elsewhere, her name appears as having found and corrected an error in a set of calculations in the Jupiter tables. Drs Iliffe and Willmoth conclude that Margaret was probably not a regular assistant. She was, however, her husband's amanuensis especially in his last frail years, copying letters and manuscripts, and was personally well-known in his astronomical world. Correspondence with other scientists shows that, independently of the observatory's work, she performed experiments with optics, and had lenses ground for her microscope by some of her husband's distinguished colleagues in the Royal Society.[xii]

She also got on well with the younger generation. A former pupil of Flamsteed's who was travelling in Europe as a tutor in a family reported his adventures in a humorous letter to Margaret in which he enclosed – "to divert you, Madam" – a drawing of an eclipse he had witnessed, and ended "I drink your health and the Doctor's". Writing to Flamsteed from Germany the same tutor described an evening in the company of mathematical men, where "Nothing but Mr. Flamsteed was our subject for a while, till I began your Lady's health, and told them That very Lady I mentioned was a well wisher to the Mathematics, and understood them perfectly well, upon which her health went round with great devotion". A letter to Margaret from an erstwhile assistant, which also survives, is characterised by a similar jocular tone.[xiii] One is left with an impression of Margaret (then in her mid-30s) as a friendly and good-humoured colleague to her husband's juniors, clever but no dreaded blue-stocking.

Margaret presumably had her duties as mistress of the home, where visitors, including important personages and astronomers from home and abroad were frequent. Few references to her exist in Flamsteed's voluminous correspondence, but one in particular throws light on her talent as a hostess. It is a letter from Samuel Pepys, the famed diarist and influential Government officer, who, though not himself a scientist, moved in the company of scientists and was a Fellow of the Royal Society (and at one time its President).[xiv] He visited the observatory in 1697, and wrote afterwards to his host expressing "thankful acknowledgement of your crowd of favours, both intellectual and culinary, both of them heightened by the conversation and kindness of your excellent lady".[xv] Margaret had wider interests. She was a founder of a charity school for poor girls, where they were taught useful skills for their future employment. The school was run by an all-women committee, with Mrs. Flamsteed as treasurer for 14 years. The Flamsteeds had no children, but from early in their marriage a niece of Flamsteed's, Anna, lived with them like an adopted daughter. She became the wife of one of his assistants.

Flamsteed, whose health was frail at the end, died at the Royal Observatory on the last day of the year 1719 at the age of 73, after 44 years in his post. After her husband's death Margaret fought hard and successfully on his behalf in the matter of the publication of his catalogue, *Historia Coelestis.* There had been controversies and complications in Flamsteed's relations with the scientific heroes of the day, including Isaac Newton himself, and quite particularly with Edmund Halley. The trouble lay to some extent in the differing temperaments of the men themselves, but the most serious disagreements had their root in lack of clarity as to the duties of the office of Astronomer Royal. Flamsteed, who provided the instruments, and paid his assistants and calculators entirely from his own income, saw his observations and reductions as his own to publish and present to the world when he was ready.[xvi] His illustrious contemporaries were impatient to have access to them sooner. After much acrimony a Board of Visitors was set up to supervise the work of the Royal Observatory and a version of the incomplete catalogue, edited by Flamsteed's bitterest enemy Halley, was published by order of a Royal Commission in 1712. Flamsteed tried to have all copies of this pirated edition destroyed while he prepared his own. Some copies had already been distributed, but he gained possession of the remaining 300 printed copies and burned them in a bonfire. His final completed catalogue was ready at the time of his death. He had also finished writing the Preface in English. What remained to be done was a Latin translation of the Preface, and the engraving of charts – as well as seeing to the actual publication.

Without his widow's efforts, the *Historia Coelestis Britannica* (1725) (The British Catalogue) and the *Atlas Coelestis* (1729) would never have come to fruition. It took Flamsteed's devoted assistants 6 years after his death to prepare the Catalogue and to have it published (with its Preface in Latin) while Margaret oversaw the work and took charge of the finances. She also saw to the publication of Flamsteed's *Atlas*, a set of beautiful maps to accompany the catalogue to which she devoted meticulous attention in the choice of engravers (Fig. 1.2). It shows 25 Zodiacal

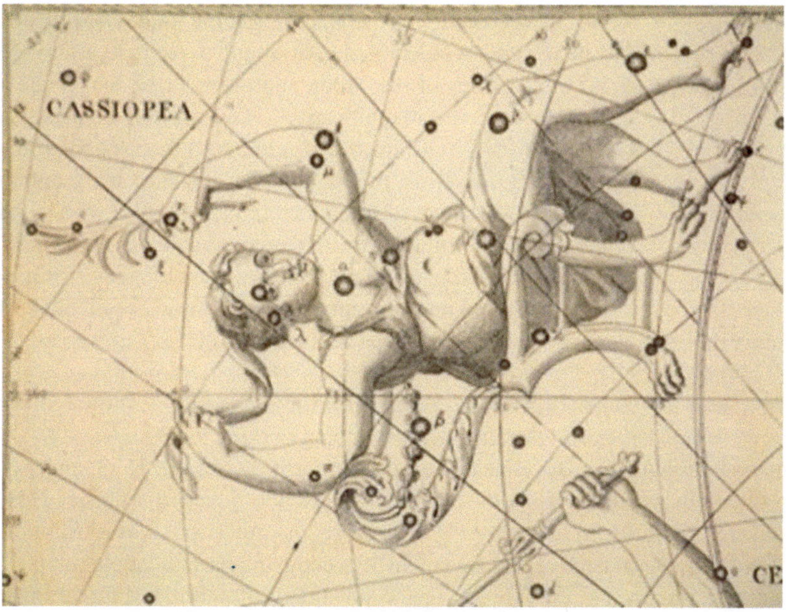

Fig. 1.2 The constellation Cassiopeia, from *Flamsteed's Atlas*. (Royal Observatory, Edinburgh)

constellations visible from Greenwich, edited by herself and James Hodgson, the husband of Flamsteed's niece Anna.[xvii] The constellation figures were drawn by James Thornhill, a famous artist of the time.

Margaret's image, according to Lesley Murdin in her investigation of seventeenth century astronomy in Britain,[xviii] was unfortunately tarnished by the fact that she left nothing in her will to Joseph Crosthwait, who had been her husband's faithful assistant and her own loyal supporter but had received no payment for his work. It is likely, however, that she was not as well-off when her task was completed as she expected; or perhaps, being herself so selflessly devoted to the great catalogue, she assumed that others felt the same.[xix] There were quarrels, too, when, on leaving the observatory to allow her husband's successor, Edmund Halley, to move in, Margaret carried away all her husband's papers and the observatory instruments, arguing – correctly – that he had paid for them.

Margaret Flamsteed died, aged 60, only one year after the publication of the *Atlas*. She was buried next to her husband beneath the chancel in St Bartholomew's Church at Burstow.[xx]

She was clearly a determined woman. She was also exceptional. No other female votary of astronomy emerged in Britain for the next half a century, and she assured for the world the indispensable and incomparable British Catalogue of stars.

There is an almost uncanny resemblance between the Flamsteeds' story and that of the most famous husband and wife team in the whole of astronomy, Johannes Hevelius and his wife Elizabeth. Hevelius (1611–1687) was a wealthy Polish astronomer, a brewer by profession, whose observatory, erected in 1640 on the roof of his house in Gdansk (then Danzig) on the Baltic, was the most magnificent and best equipped in Europe. His most celebrated instruments were his very long telescopes for the study of the surfaces of the Moon and planets (the longer the focal length of a telescope the greater the magnification). The greatest of these, all of 150 feet long, was mounted on a huge mast and required a team of men to raise it skywards. At the age of 52, Hevelius, then a widower, married his second wife Elizabeth (1647–1693), the daughter of a Dutch merchant in Gdansk who, at barely 16 years of age became his active collaborator. This remarkable young woman was highly educated and talented. She spoke and wrote Latin, was able to perform all the mathematical calculations required in astronomy, and became an excellent observer. Not only that, she also managed the family estates and the brewery, as well as raising four children.

The Hevelius couple worked in partnership for 20 years.[xxi] The large sextant required two observers; a portrait showing the pair working together with this instrument was published in Hevelius' volume *Machina Coelestis* in 1673, and is often reproduced in astronomy books (for example, on the cover of Michael Hoskin's *Cambridge History of Astronomy*[xxii]). The Hevelius' great opus was a star atlas listing 1,500 stars with a set of 54 charts which became the basis of subsequent atlases including Flamsteed's. For observing the positions of stars, Hevelius used traditional quadrants and sextants with which the separations of stars from each other on the sky are measured by naked eye sighting. He was the last great master of pre-telescopic astronomy, while Flamsteed, with telescopic sights, was the first of the new. Hevelius died while still working on the Atlas. It was completed by Elizabeth who published it in three parts: the catalogue itself, the maps, and a Prologue. In this Prologue she dedicates the volume to the King (of Poland) and signs herself Elizabeth Vidua, the widow. Elizabeth died, aged 46, six years after her husband and only two years after completing the work.

Flamsteed never met the illustrious Polish astronomer who was 35 years his senior in age and in achievement: the separation in time of their respective Atlases was also 35 years. Both died before the publication of their catalogues; in both cases, their much younger wives made themselves responsible for completing and publishing their respective three-part works, in accordance with their husbands' wishes. A final link between these devoted women was that both died relatively young, within a very few years of completing their immortal tasks.

Notes

[i] Eric Forbes. 1975. *Greenwich Observatory*: Volume 1, chapter 1. London: Taylor and Francis.
[ii] Steve J. Dick. 1991. National Observatories: an overview. *Journal for the History of Astronomy* **22**, 1–4.
[iii] Allan Chapman. 2003. *England's Leonardo. Robert Hooke and the Seventeenth-Century Scientific Revolution*. Bristol: Institute of Physics.
[iv] John L Birks. 1999. *John Flamsteed, the First Astronomer Royal at Greenwich*. London: Avon Books.

[v] W.H. McCrea. 1975. *The Royal Greenwich Observatory*. London: Her Majesty's Stationery Office. He was not always on the friendliest of terms with these great men, however.
[vi] McCrea. op. cit. p 7.
[vii] Frances Willmoth. *Oxford DNB*.
[viii] Birks. op. cit. Chapter 12.
[ix] Rob Iliffe and Frances Willmoth. 1997. *Astronomy and the Domestic Sphere*, in Lynette Hunter and Sarah Sutton (eds.) *Women Science and Medicine 1500–1700*, p 246. Stroud: Sutton.
[x] Forbes op. cit. p 47.
[xi] Iliffe and Willmoth op. cit, p 247.
[xii] Lesley Murdin. 1985. *Under Newton's Shadow, Astronomical Practices in the Seventeenth Century*. p 64–68. Bristol: Hilger.
[xiii] Quoted in Iliffe and Willmoth op. cit. p 253. The letters are dated 1706.
[xiv] Claire Tomalin. 2003. *Samuel Pepys, the Unequalled Self*, Chapter 17. London: Penguin.
[xv] Eric Forbes. Leslie Murdin and Frances Willmoth (eds.). 1995. *The Correspondence of John Flamsteed*, vol. 2. Bristol: IOP.
[xvi] Eric Forbes. 1975. *Greenwich Observatory*. Volume 1. Chapter 3. The details of these arguments and controversies are recorded in this official history of the Royal Observatory. The circumstances are also described by Allan Chapman (op. cit.) in his Chapter 5 (Hooke and the Astronomers).
[xvii] Margaret Flamsteed and J. Hodgson. 1729. *Atlas Coelestis*. London.
[xviii] Lesley Murdin. op. cit. p 126.
[xix] Iliffe and Willmoth. op. cit.
[xx] Birks. op. cit. p 115.
[xxi] Patricia Fara. 2004. *Pandora's Breeches, Women Science and Power in the Enlightenment*. London: Pimlico. A chapter is devoted to the Hevelius partnership.
[xxii] Michael Hoskin (editor). 1997. *The Cambridge Illustrated History of Astronomy*. Cambridge University Press, Cambridge.

Chapter 2
The Labyrinths of Heaven

Astronomy became a popular subject of study and interest with the general public in the eighteenth century. Indeed, scientific lecturing could be a successful career for enterprising amateurs who would advertise their courses, hire lecture rooms and charge fees. Lectures on astronomy were illustrated by means of models of various kinds. The basic equipment was a pair of globes – the familiar terrestrial globe as still used showing the territories of the Earth, and a celestial one, showing the principal stars in their constellations. Each globe had its engraved coordinates – the terrestrial one with its circles of longitude and parallels of latitude; the celestial one with the equivalent set of coordinates, right ascension and declination. Mounted in a stand, each globe was rotatable around its polar axis which was tilted at an angle of 23.4° to the horizontal, indicating the inclination of the equators of the Earth and the celestial sphere to the ecliptic, the plane of the Earth's path around the Sun. Pairs of beautiful globes – today much sought-after artistic items – were part of the furniture of well appointed houses. From the terrestrial globe one could discover, in the simplest example, the time difference between geographical locations. The celestial globe could be used to indicate the position of the Sun in the zodiac at various times of the year. With further instruction, more complicated problems relating to the appearance of the heavens could be solved. "The use of the globes" came to be a standard element in geography lessons in schools. Another topic of instruction was "dialling", or the art of making faces of sundials. Ladies, who might not consider using a telescope in the cold outdoors, could learn about the heavens in the comfort of their drawing-rooms from a celestial globe or from suitable books that explained the movements of the heavens. The more studious among them might even attend – or be permitted to attend – public scientific lectures.

The working of the solar system was illustrated by the orrery (Fig. 2.1), a mechanical apparatus that represented the planets in their orbits as balls moving by wheelwork in circles around a model Sun. One type of orrery showed the Earth, accompanied by its orbiting Moon, and could be used to explain the phases of the Moon and the occurrence of solar and lunar eclipses. A more elaborate and very beautiful instrument, sometimes also called a planetarium (though in modern times a planetarium has a different meaning) showed the known planets of the solar system, with their respective moons, all capable of performing their assigned movements by the turning of a handle. The entire interest was in the mechanics of the solar system.

M. Brück, *Women in Early British and Irish Astronomy*,
DOI: 10.1007/978-90-481-2473-2_2, © Springer Science+Business Media B.V. 2009

Fig. 2.1 An orrery by James Ferguson. (National Museums, Scotland)

The stars, at unfathomable distances, were merely a backdrop against which the movements of the Sun, Moon and planets were observed. Astronomy, in the sense of a knowledge of the nature of stars or of the wider universe, played no part.

Before the discovery of the distant planet Uranus by William Herschel in 1783, the planetary family consisted in those bodies recognised from antiquity – Mercury, Venus, Earth (and its moon), Mars, Jupiter and Saturn. Galileo's momentous breakthrough with the telescope at the beginning of the seventeenth century revealed that Jupiter was accompanied by four moons or satellites, and that Saturn was surrounded by a ring. The discovery of Jupiter's moons in 1610 proved that the phenomenon of bodies going around each other extended beyond our earth-and-moon. It was the great triumph of Isaac Newton in 1687 to interpret this in terms of Gravitation, the attraction between massive bodies. The motions of the planets around the Sun, and of satellites around their parent, all became wonderfully clear. The tides could also be accounted for by the gravitational influence chiefly of the Moon on the waters of the oceans. So could the phenomenon of precession, a slow apparent movement of the entire celestial sphere known from antiquity; so too could the rate of a pendulum's swing, and other physical facts.

Astronomy therefore had to be studied in conjunction with Newton's physics if it was to be properly appreciated. A series of special events reported in the news from time to time also aroused the curiosity of intelligent members of the public. There were two near-total eclipses of the Sun visible in Britain in the eighteenth century, much talked about in advance. The first, in 1748, was actually seen as total in the north of Scotland, while in London most of the Sun was obscured, leaving only a narrow crescent of light. The second, in 1764, was annular, that is, the Moon covered all but a narrow ring of the Sun's face. A sensational occurrence in 1759 was the appearance of a bright comet, as predicted according to Newton's laws by the British astronomer Edmund Halley in the previous century. Halley had correctly identified the bright comet of the year 1682 with comets recorded several times previously in history, at fixed intervals, and predicted its next return in 1759. Though he himself did not live to see it, the comet which was naturally given his name, Halley's comet, returned faithfully as predicted, and has continued to do so every 75 years.

Another much discussed rare astronomical event was the Transit of Venus in 1761, when that planet fell directly between the Earth and the Sun, showing up as a small dark spot traversing the Sun's bright face. The phenomenon of the Transit of Venus is also associated with the name of Halley who had pointed out the importance of observing it as a means of finding the dimensions of the solar system. There was therefore considerable excitement among astronomers in the run-up to the Transit. Such Transits occur in pairs approximately once a century – in 1761 and 1769; and again in 1874 and 1882. The first of the next pair was observed amid much excitement in 2004; its partner will occur in 2012.

In anticipation of these predicted events, courses in physics and astronomy were offered by enterprising private individuals in response to the demand among the public. Books on astronomy naturally accompanied these educational activities. Some talented teachers and demonstrators were also successful authors. One of the earliest was the erudite Benjamin Martin (1705–1782) whose wide repertoire covered natural philosophy (i.e. physics), navigation and astronomy, described in more than 30 popular books. He was also an expert on optics, an inventor and a constructor of demonstration models. His lectures involved the uses of lenses and mirrors, and mechanical apparatus such as pulleys, levers and airpumps. Women were welcomed to his courses. His book *Introduction to the Newtonian Philosophy*,[i] illustrated by copperplate drawings of instruments, was "designed for the use of such gentlemen and ladies as would acquire a competent knowledge of this science without mathematical learning, and more especially those who have, or may attend, the author's course of six lectures and experiments on these subjects". One of his books had women readers specially in mind: *The Young Gentleman's and Lady's Philosophy* (1860) took the form of a dialogue on science between a learned student and his pupil sister who is avid for knowledge but lacks opportunity.

Women were also warmly encouraged by Martin's contemporary James Ferguson (1710–1776), perhaps the most famous of the astronomical demonstrators of his day.[ii] This remarkable man began his life as a simple shepherd boy in the Scottish Highlands where his passion for the heavens was aroused on his long night vigils in charge of his flock. As a child, he also discovered a gift for mechanics and for art. He developed into a master constructor of ingenious astronomical models which were not only beautiful to behold, but were of great accuracy and intricacy. He arrived in London to seek a career as a lecturer and demonstrator, and, while not the first in the field, he was eventually accepted and admired in the highest scientific and social circles, going on to become a Fellow of the Royal Society. He travelled around the country, giving courses of lectures at up to 20 different venues throughout Britain, published numerous books and articles, as well as designing and constructing his elaborate apparatus. His most important book, *Astronomy Explained upon Sir Isaac Newton's Principles*, first published in 1756, went into 12 editions.[iii] It was an excellent textbook of classical astronomy, written in a clear style, non-mathematical, yet satisfyingly complete and beautifully illustrated. *Astronomy Explained*, which provided inspiration for such a distinguished figure as William Herschel, and *Lectures on Select Subjects in Mechanics, etc.*, were of such enduring usefulness that Sir David Brewster, the famous Scottish physicist, published updated editions of them some 30 years after the author's death.

One reviewer of *Astronomy Explained*, on it first appearance, remarked that "the ladies may now rise above the regions of vapours without running the risque of a vertigo". The ladies not only read the book but also patronised the public lectures, apparently attending in groups. In an advertisement to a forthcoming lecture course in 1759, Ferguson wrote: "It gives him great Pleasure to find that many Ladies form themselves into Companies, and attend his Lectures on the Globes and Orrery; and can now say with truth, that he has sufficient Reason, from the Quickness of their Appreciation, to ridicule a Notion which was never his own; namely that these Things are without the Sphere of Ladies Capacities, and what they have no business with". Some years later, he published *The Young Gentleman and Lady's Astronomy*[iv] (1768). It was based on lessons he gave to a young girl in real life, and took the form (already used by his rival Martin) of a dialogue between a knowledgeable young man – in this case a university student named Leander, imagined as being home from Cambridge – and his young sister Eudosia who is anxious to learn. (Benjamin Martin's fictional master and pupil had even more exotic names – Cleonicus and Euphrosyne.)

The Young Gentleman's and Lady's Astronomy lived up to its claim of being an easy introduction to astronomy. It was divided into 10 chapters, representing 10 conversations in which the brother explained various topics to his sister. The course covered all the main elements of astronomy – the Earth, the solar system, latitude and longitude, the explanation of day and night and the seasons; the Moon, its phases and the tides; eclipses, and how to predict them; solar and sidereal time. There was also a chapter on the Transit of Venus of 1761 and its scientific results. The book was illustrated by Ferguson's own beautiful diagrams.

Apart from its instructional value, *The Young Gentleman's and Lady's Astronomy* shows that already in the mid eighteenth century there were women who aspired to higher education and to a knowledge of science. At one point Eudosia interrupts her brother's discourse with a sigh. Asked why, she replies: "because there is not a university for ladies as well as gentlemen. Why, Leander, should our sex be kept in total ignorance of any science, which would make us much better than we are, as it would make us wiser?" He replies that she is "far from singular in this respect"; that many ladies think as she does. But his solution is less daring than hers. "If fathers would do justice to their daughters, brothers to their sisters, and husbands to their wives" he says, "there would be no occasion for a university for the ladies ... The consequence would be that the ladies would have a rational way of spending their time at home, and would have no taste for the too common and expensive ways of murdering it, by going abroad at card-tables, balls, and plays: and then how much better wives, mothers and mistresses they would be, is obvious to the common sense of mankind. The misfortune is, there are but few men who know these things: and where that is the case, they think the ladies have no business with them; and very absurdly imagine, because they know nothing of science themselves, that it is beyond the reach of women's capacities."

It is most likely that this conversation, or something close to it, actually took place. Ferguson's model for Eudosia was a young pupil Anne Emlyn (1753–1801), daughter of Henry Emlyn (or Emblin) (1729–1815),[v] an architect in Windsor, a man of some distinction in his profession. He was the author of a book on architecture, and executed restorations in St George's Chapel, Windsor, for King George III in 1787–1790. He is buried in the Chapel where a tablet was erected to his memory.

At the age of 25, Anne married Capell Lofft (1751–1824), a lawyer by profession, described as a "man of many accomplishments", a good classical scholar, a lover of literature especially Shakespeare, a poet, an enthusiast in music and natural history, and a skilled astronomer[vi] (the last probably acquired after his marriage to Anne). He belonged to an interesting intellectual family: his father was the private secretary of Sarah Duchess of Marlborough, and his mother was the sister of Edward Capell, an editor of the works of Shakespeare. He was himself a man of independent mind, something he could well afford to be after he inherited estates from the families of both his parents. As a lawyer he was regarded as a firebrand for his strong liberal views, and was struck off the roll of magistrates for trying (unfortunately without success) to save the life of a young girl hanged for petty theft. He was the author of many works on law, classics and politics, as well as books of poetry, most of them written during Anne's lifetime.

In 1780 Lofft published *Eudosia, a Poem on the Universe*,[vii] obviously inspired by his wife's knowledge and love of astronomy. The poem was written after their marriage in 1778, and takes the form of an educational journey by the poet, accompanied and guided by his Muse, instructress and wife "Eudosia". Ferguson's *The Young Gentleman and Lady's Astronomy* was published in 1768 when Anne was 15, from which one therefore deduces that she was no older than that age when she was Ferguson's pupil. She was clearly a favoured and very gifted one, with whom he remained in communication. In a note at the end of his published poem, Lofft tells his readers that "Some manuscript Tables, diagrams and philosophical correspondence of this heaven-taught philosopher [Ferguson] are in my hands: which were given by him to my Eudosia before our marriage: for he had no pupil whose genius and disposition he more esteemed than hers; nor can I reprove myself for the pride which I often feel in reading over his letters to Ms. Emblin, written to her before our marriage". Lofft included in his poem a charming verse in praise of the "heaven-taught philosopher":

> "Nor shall thy guidance but conduct our feet,
> O honoured shepherd of our later days!
> Thee, from the flocks, while thy untutored soul,
> Mature in childhood, traced the starry course,
> Astronomy, enamoured, gently led
> Through all the splendid labyrinths of heaven,
> And taught thee her stupendous laws; and clothed
> In all the light of fair simplicity,
> Thy apt expression."

and elsewhere described him as

"Ferguson, than whose no beam more clear
Pierces the gloom, where science is concealed".

Ferguson himself did not live to read these compliments: he died in 1776.

Eudosia is written in blank verse and divided into Books or chapters, which follow closely the chapters in Ferguson's *Astronomy Explained on Isaac Newton's Principles.* It has copious notes and references. The dedication on the title page reads: *To the person to whom he owes the greatest possible obligations. This poem is affectionately inscribed by the author, Capell Lofft.*

It is a remarkable piece of work, a combination of a scientific exposition and a love poem. The poet succeeds in translating the contents of Ferguson's work accurately into verse while adopting the style of classical poetry. He begins by asking to be enlightened by Urania, the Muse of Astronomy:

"But what reply to these Urania deigns
Well knows EUDOSIA; her sacred muse
Of true philosophy delighted sees
And with her walks familiar; not displeased
Will she resolve the arguments of truth
In strains attempered to the shepherd's reed".

and:

"Lead me, O fair and wife! point to my view
The wonders of the Heavens well known to thee;
Correct, refine, and animate my song."

The first chapter deals with phenomena associated by the rotation and revolution of the Earth. On moving on to the planets:

"Now, my EUDOSIA. to a nobler scene
Now, leaving mortal cares, the fields of Heaven
Free and sublime, and free, as is thy mind,
we range...."

This elaborate poem includes the history of astronomy and acclaims the great philosophers who shaped its progress – Pythagoras, Copernicus, Kepler, Galileo, Newton, Halley – and explains many complex ideas in verse such as Kepler's Laws. the Law of Gravitation or the Transit of Venus ("the conjunction of the Star of Eve/with the all-radiant sun"). The last Book of the poem combines the praise of scientific progress with high moral sentiments:

"Fruitful of good, by gradual advance
The golden age of wisdom and of bliss.
Thine be the praise of ever-growing worth,
O my EUDOSIA! Thus the stars are won,
Thus all the heroes of Philosophy
Have triumphed over fate
And blessed mankind."

It seems very likely that Lofft was first introduced to astronomy by his wife, and that he (or they) actually acquired a telescope and made observations. In 1818 (after Anne's death) Lofft published an observation of a transit of what he thought was a planet inferior to Mercury (i.e., nearer the Sun),[viii] which, however, was probably a sunspot. Anne, who must surely have had an interesting married life with her flamboyant husband, and had several children, died sadly at the age of only 48. In the person of Eudosia, however, she deserves to be remembered and honoured. In expressing her resentment at women's lack of access to university, Anne was well ahead of her time and ahead even of her teacher. Ferguson's ideal of women as educated companions to their enlightened menfolk in the family circle was still being extolled by Maria Edgeworth more than 50 years later (Chapter 5). For all her admirable pioneering views on female education, Maria stopped well short of recommending women to join the man's world on equal terms. Eudosia's dream of a university for women had to wait a 100 years.

"Eudosia" was not the only woman of her generation to have studied astronomy by Ferguson's method. And not all teachers were men. Margaret Bryan (fl c.1797–1816), who was perhaps about 10 years younger than Eudosia, was the author of an excellent textbook for older children, *A Compendious System of Astronomy*, first published in 1797.[ix] She wrote one also on Physics, *Lectures in Natural Philosophy* (1806), and, in 1816, *An Astronomical and Geographical Class Book for Schools*. As to the identity of the author, The Dictionary of National Biography tells us little more than that she was the wife of Mr. Bryan (but who was Mr. Bryan?) and "a beautiful and talented schoolmistress". The description is amply borne out in her portrait by Samuel Shelley, a famous and fashionable engraver of the day, which forms the frontispiece of her first book. She is shown there with her two charming young daughters posed among her astronomical apparatus (Fig. 2.2).

An advertisement at the beginning of that book announces: "Mrs. Bryan receives young ladies for the purpose of education at Bryan House, Blackheath" (perhaps a boarding school, presumably at her private house in Blackheath, near Greenwich). It is believed that she later ran a school at Hyde Park Corner, and another at Margate. Mrs. Bryan was clearly well connected. The second edition of the book (1799) gives an impressive list of some 400 subscribers who financed the original publication. The list includes eminent scientists – Rev. Dr. Maskelyne, Astronomer Royal (whose only daughter may have been one of the pupils); Alexander Aubert FRS (astronomer friend of the eminent William Herschel and his sister Caroline); William Boys FRS (surgeon and polymath scholar); John Bonnycastle, R.M. Academy, Woolwich (author of many well-known textbooks); Thomas Keith (author of *Treatise on the Use of the Globes* and one time tutor to the Royal Princesses) and Charles Hutton FRS, another of Herschel's circle, who was the author's scientific adviser. High society was well represented – the Duke of Marlborough, the famous and beautiful Lady Georgina Cavendish of Devonshire House, herself a student of chemistry and mineralogy (who took two copies); His Grace the Archbishop of Canterbury; Rt. Hon. William Windham, the Secretary-at-War. There were several members of Parliament, Army Officers, members of Cambridge Colleges and clergymen.

Fig. 2.2 Mrs. Bryan and her daughters. (Margaret Bryan's *Astronomy*, 1799)

Among the many ladies was Ms. Thrale, daughter of the famous Mrs. Thrale, friend of Samuel Johnson. Other subscribers appear to have been parents of pupils: from the book's *Dedication* to her pupils it is evident that she had been teaching the course of astronomy for some years before the book was written.

Mrs. Bryan well merited this august clientele. Her text had been approved by the eminent Charles Hutton, professor of mathematics at the Royal Military Academy at Woolwich. So pleased was she with his endorsement that she published in the Preface the letter in which he declared that he had read the text "with great pleasure: and the more so, to find that even the learned and more difficult sciences are thus beginning to be cultivated by the extraordinary and elegant talents of the female writer of the present day". Hutton was well-known in his circle. Apart from being the author of reference books on mathematics and natural philosophy, he had also distinguished himself as the collaborator of the Astronomer Royal Nevil Maskelyne in a famous experiment to find the mass of the Earth by the indirect method of observing the gravitational influence of a mountain. The experiment made use of Newton's law of gravitation which states that massive bodies exert an attraction in proportion to their mass and to the inverse square of their distance. Normally, a plumbline will hang perfectly vertically, being attracted by the Earth itself towards its centre. If, however, the line is held in the vicinity of a massive mountain it will deviate slightly from the true vertical under the mountain's gravitational influence. The mountain chosen for this difficult experiment was Schiehallion in Perthshire, Scotland. Maskelyne and his team succeeded in observing a slight deviation of the plumb line from the true vertical (determined by observations of the stars). Though their result was improved upon by later experiments, it was historic for its boldness and for the fact that it worked.

Hutton introduced Mrs. Bryan to William Herschel, the illustrious astronomer who had his observatory at Slough where she once visited him, and met his wife and his astronomer sister Caroline (Chapter 3). Hanging on the wall in Herschel's study was a copy of her portrait, as she reminded him in a letter written in 1811 when she asked him to send her copies of his astronomical papers.[x] It is exciting to reflect that Herschel's son, the young prodigy John, who was born in 1792, may have used Mrs. Bryan's delightful book as an elementary introduction to the science in which he was soon to shine.

Mrs. Bryan's book was a set of 10 lectures in a well-planned course covering the usual subjects treated by Ferguson and others in their books – astronomy, optics, Newton's laws, gravity, planetary orbits, motion of the Moon, eclipses, transits – in a sentimental conversational style to suit her young lady pupils. New material since Ferguson's time included the discovery of the planet Uranus by William Herschel in 1780, still called Georgium Sidus, the name given to it by Herschel in honour of his patron King George III. Herschel also discovered (in 1789) two attendant moons around his planet and (in 1798) believed he caught glimpses of four more which, however, turned out to be illusory. Mrs. Bryan recorded the discoveries of the planet and its two moons in the first edition of her book (1797) and added a footnote increasing this number to six in the second edition of 1799 – an example of her care in keeping up to date with observational advances. More importantly, she included a chapter on the fixed stars and the universe, with an account of the researches of Herschel, a "superior genius", into the distribution of stars in the sky, and his discovery in 1784 that "the galaxy, or milky way, as it is called, ... consists of an innumerable quantity of fixed stars". There was also instruction on elementary plane geometry and trigonometry, and a set of problems with solutions.

The book was written in clear and simple language, and illustrated with the author's own beautifully precise drawings and diagrams. In the practical part of Mrs. Bryan's curriculum, models such as the orrery were demonstrated and authentic instruments were shown but not put to actual use on celestial objects. A reflecting telescope was recommended for "common purposes of celestial observations" for those who might wish for such "amusement", but night sessions out of doors were not part of the taught course for her lady pupils. She herself, however, was an observer. In her letter to William Herschel in 1811 she mentions her efforts to observe the much-discussed comet of that year. (It was not one of Caroline's; it was discovered by the famous comet hunter Jean Louis Pons.)

Margaret Bryan's second book, on Natural Philosophy[xi] (i.e. Physics), followed a similar pattern to the first. It was dedicated, with expressions of admiration and homage, to "Her Royal Highness The Princess Charlotte of Wales". Princess Charlotte (1766–1828) was the eldest daughter of King George III, who in 1797 married a German prince, Frederick, future King of Württemburg. "May the blossoming genius of Your Royal Highness' mind ripen into maturity with that vigour and sweetness may render it a public blessing and delight, adorning your elevated station, and diffusing a salutary and benign influence through your native land". A further dedication paid tribute to Dr. Charles Hutton who had encouraged her as he had done with the first book. There was a beautiful portrait of herself by Kearsley who was also her publisher and bookseller and, as before, a long list of socialite subscribers. These included, once again, Dr. Maskelyne, the Astronomer Royal, Charles Hutton and various members of his family. A new name was Charles Burney, musician and scholar, father of the authoress Fanny Burney and friend of the Herschel family.

In her Preface the authoress explains that, having studied her subject for eight years and taught it for seven, she was "encouraged by the very honourable and distinguished patronage" which her lectures had received to have them published. She also declared that she had followed "the very excellent divine, Dr. Paley, in his Natural Theology". Here was a new element in her teaching. William Paley, an eminent Anglican theologian and philosopher, studied and published work on moral philosophy and on the connection between science and theology. His last and most influential book was *Natural Theology*, published in 1802, which puts forward the argument that the world, as revealed by scientific observation, was designed by a benevolent Creator. This work had not yet been published when Margaret Bryan wrote her first book, but it now clearly had a profound influence on her. All the lectures in her course follow the same pattern: an account of the wonders of the Almighty as revealed in the particular phenomenon under discussion, a discursive scientific explanation, and an account of the usefulness of these findings to the human race.

The book contains 13 chapters – each a self-contained Lecture – dealing with Properties of Matter, mechanics, pneumatics (properties of gases), acoustics and sound propagation, hydrostatics (properties of liquids), magnetism, electricity. The last few chapters deal with optics, lenses, mirrors, telescopes, microscopes, the spectrum and spectroscope. The very last was on astronomy. No mathematics whatever were used, but one must admire the clarity with which Mrs. Bryan could explain the principles accurately in words, aided by copious meticulously drawn diagrams, some from her own hand. There were also numerous tables, lists and simple problems.

The book ends with a "Concluding Address to my Pupils". "In these lectures, I have had the delightful satisfaction to delineate some of the most striking evidences of the power, wisdom and benevolence of the Great Creator.... To seek for evidences of the Deity in the operations of his power, and the arrangement of his plans, is no impeachment of our faith." Her address continues with exhortations to the pupils to be "obedient to your parents ... faithful to friends ... as sisters, affectionate and generous." On a subject that was the hoped – for fate of young women of that age, marriage, she gives gentle uplifting counsel: "When wives, consider the solemn oath pledged before God, and strictly obey its mandates. Let cheerful acquiescence evince your affection towards your husband ... As mothers, remember you were once young", amplifying the theme of family life at some length, and, in general, of their duty to God and man.

Margaret Bryan comes across to the reader through her books as a person of great charm who genuinely loved her young charges and saw it as her duty to form their characters as well as to enlighten them in science. Though she did not make use of mathematics, and was perhaps not versed in it, she was clearly a person of considerable talent. The friendship of such distinguished scientists as William Herschel and the Astronomer Royal, the patronage of society luminaries who evidently entrusted their daughters to her care, and the acquaintance with Royalty itself, distinguishes her as an unusually erudite and charming lady. One longs to know more of her family – her husband and her delightful daughters.

Mrs. Bryan was confident – as who would not be, with the example of her contemporary Caroline Herschel before her – that in teaching astronomy to female pupils she had the support of those "whose extensive learning and liberality leads them to judge impartially; for they, rising superior to the false and vulgar prejudices of many, who suppose these subjects too sublime for female introspection (ascribing to mental powers the feebleness which characterises the constitution,) invalidate the idea by affording all laudable exertions their avowed patronage, – acknowledging truth, although enfeebled by female attire."

In 2004 a miniature portrait of Margaret Bryan and her children was by good fortune discovered and acquired by the Herschel Society.[xii] It is the same portrait (or a copy of it) that once hung in William Herschel's observatory in Slough, and was reproduced as an engraving in the *Compendious System of Astronomy*. Now, two centuries later, it fittingly and delightfully adorns the Herschel Museum in Bath. The museum also contains an exhibit relating to Margaret Bryan's life and work.

Lectures and demonstrations, given by itinerant teachers such as James Ferguson did not normally include actual viewing of the heavens, one reason being that they were held in hotel rooms or hired halls, where the speaker spoke and listeners sat and listened. There was also the practical problem of the cost of even moderately sized telescopes, and of the weather: it was much more effective for a lecturer to provide globes and orreries. There were, however, exceptions.

> "What crowd is this? what have we here!
> we must not pass it by.
> A telescope upon its frame, and pointed to the sky:
> Long is it as a barber's pole, or mast of little boat,
> Some little pleasure-skiff, that doth
> on Thames' water float.
> The showman chooses well his place.
> 'tis Leicester's busy Square,
> And is as happy in his night,

for the heavens are blue and fair;
Calm, though impatient, is the crowd;
each stands ready with the fee,
And envies him that's looking – what an insight must it be!"

wrote the poet William Wordsworth in 1802.[xiii] And Dr. Allan Chapman in *The Victorian Amateur Astronomer* tells the tale of Mr. Treford's kerbside shows where members of the public could view the heavens at "a penny a peep".[xiv]

A considerably more sophisticated venture could be found in Edinburgh where a plucky lady called Maria Short (c. 1788–1869) owned and operated an observatory for the paying public in the mid-nineteenth century. Maria had a good astronomical pedigree. Her father, Thomas Short, was a scientific instrument maker in Edinburgh. Thomas was a brother of James Short, a renowned telescope maker who flourished in the period before William Herschel began constructing his own magnificent instruments. Like Herschel, his telescopes were of the reflecting kind. James began his career in Edinburgh, instructed in mathematics under the influence of the university's great mathematicians, and moved his workshop to London where he carried on a very successful business for over 35 years. His telescopes made use of speculum (metal) mirrors which were figured and polished to such perfection as to surpass all competitors. He was admired by leading scientists in Edinburgh and London and was elected a Fellow of the Royal Society. He made telescopes of various sizes up to 18 in. in diameter, and could command high prices – many hundreds of pounds for the large ones – from wealthy customers who were for the most part aristocratic dilettanti, even foreign royalty.[xv] When he died, unmarried, in 1768, his brother Thomas came to London to manage the workshop and acquired several mirrors and unfinished telescopes, including one of 12 in. aperture valued at thousands of pounds, originally intended for the King of Denmark. Thomas returned to Edinburgh and formed the idea of establishing an observatory in the city where this large instrument would be installed as a commercial venture and used by the university and others on payment of fees.[xvi] He meantime set up telescopes in daytime where visitors could, for a charge, view the city and the surroundings. When an eclipse of the Moon occurred, he set up the large telescope in the university, and sold tickets to watch the event. It was a great success. It was reported that "a very numerous and genteel company of ladies and gentlemen convened in the New Library Room of the university, observing the Lunar Eclipse. They were all exceedingly entertained with the different appearances of the moon exhibited, and which were beheld with great advantage through Mr. Short's telescope".

The plan for a permanent observatory was less successful. The city of Edinburgh was to provide a site and promised further aid in the future. In return, the city was to have the rights to the building and the instruments, Short himself to be allowed to charge for their use. The university, which had long wished to have an observatory for teaching and research, also became entangled in the scheme. The foundation stone for the building was laid on Calton Hill, that conspicuous landmark in the centre of the city, in 1776. At the time of his death in 1788 Thomas Short was left in possession of an observatory still unfinished, after 20 years of confusion and exhausted funds. The building, with living accommodation, was completed by the city some years later. The instruments were inherited by a grandson, but not without a family feud and an unsuccessful attempt by Short's widow, his second wife Jacobina, to take possession by force of the premises from which she had been evicted, helped by accomplices with pistols, cutlasses and a blunderbuss. The observatory eventually came into being in 1792: one of the first visitors was William Herschel who was interested in examining James Short's large reflector. The project was not a commercial success, and gradually ceased to function.

A new start was made in 1818 which resulted in the erection of the university (later Royal) observatory, built quite independently of the earlier one on a site nearby. Now known as the City Observatory, it is still in active use by the Astronomical Society of Edinburgh.

In 1827, long after the demise of the original Short observatory, Maria,[xvii] the youngest of Thomas Short's nine children by the fiery Jacobina, arrived in Edinburgh from abroad where she had been living, probably in the West Indies and later in Ireland with a married sister. Maria, who was by this time at least 39 years of age (her father having died in 1788), set about establishing a legal claim to the Great Telescope. Exploiting the reputation of her illustrious forebears, she wrote to influential citizens of Edinburgh and succeeded in collecting a long list of supporters. Some were sceptical of her authenticity. William Wallace, Professor of Mathematics and former tutor of Mary Somerville (Chapter 6), who could claim to know all the local opticians and telescope makers, wrote to the Lord Provost warning him of "a person calling herself the daughter of the late Thomas Short" though he had never heard of his having a legitimate daughter. Maria, however, won the day, and was awarded possession of the telescope by the city in recognition of her father's efforts to establish the first public astronomical observatory in Edinburgh.[xviii] A list of about 200 names of distinguished individuals who responded to her appeal included the Duke of Buccleugh, Lord Jeffrey and other lawyers, several university professors, Robert Stevenson the lighthouse builder, Alexander Nasmyth the artist, Sir George Clerk of Penicuik, and Ms. Susan Ferrier the famous novelist. Even William Wallace succumbed, and signed a message of goodwill from the University of Edinburgh.[xix]

Maria erected a wooden building on Calton Hill and in May 1835 opened "Short's Popular Observatory" which ran successfully for 15 or more years. The scientific collection comprised the Great Telescope which Maria had repaired in London and re-mounted, various smaller ones, and other instruments. A printed prospectus declared that "the sublime truths of science are no longer confined to the wealthy and the learned". The observatory figured in tourist pamphlets of the time. It was open every day from 9 a.m. to 9 p.m. One presumes that, for a charge, members of the public were shown the planets and interesting celestial objects – a forerunner of the twentieth century's Planetarium. Maria also installed a camera obscura – an apparatus that gave an impressive panoramic view of the surrounding skyline in daytime.

The camera obscura was at that time a popular device among optical experimentalists, widely used in Edinburgh's lively artistic circle. In essence it was a convex lens mounted in a dark box which projected an image of surrounding illuminated objects on a screen within the box. Such an arrangement had its use as an aid to artists who could trace the scene directly, such a drawing having the advantage that the proportions of the objects in it were correct. The arrangement was in a sense the predecessor of the photographic camera. When the lens, together with a sloping mirror, was mounted at the top of the box, the image formed by the lens could be projected downwards on to a table underneath. In this form the device was useful for serious students of perspective, and was also popular as a sort of toy by amateur artists. As the mirror was rotated, a panoramic scene was generated, to the delight of the viewer. The first such panorama display for the public was that of the Edinburgh skyline from Calton Hill, demonstrated by Maria Short.[xx] According to newspaper reports, the observatory attracted thousands of visitors.

In spite of the venture's commercial success, Maria Short appears to have been on less than good terms with the Town Council which owned the site. The observatory on Calton Hill closed, or was forced to close, in 1851. Five years later Maria reopened the Camera Obscura in a house at the top of Castle Hill (near Edinburgh Castle). (A Camera Obscura still operates on the same site, and is still one of the city's tourist attractions). Maria Short and her husband Robert Henderson whom she married in 1849, lived at Castle Hill and operated Short's Observatory there until 1861.

The ultimate fate of the large telescope is strangely unknown. Nothing more is recorded, either, of the personal life of its final owner and spirited "astronomess". Indeed, there is some mystery about it. Veronica Wallace's research[xxi] has revealed that in the census of 1861, Robert Henderson and Maria are described respectively as "Keeper of the Observatory" and "his wife, conductress of the Observatory". Their ages are given as 59 and 58 – an impossibility in Maria's case, if she was truly Thomas Short's daughter whose age in 1861 would have been at least 73.

But whatever her age and lineage, this "resourceful and ambitious woman", "Maria Obscura" (as Veronica Wallace so aptly dubs her), brought for 20 years of the nineteenth century the magic of astronomy and a taste of technology to an appreciative public in her native city.

Notes

[i] Benjamin Martin. 1754. *A Plain and Familiar Introduction to the Newtonian Philosophy.* London: W. Owen.

[ii] John R. Millburn, in collaboration with Henry C. King, 1988. *Wheelwright of the Heavens, the Life and Work of James Ferguson FRS.* London: The Vade-Mecum Press.

[iii] James Ferguson FRS. 1778. *Astronomy Explained upon Sir Isaac Newton's Principles.* (Sixth Edition). London: W. Strahan and others.

[iv] James Ferguson F.R.S. 1779. *The Young Gentleman's and Lady's Astronomy. An Easy Introduction to Astronomy for Young Gentlemen and Ladies.* (Fourth Edition; first edition 1768). London: T. Cadell.

[v] *Dictionary of National Biography.* 1975. Compact edition. Oxford. Lofft spells his wife's name "Emblin" in his published poem. The DNB has "Emlyn".

[vi] idem.

[vii] Capell Lofft. 1781. *Eudosia: or a Poem on the Universe*. London: W. Richardson and C. Dilly.

[viii] The entry on Lofft in the DNB states that the observation was published in *Monthly Notices of the Royal Astronomical Society*. This is erroneous, as that journal did not exist in 1818. The correct reference has not been located.

[ix] Margaret Bryan. 1799. *A Compendious System of Astronomy*. (Second Edition). London: J. Wallis, and Wynne and Scholey.

[x] Letter from Mrs Bryan to W Herschel, RAS archives.

[xi] Margaret Bryan. 1806. *Lectures on Natural Philosophy, the result of many years' practical experience of the facts elucidated, with an Appendix containing a great number and variety of Astronomical and Geographical problems and some useful tables*. London: George Kearsley.

[xii] Debbie James. 2004. *Speculum, the Journal of the Herschel Society*, Vol. 3, no. 1, 2–3.

[xiii] William Wordsworth. Stargazers reproduced in the delightful collection of poems about the stars compiled by M.G.J. Minnaert. 1959. *Dichters Over Sterren*. Arnhem: van Loghum Slaterus. Henry C. King uses the first line of this poem as an opening motto for his *History of the Telescope* (reference below).

[xiv] Allan Chapman. 1998. *The Victorian Amateur Astronomer*, p 174. Chichester: Praxis Publishing Ltd. and John Wiley.

[xv] Henry C. King. 1979. *The History of the Telescope*, p 84. New York (Dover edition).

[xvi] D.J. Bryden. 1990. The Edinburgh Observatory 1736–1811: a story of failure. *Annals of Science* **47**, 445–474.

[xvii] Veronica Wallace. 1992. Maria Obscura. *Edinburgh Review* **88**, 101–109.

[xviii] D.J. Bryden. 1968. *James Short and His Telescopes*. Edinburgh: Royal Scottish Museum.

[xix] Veronica Wallace. op. cit. The complete list of patrons is published in this paper.

[xx] Alison Morrison-Low and Allen Simpson. A New Dimension: a context for photography before 1860. in Sara Stevenson. 1995. *Light from the Darkroom*. Edinburgh: National Galleries of Scotland.

[xxi] Veronica Wallace. op. cit.

Chapter 3
Martyr to Astronomy

While Eudosia's dream of university for women was still far in the future, one fearless hardworking woman of a very different temperament appeared on the astronomical scene in Britain. She was Caroline Herschel (1750–1848) (Fig. 3.1), sister and collaborator of William Herschel, one of the greatest observational astronomers of all time, pioneer of the study of the universe of stars and father of modern cosmology. Unlike traditional astronomers who busied themselves with measuring star positions and refining their knowledge of movements of the planets, Herschel turned his eyes towards the wider universe and aimed at unfolding "the construction of the heavens". He discovered the existence of double stars (pairs in orbit about each other), and the movement of the Sun in space. He investigated the structure of the Milky Way system of stars and listed thousands of nebulae, recognising many of them as unresolved clusters of stars at great distances and suggesting the possibility of other "island universes" beyond our own.[i] "He broke the barriers of the heavens" was the epitaph inscribed on his tomb; and Caroline, his obedient and unquestioning assistant, helped him in that breakthrough. Caroline also attained renown in her own right as a leading discoverer of comets. Admired in her lifetime, she remains to this day the undisputed heroine of women astronomers.

William and Caroline Herschel came originally from Hanover, in Germany. Their father Isaac was a professional musician, an oboist in the Guards and afterwards a teacher of music and a performer. Frederick William (to give him his full name) was the third and Caroline the fifth in the family of six surviving children. The father raised all his four talented sons in his own profession. William joined the Hanoverian Guards as an oboist (he also played the violin) at the age of only 14, but four years later fled to England as a refugee from the Seven Years' War.[ii] There he decided to try to make a career for himself as a musician. He worked hard as a concert performer, teacher and composer, and gave great satisfaction to his patrons. The outcome was his appointment in 1766 at the age of 28 to the post of organist at the Octagon Chapel in Bath. That beautiful town was a fashionable resort of the wealthy in the eighteenth century, with many opportunities for private recitals and for giving music lessons. His reputation grew, and William, knowing that his sister was unhappy at home, decided to invite her to join him in Bath and become, perhaps, a member of his choir.

M. Brück, *Women in Early British and Irish Astronomy*,
DOI: 10.1007/978-90-481-2473-2_3, © Springer Science+Business Media B.V. 2009

Fig. 3.1 Caroline Herschel in old age. (Royal Observatory, Edinburgh)

In 1772 William arrived in Hanover to fetch her. She was now 22 years of age and had never been away from home where her situation was far from easy. Her much older only sister was long married. Her beloved father had been dead for five years. Her oldest musically-dedicated brother Jacob was overbearing and demanding, while her overwrought mother, long oppressed with worry, had little sympathy with Caroline's hope to acquire some skill by which she might earn her own living. She had learned to read and write at school. She had also been taught to knit, and attended – though only for a few months, as her mother could not spare her for longer – a dressmaking class run by "an elegant lady" for girls from "genteel families". While her father lived, the quick-witted young Caroline was able to pick up some musical knowledge and was "frequently called to join the second violin in an overture, for my father found pleasure in giving me sometimes a lesson before the instruments were laid by, after practising with Dietrich (the much loved youngest brother and family favourite), for I was never missing at those hours, sitting in a corner with my knitting and listening all the while".[iii]

Caroline had fond memories of their father's interest in learning and of his scientific discussions with the young William. She remembered how he took her out on a clear frosty night to point out the constellations and to watch a comet, and how, on the occasion of an eclipse of the Sun, he assembled the family around a tub of water and showed them how to observe it safely by reflection. Now, however, her life was one of endless drudgery, more miserable even, in Dr. Hoskin's opinion,[iv] than depicted in her memoirs – until William's offer promised an escape.

Before taking Caroline away, William made provision for their mother to hire help in the house in place of her daughter. Caroline for her part, in anticipation of her departure, had knitted two years' supply of stockings for her brothers, the plan being that she should join William for two years' trial in the first instance.[v] She left home in 1772, never to see her mother or sister again. Nor did she return to Germany until after William's death 50 years later.

Caroline has left a lively account of the vicissitudes of the brother and sister's journey from Hanover to Bath. They moved house in Bath more than once during the following years, but their last home, made famous by William Herschel's astronomical discoveries, was 19 King Street. It is today the William Herschel Museum.

It was a big challenge for Caroline to adjust to her new life in a foreign land. She spoke no English and felt lonely and homesick. She had to learn a new language, to run the household, and to get used to marketing for unfamiliar food. Her brother was too busy to instruct her in these arts, or even to offer her companionship in the evening, because, as she soon discovered, every free moment was taken up with his new consuming interest – astronomy. It was helpful, however, that their brother Alexander was often with them: he too established a career in music as a cellist, and spent most of his life in Bath.

Caroline's musical training began almost immediately. William was well satisfied with her voice and gave her singing lessons in preparation for the "season" which began in October when visitors came for the winter and there was ample work for musicians. She met his pupils and his solo performers. As soon as she could pronounce English satisfactorily, she joined the Octagon chapel choir.

With her considerable natural talent and intelligence, Caroline made rapid progress as a singer and as helper in her brother's busy musical schedules. She copied scores and trained the treble chorus. In 1777, at the age of 27, after five years in Bath, Caroline appeared for the first time as soloist in the Easter oratorios, with William conducting. Now principal singer at William's concerts, she was offered, after a performance of the Messiah, a professional engagement in Birmingham. "I must have acquitted myself tolerably well", she recorded modestly in her memoirs. She declined, not willing, she said, to work under any conductor but her brother. It might have been her dream of independence come true; but, as Michael Hoskin remarks, it would not have occurred to William to give her the choice.[vi] She continued with her work for the choir, and sang solo at concerts and at private functions; but her own career could advance no further; as she herself wrote, "the interruption [by astronomical activities] in my practice, besides accumulation of copying music, etc., left me no time to take care of myself or stand upon nicetys".

Despite the hardships of his early life, William Herschel was more than a simple army bandsman and amateur astronomer. He was a well read thoughtful scholar, and a man of wide cultural aspirations. As a young musician in England, aiming to excel in composition, he studied the theory of music and harmony, a subject that brought him in touch with mathematics. By 1766 – the year he came to Bath – he was already spending his leisure hours reading advanced treatises such as those of the Scottish mathematician Colin Maclaurin and Robert Smith's *Harmonics or the Philosophy of Musical Sounds*.[vii] The *Harmonics* was to lead him in turn to the same author's *A Complete System of Optiks*, containing the theory of lenses, mirrors and telescopes. When he read of the "many charming discoveries" made by means of the telescope, he "wished to see the heavens and the planets with my own eyes through one of those instruments".[viii]

Caroline was to witness the transformation of her brother from musician to full-time astronomer. Not long after her arrival (Spring 1773), William bought some books, among them James Ferguson's popular *Astronomy Explained on Isaac Newton's Principles* which had an important influence on him. He acquired lenses from London to make small telescopes for looking at the planets. Caroline, too, was recruited to make tubes for them out of cardboard. William soon found that such instruments, using lenses mounted in long tubes, were difficult to handle, so he thought of making his own reflecting telescopes, following the description in Smith's book on *Optics*. Having acquired the necessary tools from a local amateur astronomer, William and Alexander soon taught themselves to grind and polish mirrors. By the summer, Caroline was dismayed to find "every room turned into a workshop", her "handsome drawingroom" occupied by a cabinetmaker, and the mechanically-minded Alexander erecting a machinery in a bedroom. The brothers had several mirrors cast for them, and later progressed to casting their own – a hazardous task for which they built a furnace for melting their metal alloys.

William Herschel's labours in casting mirrors and his skill in polishing and mounting them, constitute one of the most remarkable achievements in the history of astronomical technology. Grinding and polishing had to be done by hand, and while William worked (on one occasion he polished without pause for 16 hours), Caroline kept him amused by reading to him from the *Arabian Nights* or *Don Quixote*, and fed him by putting pieces of food in his mouth. At mealtimes he occupied himself making drawings or doing calculations. At bedtime, he retired with a bowl of milk and his favourite books on astronomy. There was no longer time for the extra music lessons that Caroline would have appreciated.

Extraordinarily, throughout all this activity, William Herschel continued to conduct rehearsals in the house, and to give recitals. He juggled his two lives: one pupil remarked that the room in which he had his music lessons resembled an astronomer's more than a musician's, with globes, maps and telescopes piled on top of the piano.[ix] Caroline carried out her own musical obligations conscientiously, while putting every spare moment at her brother's disposal. "I became in time as useful a member of the workshop as a boy might be to his master in the first year of his apprenticeship ... So we lived on without interruption".

In the course of a few years William became highly skilled in making beautiful mirrors of increasing size which he mounted himself. The activities of this unusual musician-astronomer, not surprisingly, attracted attention in the town. William Watson, a well-connected local medical doctor and Fellow of the Royal Society introduced Herschel to a group of other devotees of science in Bath and further afield. Visitors who called to see his instruments included the Astronomer Royal,

Nevil Maskelyne in person (Caroline was quite unaware of his identity), and Thomas Hornsby, professor of astronomy at Oxford University. Another visitor was Charles Blagden, secretary of the Royal Society. Alexander Aubert, an amateur with his own observatory became a special friend of Caroline as well as of her brother. These distinguished men recognised Herschel's remarkable talents, and became active friends, correspondents and encouragers.

One of a number of instruments that William Herschel made for himself in Bath was of 7 ft focal length with a mirror of 6.2 in. aperture. With this telescope he embarked on what he called a "review" of the heavens. Michael Hoskin, the leading authority on the life and work of both William and Caroline, points to the likely inspiration here of Ferguson's account of the universe of stars – "thousands and thousands of suns" at immense distances, accompanied by tens of thousand upon thousands of Worlds "all in rapid motion, yet calm, regular and harmonious" – to search for that harmonious structure.[x] His plan was to scan the entire sky systematically, initially for double stars (or, to be exact, pairs of stars that appeared close together in the sky, a line of enquiry that led to his discovery of real orbiting systems). On the evening of 13 March 1781, as he scanned his current area of sky, he came across a small object with a noticeable disk which he reckoned was either "a nebulous star or a comet". Four days later he returned to the strange object and found that it had moved – indicating, he thought, that it was a comet. His friend Watson, the first to whom it was shown, reported the discovery immediately to the Royal Society in London. When its motion was examined by expert mathematicians, it transpired that the object was not a comet but a planet, located beyond Saturn, the first new planet ever to be discovered in the history of mankind. It was a momentous discovery. It shot Herschel to fame overnight and earned for him not only Fellowship of the the Royal Society but its highly prized Copley medal.

Life was now hectic for the Herschels – as regards both astronomy and music. William did not interrupt his observing programme by night nor did he and Alexander slacken by day in the task of casting new and larger mirrors. "I saw nothing else and heard nothing else but about those things when my brothers were together", recalled Caroline. But Alexander had to look after his own musical affairs, so William had to call on his sister's help which was gladly given. While he laboured at the telescope at night she stayed up copying catalogues and scientific papers for him, and kept herself ready to lend a hand "when any particular measures had to be made or a fire kept in or a dish of coffee [provided]". She "undertook with pleasure what others might have thought a hardship".

Throughout, the Herschels contrived to fulfil their musical engagements. Easter 1782 saw them performing oratorios in Bath or Bristol every night of the week, followed after Easter by two full-scale performances of the Messiah. Whit Sunday was the last date in their musical diary, when one of William's anthems was sung at St Margaret's Chapel and he performed on the organ for the last time. Two days later, carrying with him his 7 ft telescope, he went to London on the advice of his scientific friends, to demonstrate his skills to the great and good.

It was the end of the Herschels' musical careers. Caroline, 32 years of age, had been just 10 years in that profession.

William's influential admirers in London looked for a way that would allow him to devote himself full-time to astronomy. His supporters included the most prestigious scientist in the land, Sir Joseph Banks, President of the Royal Society. Banks, distinguished botanist and naturalist explorer, had accompanied Captain Cook on his famous voyage of discovery to the South Seas in the ship Endeavour between 1768 and 1771. On that voyage he was in the company of the astronomers sent officially by the Government to observe the transit of Venus of 1769 in Tahiti. Banks led the move to find Royal patronage for Herschel. The outcome was a pension from King George III of £200 a year. "Never bought monarch honour so cheap", was Watson's reaction to this figure,[xi] though William himself did not disclose the sum to anyone else. It was by some standards reasonable enough – the Astronomer Royal's salary was £300 – but the realistic Caroline who was responsible for the household purse was of the same mind as Watson.

In London Herschel demonstrated his telescope at the Royal Observatory at Greenwich, and was then summoned to repeat the demonstration at the Royal Palace at Windsor Castle (near London, where the Royal family usually spent weekends). He showed the heavens in a clear sky to the King and his entourage. The following evening it was the turn of the young princesses, but as the sky was cloudy, Herschel, well prepared for this eventuality, was able to show them an artificial Saturn made out of card which he pinned on the garden wall. The ladies were much pleased with what they saw. Sir Joseph Banks suggested that it would be a gracious gesture to dedicate the new planet to the King. This Herschel did, giving it the name Georgium Sidus (the Georgian Star). That name was used for a long time in Britain, and always by Herschel himself and by Caroline. The continental astronomers preferred a traditional classical name, and eventually the name Uranus, proposed in Germany, became universal.

In his new capacity as King's astronomer, Herschel was expected to reside near the Royal residence and to be available to show the stars to the Royal family and their guests. He rented a house with a suitably large garden in which to erect his telescopes at a place called Datchet not far from Windsor. At the end of July 1882 he arrived back in Bath. Everything was settled. Within days the family moved to their new home. Caroline had not been consulted, and the shock of leaving her old life was all the harder when she realised that Alexander would not be able to stay with them permanently. Until then she "had not had time to consider the consequence of giving up the prospect of making myself independent by becoming (with a little more uninterrupted application) a useful member of the musical profession. But besides that my brother William would have been very much at a loss for my assistance, I had not spirit enough to throw myself on the public after losing his protection".

William, obsessed with astronomy and full of energy, did not waste much time before directing Caroline into useful work. "I found I was to be trained for an assistant-astronomer", she recalled with more than a hint of resentment in her old age. "By way of encouragement" she was given a little telescope with which she was to scan the sky and make a note of unusual objects,[xii] her first feeble attempt being made within weeks of the move from Bath. On the domestic side, the house she took over was in an utterly neglected state, and the garden overgrown with weeds. She could not find a servant and prices of goods in the shops appalled her. Yet William expected her to work nights as well! No wonder she was "anything but cheerful". The Herschels lived and worked through some exceptionally cold years in this uncomfortable place, where every room in the house leaked the rain, before moving, first to Old Windsor, and then to their final home at Slough where William spent the rest of his life.

Caroline was instructed to record all unusual objects – double stars, clusters (crowded groups of stars) or fuzzy nebulae – and to note their positions with respect to nearby recognisable stars. She would then check them against existing catalogues to see if they were already known. The objects that were to yield the greatest rewards were the nebulae. Nebulae were frequently mistaken for comets, and for this reason the French astronomer Charles Messier, the world's most successful discoverer of comets at the time, compiled a list of diffuse objects which would help observers distinguish genuine comets from permanent nebulae. His famous catalogue of 103 objects (nebulae and clusters) was published in 1781. The objects in this useful list are still referred to by the letter M and their catalogue number. Messier's catalogue was obviously by no means exhaustive. Caroline's first discoveries, made in February and March 1783 with her small refractor, were four previously unknown nebulae and a star cluster. One can imagine her satisfaction when she could record in her notes: "Messier has it not".[xiii] It was an excellent start; Michael Hoskin, who has studied the observing notes of both Caroline and William, finds that William was impressed by his sister's findings, and paid her "the sincerest form of flattery" by commencing his own sweeping of the sky for nebulae. He also decided that Caroline deserved a better instrument. In 1782, after just a few months with her original little refractor, William made her a small reflector specially designed for the job. It had a 4 in. (10 cm) diameter mirror and a focal length of 27 in. (69 cm) in a Newtonian mounting, which meant that the observer viewed through an eyepiece close to the top of the instrument. Its field of view was relatively wide (over 2°), suitable for searching the sky for unknown objects. Caroline called it her "sweeper" (its optics are preserved in the Museum in Hanover), and she used it in the same manner that William used his large instruments, that is, she combed the sky in vertical scans or sweeps. The telescope was cleverly mounted in a frame on a stand, with strings attached for easy vertical movement, and in such a way that she could operate it conveniently while sitting in a chair. Perhaps she would sometimes have a quick sweep before William started observing, because, when he was working, she was "attached to the writing desk". Some years later he made her a bigger one. However, her brother's programme always took precedence over her own.

Caroline discovered her first comet on 1 August 1786, while her brother was absent in Germany (Fig. 3.2). She recorded in her notes: "This evening I saw an object which I believe will prove tomorrow night to be comet". Next evening she could write: "The object of last night *is* a comet". She wrote immediately describing its position and appearance to her amateur friend Mr. Aubert and also to Dr. Blagden at the Royal Society asking him to inform her brother's astronomical friends. The announcement came at a fortunate time: it was Visitation Day at the Royal

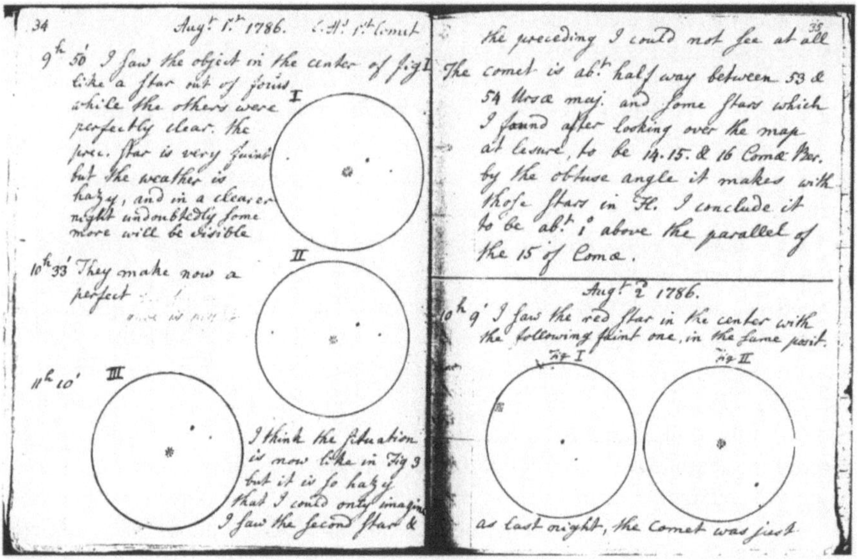

Fig. 3.2 Caroline's observation and drawing of her first comet, 1 August 1786. (Royal Astronomical Society)

Observatory (the annual meeting when a report on the year's progress was presented to a distinguished committee), and Caroline's discovery was thus made known to the country's leading astronomers. Blagden furthermore spread the tidings to France and Germany.[xiv] A few evenings later, he and Sir Joseph Banks, President of the Society, came to Slough to see the comet for themselves through Caroline's instrument. "You have immortalised your name", wrote Aubert, who had by now picked up the comet in his own telescope.

When William returned, he was summoned to Windsor by the King to show the comet to him and the Royal Family. Fanny Burney, the writer, who was at that time lady-in-waiting to the Queen, had the good fortune to be present and to be included in the demonstration. To her eye the comet had "nothing grand or striking in its appearance", but it had the distinction of being "the first lady's comet". On that occasion Herschel also showed the company some of his "new discovered universes" – the nebulae.[xv] It is interesting that Caroline was not of the party: William, not she, was the King's astronomer.

William's work was done mainly with the larger of his 20-ft telescopes which was mounted on a wooden frame facing due south like a transit instrument, and movable in altitude. It produced an image at the upper end of the tube where the observer (William himself) sat aloft on a platform. To "sweep" the sky, a workman, using pulleys, moved the telescope tube up and down vertically by a few degrees while the observer had time to note the interesting objects in that particular strip of sky. The slow rotation of the earth brought new strips of sky progressively into view, each to be examined in the same way.

The procedure was rendered faster and more efficient when Caroline joined in. She kept tally of which parts of the sky had been swept, and which remained to be examined. She sat at a desk in a hut at ground level, with a star chart in front of her and a sidereal clock in view, and recorded her brother's observations as he dictated them. Knowing the position in which the telescope was pointing, and the sidereal time, she knew effectively where the object was, and could locate its position on the chart. Also beside her for reference was Messier's list. On her desk were sheets of paper ruled into squares representing quarters of a degree of sky, and as each area of sky was observed, she marked it off with a cross.[xvi] Her brother did not need to take his eye away from the telescope, and his assistant Caroline was able, in her own words, "to execute his commands with the quickness of lightning".

Next day, after each night's observing, Caroline copied out the observations in a fair hand, keeping double stars and nebulae in separate "Register Sheets". A stabling in the grounds had been converted into a little apartment for her where she had her own study and where she kept her own telescope and her records. It had a flat roof, reached from inside, where she could "mind the heavens", as she called it, when not needed by her brother.

William's ambition from the moment he received his Royal appointment had been to construct a very large telescope to probe as deeply as possible into space, for which he received a grant from the King. The construction of the giant telescope was a formidable task. The grounds of the house came to resemble a factory. Outbuildings were converted into workshops and optical rooms, with a large staff of workers. The 40-ft long telescope with a mirror of 48 in. was completed and erected in 1789 (Fig. 3.3). Mounted on a huge wooden structure, with ladders and platforms, the great instrument was the largest and most powerful in the world, and a massive landmark which travellers looked out for as they passed on the London coach. The observer William sat aloft in a cage. An extra ingenious element was a speaking tube through which he could communicate his observations with his sister without raising the voice.[xvii] "Their manner of working together is most ingenious and curious", wrote their friend Fanny Burney, one of many visitors who described them in action. "While he makes his observations out of doors, he has a method of communicating them to his sister so immediately, that she can instantly commit them to paper, with the precise moment at which they are made".[xviii]

Herschel published his first catalogue of 1,000 new nebulae and clusters in 1786 and the last in 1802. Altogether, over a period of 20 years, with Caroline's assistance, he recorded, numbered and classified 2,510 objects. Those among them already found by Caroline were credited to her by the initials C. H. The most interesting of these, later admired, were the companion of the Andromeda Galaxy – a dwarf elliptical galaxy (NGC 205 in the later New General Catalogue) – and a beautiful spiral in Sculptor (NGC 353) which her nephew John Herschel bracketed with the Andromeda Galaxy as the "largest and finest" of its kind.[xix]

William Herschel's marriage, at the age of 49 in 1788, came as a heart-wrenching shock to Caroline who had been his housekeeper and companion for the past 15 years. Caroline moved out of the Observatory house. She later destroyed her private journals covering that period of her life; but she bottled up her feelings and

Fig. 3.3 William Herschel's 40-ft telescope with huts for Caroline and the mechanic. (Royal Astronomical Society)

continued to work, being now, as Patricia Fara puts it, "conditioned to servitude".[xx] In the course of time she warmed to her graceful and gentle[xxi] sister-in-law, especially after the birth in 1792 of John, the only child of the marriage, whom she adored and whose progress she proudly followed until the end of her life.

The star catalogue used by Herschel for the surveys was that of John Flamsteed, the first Astronomer Royal. That great catalogue, also known as the British Catalogue, contained the positions of 3,000 stars observed by Flamsteed in the course of many years' labour at the newly founded Royal Observatory at Greenwich. The final version, published only after his death by his devoted wife in 1725 (Chapter 1) was the accepted authority on star positions by Herschel and his contemporaries.

However, after some years' experience, Herschel suspected that the catalogue was not "faultless" after all. There were numerous mistakes which could not all be due to genuine variations in star brightnesses. Some stars appeared to be missing, some recorded in places where there were none. It was essential, he thought, that the catalogue should be rectified by checking against Flamsteed's original observations which had been published in a separate volume. He recommended the "arduous task" of tracking down the errors to his sister. Caroline completed the work in 20 months. She found 560 stars observed but not catalogued, and 111 catalogued but not observed.[xxii] She arranged these in "An Index to Mr. Flamsteed's Observations of the Fixed Stars" which William communicated on her behalf to the Royal Society in 1797. The Index was published at the expense of the Royal Society in 1798. She performed a similar check on a catalogue of circumpolar stars by another observer, Francis Wollaston, and produced a cross reference between his omitted stars and Flamsteed's. Her Index, involving many hundreds of stars, was of great practical use to astronomers at Greenwich and elsewhere. Her corrections were incorporated in celestial charts: for example, a beautiful celestial globe produced by John and William Cary in 1899, records her name on its label.[xxiii]

In 1791, William made Caroline a bigger sweeper. It had an aperture of 9 in. (23.4 cm) and a focal length of a little over 5 ft (160 cm). Between her two instruments, she discovered eight comets in the course of 11 years from 1786 (of which three were not first observations).[xxiv] On each occasion she alerted Maskelyne, the Astronomer Royal (usually by letter, though she once rode on horseback all the way to Greenwich to bring him the news), and one or two other scientist friends, leaving the matter of the comet's subsequent movement to them. Maskelyne, in turn, spread the tidings, never failing to inform the specialist comet observer, André Méchain of Paris.

Of all the comets discovered by Caroline, the most interesting was that which she found in 1795, later known as Encke's comet. That famous object, only the second instance of a periodic (i.e. recurrent) one (Halley's was the first) has a period around the Sun of only 3.3 years, but its reappearances were not recognised until the talented astronomer and mathematician, Johann Encke at Berlin, calculated its exact orbit and correctly predicted its return in 1822. "Your" comet, Caroline's nephew John called it, when he informed her that it had been observed from South Africa in 1836.[xxv] It continues to be a regular visitor to our Earth's neighbourhood.

Caroline's fourth comet, of January 1790, was celebrated by an admirer, the French astronomer Jerôme de Lalande, who bestowed the name Caroline on his infant god-daughter. Lalande, of the French Academy, had visited the Herschels in 1788 and was ardently devoted to both brother and sister. Caroline discovered her eighth and last comet on 14 August 1797. It was commemorated 25 years later in a poem by Mary Shelley, author of the famous story, Frankenstein, who was born in the very month of its appearance: "And thou, strange star, ascendant at my birth…". It had been seen as a good omen by her parents, Mary Wollstonecraft, early champion of women's rights, and her husband William Godwin, as they awaited her birth. Tragically, the mother died a few days later as a result of childbirth complications and medical ignorance.[xxvi]

It is a tribute to Caroline's gift as an observer that, despite being restricted by her brother's prior needs, she made so many discoveries. Her experience at his side had its advantages. It made her completely familiar with the heavens. She knew every object in Messier's catalogue and all the brighter stars by sight, and could quickly spot an interloper. In the comet hunting league, counting to the end of the eighteenth century, she took third place after her two famous French competitors, Messier and Méchain, with 14 and 10 each respectively.[xxvii] There was no jealousy among them, however. Mechain, who met Caroline when he visited Slough, had the greatest admiration for her. "Sa celebrité sera en honneur dans tous les siecles, mais ceux qui ont l'honneur de la connoitre savent combien ses respectables qualites du coeur ajoutent a son metier"[xxviii] As a comet-huntress, Caroline "enjoyed the chase and prized her comets as trophies", but took no interest in their scientific significance.[xxix]

To discover a comet brings celebrity. However, Caroline was known to her brother's astronomical associates well before her first discovery of 1786. Since the early days in Bath, visitors who came to see William and his telescopes were sure to meet Caroline as well. She knew the Astronomer Royal Nevil Maskelyne, and the distinguished amateurs William Watson and Alexander Aubert, from the very beginning of her brother's scientific career. All became lifelong friends. Similarly she knew the Royal Society luminaries, its President Sir Joseph Banks and Secretary Sir Charles Blagden – acquaintances of the highest possible prestige.

These astronomers treated Caroline respectfully as a professional colleague (and also perhaps, Dr. Hoskin surmises, as something of a curiosity). Maskelyne's response to the announcement of her second comet in 1887 was a report of his own observations of it, its coordinates and its motion, and his hopes of tracking its orbit. He went on to tell her how useful her positional observations were likely to be in this regard. "You cannot, my dear Ms. Herschel, judge of the pleasure I feel when your reputation and fame increase; everyone must admire your and your brother's knowledge, industry and behaviour", wrote Aubert after Caroline informed him of her comet of 1890. "I am always happy to hear from you" wrote Sir Joseph Banks on the same occasion, "but never more so than when you give me an opportunity of expressing my obligations to you for advancing the science you pursue with so much success". When she discovered her last comet in 1897, in her brother's absence, and her customary message to Banks was delayed, her words reveal how natural was her relationship with this great man. "This is not a letter from an astronomer to the President of the Royal Society announcing a comet", she wrote, "but only a few lines from Caroline Herschel to a friend of her brother's, by way of apology for not sending intelligence of that kind immediately where they are due".

Maskelyne in particular emerges from his letters to Caroline as a person of great charm whose wife and only daughter also held her in warm affection. "My worthy sister in astronomy", he called her. Her *Index to Flamsteed's Catalogue* was of practical interest to him, while she appreciated his part in having "the little production of [her] industry" published. "I thought the pains it had cost me were, and would be, sufficiently rewarded in the use it had already been, and might be in the future, to my brother. But your having thought it worthy of the press has flattered my vanity not a little. You see, sir, I do own myself to be vain, because I would not wish to be singular; and was there ever a woman without vanity? or a man either? only that with this difference, that among gentlemen the commodity is generally styled ambition". Only a close friend could banter in this way with the Astronomer Royal.

Caroline was invited many times to stay with the Maskelynes at Greenwich. She at last was induced to accept. She had made some further amendments to the Index which she wished to enter in the Greenwich copy of the Flamsteed Atlas. She spent a week with the Maskelynes at the Royal Observatory in the summer of 1799; but with so many excursions and diversions arranged for her, she did not manage to complete her mission. They took her up to London and at Christmas sent her a present of a pair of binoculars. The Maskelynes' little daughter Margaret (Fig. 3.4), though not more than about 10 years old, accompanied her parents everywhere. Caroline's journal records other occasions in their company, both in her brother's home, and in the homes of other friends.

In fact, social life was not lacking at Slough. Foreign astronomers came to meet the celebrated Herschel and to see the giant telescope. Caroline in her diary mentions some of them: Giuseppe Piazzi of Sicily, discoverer of the first minor planet in 1801; Jean Dominique Cassini, Director of the Paris Observatory; André Méchain, her fellow comet-hunter; Marc August Pictet, renowned Swiss physicist. There was also Jerôme de Lalande of Paris who addressed her as "astronome célèbre", and the German astronomer Karl Felix Seyffer who referred to Caroline as "the noble

Fig. 3.4 Margaret Maskelyne, daughter of the Astronomer Royal, aged 10. The portrait shows the Royal Observatory in the background. (Vanda Morton, *Oxford Rebels*)

priestess of the Temple of Urania". These distinguished guests stayed at the Observatory House with William and his wife, where Caroline would meet them. There were family musical gatherings (presumably on cloudy nights) when William played the piano, though Caroline does not appear to have joined in music making. Her life ought not to have been entirely gloomy.

One non-scientific friend was Fanny Burney. Fanny's father, Dr. Charles Burney, was a distinguished historian of music who on that account was a good friend of William Herschel. He often visited the Herschel household at Slough. Among his undertakings, Dr. Burney composed verses on the subject of the stars which William, ever gracious, had the patience to listen to.[xxx] Fanny had met William at Windsor when he demonstrated Caroline's first comet to the Royal family; she met Caroline for the first time when taken to Slough by the famous literary hostess Mrs. Delany. She found her "very gentle, very modest", "unhackneyed and unawed by the world". She described her brother as "a delightful man, so unassuming, with his great knowledge, so willing to dispense it to the ignorant, and so cheerful and easy in his manners." It was the same charming, unaffected, impression that all members of the Herschel family made on those who met them. William was at that time involved in observing what he thought was a volcano on the Moon, but "left to his sister to sweep the heavens for comets".[xxxi] "Their own pursuit", Fanny remarked, was "all-sufficient to them".

As the King's astronomer it remained a duty on William Herschel to entertain the Royal family or their guests when requested. This was accepted as an honour and a pleasure, but Caroline was "unawed" (as Fanny Burney put it) by celebrities, even royal ones. Her diary contains unembellished entries such as: "Prince Charles (Queen's brother), Duke of Saxa Gotha and the Duke of Montague here this morning. I had a message from the King to show them the instruments"; "The King had been at the Observatory"; "The Prince of Orange stepped in to ask some questions about planets, etc." On one occasion, when the elderly William was away from home recouping his health, Caroline and the 15 year-old John found "the Duke of Kent, with the Dukes of Orleans, etc." waiting to be shown round; so they showed them "Jupiter, the Moon, etc. in the 7 ft". The Duke of Cambridge (one of the Royal princes) once came in to shelter from the rain. There were special occasions, such as the visit of the Archduke Michael of Russia "with a numerous attendance", and many visits of members of the German aristocracy. A famous Royal visit took place in 1787, before the tube of the great 40-ft telescope was hoisted into place, when one could walk right through it. Among those who had that experience was a royal party and the Archbishop of Canterbury. Caroline recalled it in old age: "The Archbishop, following the King and finding it difficult to proceed, the King turned to give him a hand, saying, 'Come, my Lord Bishop, I will show you the way to Heaven' ".[xxxii]

The royal connection provided another, happy, element in Caroline's life. It introduced her to the ladies of the household of the amiable Queen Charlotte and to the charming royal ladies themselves. One of the Queen's German entourage, Mrs. Beckedorff, had been a classmate of Caroline's at the dressmaking school that she attended as a young girl in Hanover. Caroline often visited her at Windsor Castle and would sometimes meet the young princesses (there were six of them, most still in their twenties). Her diary records an evening when, after dinner with Mrs. Beckedorff, the Princesses Augusta and Amelia and their brother the Duke of Cambridge came to see her; another, when the Queen and Princesses Elizabeth, Augusta and Mary dropped in. On a trip to London, to visit her friend at Buckingham House, she met and conversed at length with the Queen and three of the princesses. Caroline was a favourite of Sophia, known to be "a very clever, well-informed woman", though destined to an unhappy life,[xxxiii] who invited her to the Castle for long conversations and borrowing of books.

When Queen Charlotte died in 1818 her household was dispersed (the King himself died in 1820). Mrs. Beckedorff returned to Hanover, where Caroline, in retirement, was to renew their friendship.

Caroline's comet-searching efforts diminished in later years. For reasons that are not clear, she was obliged to move her lodgings (which were not always congenial) several times.[xxxiv] Her observing came virtually to an end.[xxxv] She refers more than once to her "solitude" and described the last months of the year 1813 as being "spent mostly in solitude at home, except when I was wanted to assist my brother at night or in his library". Her nephew John took the Tripos examination at Cambridge with highest honours in 1814, and after some hesitation decided in 1816 to devote himself to continuing his father's work at Slough. Caroline may have felt thereafter less wanted; she noted in her diary: "Nothing particular happened, my nephew remaining at home working with his father, and I took the opportunity of working on my mss Catalogue at those times when I was left without employment".

Her main anxiety, however, was her brother's health. She put the beginning of his decline down to his exertions in October 1807 when he was overwhelmed by visitors who came to see a comet (not one of Caroline's). One evening in particular, having laboured all day with his team of workers polishing the 40 ft mirror, he stood "from dark until midnight on the grass plot surrounded by between 50 and 60 persons, without having had time for putting on proper clothing, or for the least nourishment passing his lips". The thoughtless throng included royal and society folk, officers and ladies. William was in his 70th year. From that time onwards, Caroline's journal is a heart-rending record of her increasing worries, though they both kept working, and William continued to publish papers until 1817, when he was almost 80.

William Herschel died on 25 August 1822. Caroline, at the age of 72, felt that her life's work was done. She had been, as Michael Hoskin put it, "his photocopier, his word processor, his calculator, his collaborator at the telescope".[xxxvi] She decided to return to her native Hanover and live with her youngest brother Dietrich, the last remaining sibling, and his family. She packed her books and her small "sweeper", sold her few belongings, and travelled in October with Dietrich who came to fetch her. Before departing she visited the royal princesses and took "an everlasting leave" of her own and her brother's special friends. The last, parted from with just a pressure of the hand, was her nephew's friend and her dear brother's favourite, Charles Babbage.

Caroline's brother and family welcomed her, but their company was very different from what she was used to in Slough. Dietrich's health was poor – he died in 1827 – and his wife's interests (playing cards and gossiping with her cronies) were not hers. The disenchanted Caroline confessed to her nephew in 1826: "From the first moment I set foot on German ground, I found I was alone...". Her thoughts turned constantly to the family she left behind, and she cherished the affectionate letters from her nephew John and his mother. In 1829 another correspondent was added, when John married his young wife Margaret, a wonderful letter-writer who regaled their aunt with cheerful tidings of their family of little ones. The many dozen Herschel letters, still preserved, are an extraordinary chronicle of unalloyed family devotion.

Though Caroline was "alone", in the sense of missing her life in England, she had in fact numerous friends and well-wishers in Germany. Her niece Anna became very close to her, and she had once again the company of Mrs. Beckedorff who was back in Hanover with the Royal household. The Royal persons in Hanover included the Duke of Cambridge, son of King George III, whom she knew in England, who showed her many favours and clearly enjoyed her company. She was one of the revered institutions of the city. She went to concerts, and received visits. She kept herself cheerful and lively, and above all, kept her mind occupied and her astronomy exercised.

Among the papers carefully kept by Caroline were the working records of the observations which provided William's catalogues of nebulae, in which each nebula's position was given relative to a nearby star. In 1799 William had asked her to re-order these reference stars by zone of north polar distance, a task she finished in 1818. When she settled down in Hanover she decided as her "winter's amusement" to re-order the nebulae in the same form, which would make it more convenient for a future observer to use. She divided the sky into zones of polar distance, and listed the nebulae in order of right ascension (one of the co-ordinates) within the zones. The task involved 2,500 sets of calculations. Her Zone Catalogue was completed in 1825 and sent to John. He was about to begin his planned re-observation and extension of that same catalogue of nebulae, and Caroline's Catalogue provided him with the information he would need when sweeping to check every one of William's nebulae. This major task occupied him at Slough until 1833. The Zone Catalogue, in Michael Hoskin's judgement, was arguably Caroline's greatest achievement.[xxxvii] It was John's intention that it should be eventually published, but regrettably that never came about. However, her contribution was recognised by the Royal Astronomical Society which awarded her its Gold Medal in 1828. She was the first woman to receive a prize from that all-male society, and indeed the only woman to be honoured with a prize until Jocelyn Bell Burnell, discoverer of pulsars, was awarded the Herschel medal in 1989.

John Herschel was at the time President of the Royal Astronomical Society and the address on the occasion was left to the vice-President, Sir James South. South paid an eloquent tribute to Caroline before asking John to accept the medal on his aunt's behalf.[xxxviii] John informed Caroline that the decision to award the medal had been taken by the society's Council, lest she should think that he had engineered it ("It was none of my doings. I resisted strenuously. Indeed, being in the situation I actually hold [i.e. the Presidency], I could do no otherwise".); but unfortunately she completely misunderstood his meaning (or perhaps chose to misunderstand it).

"I felt from the first more shocked than gratified by that singular distinction ... And the little pleasure I felt ... was entirely owing to the belief that what was done was both with your approbation and according to your recommendation. Throughout my long-spent life I have not been used or had any desire of having public honours bestowed on me". She was embarrassed by the Vice President's praises: "Whoever says too much of me says too little of your father". John – who was probably well familiar with his old aunt's prickliness – assured her that it was a well-merited distinction, most honourably conferred and agreed by all. South's speech was quoted in the Royal Irish Academy's notice of Caroline's death. "We scarcely know", he had said, "whether most to admire the intellectual power of the brother, or the unconquerable industry of the sister."[xxxix]

Caroline Herschel's lively writings during her retirement are a unique source of information on the extraordinary Herschel family, on William's career and on John's rise to eminence. Encouraged by John and his wife she recorded her memoirs, and wrote delightful letters to them, witty, self-deprecating and uninhibited. She was completely lacking in pomposity – as were all the Herschels – but had seen enough of the world to take an occasional ironic view of its inhabitants. Her memory was phenomenal. She recalled details not only of her youth, as many older people do, but of her entire life, especially anything that impinged on William's astronomical activities.

Caroline was as well known among astronomers in Germany as she had been in England. She received the astronomical newsletter *Astronomische Nachrichten* regularly from its founder and editor Heinrich Christian Schumacher, and acted as a link between John and such colleagues as Encke, Olbers, Bessel, and Gauss – all scientists of the highest rank – for whom he would send scientific papers, or items of news for her to distribute.[xl] Acknowledging one such delivery of a parcel from John, Encke – who had deciphered the orbit of one of Caroline's comets – took the opportunity of paying his "respects to a lady, whose name is so intimately connected with the most brilliant astronomical discoveries of the age, and whose claims to gratitude of every astronomer will be as conspicuous as your own exertions for extending the boundaries of our knowledge, and for assisting to develop the discoveries by which the name of your great brother has been rendered so famous throughout the literary worlds". Gauss, brilliant mathematician at Göttingen, who visited her in Hanover and to whom she gave an annotated copy of Flamsteed's Atlas and her own Catalogue, declared that these would be considered "the greatest ornament of the library of the Observatory". "I cannot express", he wrote, "how much I feel happy of having made the personal acquaintance of one whose rare zeal and distinguished talents for science are paralleled by the amiability of her character".

John Herschel, accompanied by his family, spent 4 years (1834–1838) in South Africa surveying the southern heavens. Caroline took a great interest in the project. If only she were 30 or 40 years younger, would she not go too! He reported many interesting finds to her – globular clusters, nebulae, the Greater Magellanic Cloud, with drawings. She was also interested in the satellites of Saturn, a favourite object of William's; John duly sent her a full account of his observations, complete with sketches and co-ordinates.[xli] The fascinating letters between John Herschel at the Cape and his aunt reveal Caroline in her eighties as the experienced old hand.

In 1835 the Royal Astronomical Society made Caroline, aged 85, an Honorary member. To elect women as Fellows of the Society would have contravened its statutes; the Council therefore devised a new rank, Honorary member, specifically as an honour for women, and conferred it for the first time on Caroline Herschel and Mary Somerville (Chapter 6). The election put her in touch with Mrs. Somerville, aged 55, friend and protegée of John, who sent her a copy of her book, *Connexion of the Physical Sciences*, accompanied by a reverential note expressing her honour at being bracketed with a person of such distinction. Caroline reciprocated with similar compliments.

A few years later, in 1838, Caroline was elected an Honorary member of the Royal Irish Academy on the nomination of the President, Sir William Rowan Hamilton, who sent the formal Diploma to her nephew to transmit to her. John, a good friend of Hamilton's, was himself since 1824 an honorary Member. Hamilton's accompanying letter to Caroline expressed his own "high sense of your services to astronomy, and of the eminent degree in which you have deserved the present testimonial". She requested her nephew to reply to Hamilton on her behalf. "It has given me pleasure", she wrote to John, despite "having lived 18 years without discovering so much as a single comet".

Caroline's last honour was a Gold Medal from the King of Prussia, to mark her 96th birthday in 1846, "in recognition of the valuable services rendered to Astronomy by you, as the fellow-worker of your immortal brother, Sir William Herschel, by discoveries, observations, and laborious calculations". It was transmitted to her by the 77 year-old Baron Alexander von Humboldt, the famous naturalist and traveller and the revered doyen of German science.

The short letter in which Caroline informed John's wife of the medal was one of the last from her own hand. Thereafter her niece wrote her letters for her. The final important event in her astronomical life was the delivery of her nephew's monumental "Cape Observations", the result of his four years of observing in South Africa. In this catalogue, John had extended to the southern hemisphere of the sky the work done in the northern hemisphere by his father and aunt in Slough. The entire sky, from pole to pole, had now been explored by the Herschel family. In sending the volume to her, in July 1847, John wrote: "You will then have in your hands the completion of my father's work – "The Survey of the Nebulous Heavens".[xlii] No gift could have been more appropriate.

Caroline died peacefully in her 98th year on 9 January 1848. She was buried next to her parents in the Gartenkirchhof in Hanover. Her grave is marked by a flat stone engraved with a moving epitaph of her own composition.

The epitaph on the grave is in German, but, as Michael Hoskin has discovered, Caroline's original version was written in English. (The English version found in the Herschel Chronicle[xliii] is a translation from the German by Constance Lubbock who mistakenly thought it was the original). Caroline's unpublished original version (which had spaces for data of her death), generously made available by Michael Hoskin, is quoted here.[xliv]

"Here rest the earthly remains of Caroline Herschel, born at Hanover, March 16th 1750, died January 9th 1848. The gaze of the deceased while here below, was turned to the starry heavens; her own discovery of comets and her share in the immortal labours of her brother, William Herschel will testify hereof to future generations. The Royal Irish Academy of Dublin and the Royal Astronomical Society of London counted her among their members. At the age of 97 years 10 months she fell asleep in calm and peaceful possession of all her powers of mind, following to a better world her father Isaac Herschel who after attaining the age of 60 years 2 months 17 days was buried at this place 25th March, 1767."

In this short epitaph, Caroline defines the achievement and essence of her life. She remembers the two people whom she loved most and to whom she owed most, her father and her brother. (It is sad and a little shocking, however, that no mention is made of her mother Anna, who lies in the same grave.) She recognises the scientific work of her brother as immortal, her own share being complete submission to his wishes. She was proud of her cometary discoveries; and though in life she disowned "any desire of having public honours bestowed" on her, it is clear that in her heart she appreciated the recognition of the world of learning, and well deserved it.

She has a unique place in the history of astronomy.

Notes

[i] Michael Hoskin. 1963. *William Herschel and the Construction of the Heavens*. London: Oldbourne.

[ii] Michael Hoskin. 2003. *The Herschel Partnership*. Cambridge: Science History Publications. p 5. Dr Hoskin is the foremost Herschel scholar and expert on the lives and work of William and Caroline. This brilliant work is an indispensible source of enlightenment of Caroline's achievements and personality.

[iii] Mrs John Herschel. 1879.*Memoirs and Correspondence of Caroline Herschel*. London: John Murray. (Reprinted by the Herschel Society, Bath, 2000). Unattributed references to Caroline's life that follow come from these memoirs, hereafter *Memoirs and Correspondence*. Caroline's memoirs have been edited and published in full by Michael Hoskin 2003.*Caroline Herschel's Autobiographies*. Cambridge: Science History Publications.

[iv] Michael Hoskin. 2005. Caroline Herschel as observer. *Journal of the History of Astronomy* **36**, 373–406.

[v] ibid.

[vi] ibid.

[vii] Constance A. Lubbock. 1933. The *Herschel Chronicle*. p 35. Cambridge University Press. (Reprint edition, Herschel Society, Bath 1997). Hereafter *Chronicle*.

[viii] Hoskin. 2005.

[ix] Lubbock. *Chronicle*. p 73.

[x] Hoskin. 1963. p 23.

[xi] *Memoirs and Correspondence*. p 57.

[xii] Hoskin. 2005.

[xiii] Hoskin. 2005.

[xiv] *Memoirs and Correspondence*. 63–68.

[xv] Lubbock. *Chronicle*. p 169.

[xvi] Henry C. King. 1979. *The History of the Telescope*, p132. New York: Dover (reprint edition).

[xvii] Lubbock. *Chronicle*, p159.
[xviii] Frances Burney. 2001. *Journals and Letters*. p 252. London: Penguin Books.
[xix] John Herschel. 1887. *Outlines of Astronomy*. p 643. London: Longman Green.
[xx] Patricia Fara. 2004. *Pandora's Breeches. Women, Science and Power in the Enlightenment*. p 161. London: Pimlico.
[xxi] Lubbock. *Chronicle*. p 177.
[xxii] A. M. Clerke. 1910. Article on Flamsteed. *Encyclopaedia Britannica*, 11th edition.
[xxiii] A beautiful example of this globe was on show at an exhibition of Treasures of Edinburgh University in 2003. It is labelled: Celestial Globe made by J & W Cary, 1799, and inscribed: "Cary's new & improved Celestial Globe on which is carefully laid down the whole of the stars and the nebulae contained in the astronomical catalogue of the Rev. Mr. Wollaston FRS, compiled from the authorities of Flamstead, de la Caille, Hevelius, Mayer, Bradley, Herschel, Maskelyne & with an extensive number from the works of Miss Herschel, the whole adapted to the year 1800."
[xxiv] David W. Hughes. 1999. Caroline Lucretia Herschel – comet huntress. *Journal of the British Astronomical Association* **109**, 78–86. The designations and dates of these comets are given here.
[xxv] J. F. W. Herschel. 1836. Letter to Caroline Lucretia Herschel. In David S Evans, Terence J Deeming, Betty Hall Evans, Stephen Goldfarb. 1969. *Herschel at the Cape. Diaries and Correspondence of Sir John Herschel 1834–1838*. p 192. Cape Town: A.A. Balkema.
[xxvi] Miranda Seymour. 2001. *Mary Shelley*. p 27. London: Picador. Mary Wollstonecroft's death is one of the saddest events in the history of British intellectual women.
[xxvii] Hughes, op. cit.
[xxviii] Lubbock. *Chronicle*. p 219. Letter of Mechain to William Herschel 1789.
[xxix] A.M. Clerke. 1895. *The Herschels and Modern Astronomy*. p 139. London: Cassell and Co. Ltd.
[xxx] Lubbock. *Chronicle*. 291–5, quoting extracts from Dr Burney's Diaries. The astronomical verses have not survived.
[xxxi] Fanny Burney, op. cit. *Journals and Letters*, September 1787. p. 252.
[xxxii] *Memoirs and Correspondence*. p 309, from a letter from Caroline to Lady Herschel, 1840.
[xxxiii] Flora Fraser. 2004. *Princesses, the six daughters of George III*. p386. London: John Murray.
[xxxiv] Hoskin. 2003. p115.
[xxxv] Hoskin. 2005.
[xxxvi] Michael Hoskin. 1998. Heavenly Siblings: the partnership of William and Caroline Herschel. (lecture to the Royal Astronomical Society). *Observatory* **118**, 260–61.
[xxxvii] I thank Michael Hoskin for pointing the important fact out to me that without the Zone Catalogue, John Herschel would not have been able to produce what was the basis of the famous NGC catalogue.
[xxxviii] The address is given in full in the *Memoirs and Correspondence*. 222–226.
[xxxix] Proceedings of the Royal Irish Academy **6**, 97–8, 1847–50.
[xl] *Memoirs and Correspondence*. p 373.
[xli] Evans et al. op. cit. 300–303.
[xlii] Memoirs and Correspondence. p 342.
[xliii] Lubbock's *Chronicle, Appendix III*.
[xliv] Caroline wrote two versions in English, this being the later one which was translated and which survives in Caroline's own hand. I thank Dr Hoskin for sending me a copy of this and for giving me permission to use it.

Chapter 4
The Art of Navigation

The problem of determining a ship's position at sea – that which the Royal Observatory was committed to solve – was essentially one of determining geographical longitude. Geographical latitude is, in principle, easily found. The altitude (height above the horizon) of the pole of the sky, is equal to the latitude of the observer. The celestial pole is that point which stands directly above the pole of the Earth and, conveniently for inhabitants of the northern hemisphere, this point is marked by an easily recognisable star very close by, Polaris (the Pole Star), which has been the seafarers' guide for centuries. This friendly guide also points to the north.

The Sun by day is capable of providing the same information. The direction of the Sun at noon – the moment when it is at its highest in the sky – marks the south. Its altitude at that moment depends on the time of year and also on the observer's latitude. The Sun's path in the sky in the course of the year has been well studied and the mariner is equipped with tables giving that information. To find the latitude from the Sun's noonday altitude was a matter of a simple piece of arithmetic. These were the basic methods of finding latitude, though in practice more elaborate observations were made in order to get that information. As nautical instruments became more refined, the accuracy of the result was improved.

Longitude, however, was a different matter. There was absolutely no way of finding longitude simply by looking at the sky. Longitude difference between two places on the surface of the earth shows itself as a time difference, brought about by the rotation of the Earth on its axis. The only way to find the difference in longitude between two places is to know the difference between their local times. No clocks existed in the seventeenth century capable of being transported from land to sea without losing track of time. Until such a clock was invented (as was achieved in 1759), the challenge to the astronomers was to find an alternative way of transmitting time across large distances, even across oceans.

The answer was an astronomical time signal: an event in the sky that could be seen simultaneously by the mariner at sea and the observer on land. Galileo who discovered Jupiter's satellites, had long ago suggested the timing of the movements of these bodies for this role. The most promising candidate, however, was our nearest neighbour, the Moon, which goes around the Earth once a month and appears to move through the starry sky like the hand of a giant clock at the average

M. Brück, *Women in Early British and Irish Astronomy*,

DOI: 10.1007/978-90-481-2473-2_4,

rate of 13° per day. If the mariner were provided with a pre-calculated list of the instances (Greenwich time) when the Moon reached certain recognisable points in its itinerary, he could compare these with his own local time (found by the Sun) and note the difference.

The astronomer who succeeded in putting this method into practice was the fifth Astronomer Royal in that office, Nevil Maskelyne (1732–1811),[i] appointed 40 years after the publication of Flamsteed's catalogue. Maskelyne has been called "the seaman's astronomer" on account of the practical improvement he brought to the science of navigation by providing tables of coordinates of the Sun, Moon and planets, calculated in advance. Of these celestial bodies, the Moon was that principally intended for the purpose of finding longitude; but though the Moon is the nearest body to the Earth, its orbit is exceedingly complex and its rate of movement through the sky is far from uniform. Since Flamsteed's time, a great deal of effort, by mathematicians as well as by observational astronomers, had gone into recording and unravelling its orbit. It was only in 1756 that tables were published (by a German astronomer and mathematician, Gustav Mayer) from which the Moon's precise position could be worked out for dates into the future.[ii]

Maskelyne, a Cambridge mathematician and an experienced astronomer, had observed the Transit of Venus of 1761 from the island of St Helena, where he had been officially sent by the Royal Society. On this voyage he had made a study of the problems of navigation and – though not the very first to use it – had tested, successfully, what was known as the method of "lunar distances", to determine longitude.[iii] The method entailed measuring the angular distance on the sky between the Moon and a star from a list of selected stars with known coordinates. After his return Maskelyne published *The British Mariners Guide* (1763) which contained in convenient form the information needed for the computations. As Astronomer Royal, he continued to publish this guide annually under the title *Nautical Almanac*. The first volume with data for the year 1767 came out in 1766, one year after he took office. The soon famous *Nautical Almanac* also supplied the coordinates of the planets in advance and was useful for observers on land as well as on sea. It was published annually until 1960 when it was united with the American equivalent publication to continue as the *Astronomical Ephemeris*.

The preparation of the *Nautical Almanac* was no trivial task. The path of the Moon does not repeat itself month by month or year by year, and its position in the sky had therefore to be calculated day by day for each year, making use of the theoretical tables of its motion and auxiliary mathematical and trigonometrical tables. In the case of the planets, their orbits around the Sun had been worked out, but to find their positions at a future date as seen by an observer on the orbiting Earth involved a substantial amount of computation. Maskelyne employed human "computers" to carry out these calculations, and worked out routines for them to follow. His computers were educated people like clergymen or schoolmasters, who worked in their own homes. The Almanacs were prepared three or four years in advance, and four computers were usually needed at any one time to keep up with the job. Each set of computations was carried out independently by two people and cross-checked by a "comparer". Among those employed by Maskelyne in the course of many years was one woman, Mary Edwards (c. 1750–1815), whose name, but without further details, appears in the history of the Royal Observatory.[iv]

The title of Britain's first professional woman astronomer is usually reserved for the great Caroline Herschel, who was granted a salary as assistant to her illustrious brother by King George III in 1787 (Chapter 3). Mary Edwards' much less spectacular appointment predated Caroline's by three years. The story of this shadowy figure has been brought to light by Mary Croarken, a historian of computing who has made a special study of the early days of the *Nautical Almanac*.[v]

Little is known of Mary Edwards' childhood or education, even of her maiden name, but, from her record as a respected computer for the *Nautical Almanac*, it is clear that she understood the mathematics and astronomy behind the calculations, and was later an instructor of younger computers. Mary was the wife of John Edwards (1748–1784), a Church of England clergyman who was a curate in Ludlow, Shropshire. Edwards was an amateur astronomer and telescope maker whose chief interest was in experimenting with metal alloys for astronomical mirrors. Isaac Newton was the chief pioneer of the reflecting telescope; by Edwards' time, such telescopes were being manufactured and improved upon by various opticians and instrument-makers. Mirrors had the advantage over lenses that they could be made larger. Their drawback was that they were made of metal alloys which were inefficient reflectors of light. The chief ingredients in their make-up were usually tin, copper and brass, and telescope makers experimented with the proportions of these in their mixtures with the aim of improving the reflecting qualities of the final product. It was laborious work. The ingredients had to be melted together, cast in a mould, and the mirror then ground and polished to the desired shape.

Telescope making was an expensive hobby, and Edwards supplemented his modest income as a curate by taking in pupils as boarders in his home and by acquiring computing work from the Astronomer Royal whom he got to know through a Cambridge mathematician friend. Edwards was hired as a computer in 1773. His assignment of six months' worth of computing a year earned him more than his official salary. He was married by this time, and his wife helped him with the computing from the start. In fact, from her own statement and from the testimony of a former pupil of her husband, it would appear that she did most of it.[vi]

John himself pressed on with his experiments. He tried over a hundred different mixtures and proportions of metals in his search for better reflecting quality. He also varied the methods of casting and devised tools for grinding and polishing his mirrors. His results were later published by Maskelyne in the *Nautical Almanac*. He made several small telescopes, none of which unfortunately survives.[vii]

One can imagine that the Edwards' busy life was not unlike the Herschels' on a smaller scale, with furnaces of boiling metal and mirror moulds intruding on the domestic scene. William Herschel, with Caroline in attendance, began making his own telescopes in 1773, the same year in which the Edwards' joint astronomical activities began. Mary and Caroline were close contemporaries in age. Did they ever meet? Probably not. Ludlow was remote from the Herschel homes in Bath and Slough, and Mary had children to look after. She did, however, know Maskelyne whom she was to describe as "a kind friend", and was undoubtedly acquainted with her husband's circle of opticians and telescope-makers. Another probable link was Maskelyne's sister, Lady Clive, widow of the legendary empire-builder Clive of India, whose home was near the Edwards' and with whom Mary appears to have been acquainted.

After little more than a decade of this shared life, in 1784, John Edwards died tragically and somewhat gruesomely, from inhaling arsenic fumes during one of his experiments. Arsenic, which Isaac Newton had also favoured, was included in some of Edwards' recipes for its reputed brilliance-enhancing properties. Mary, still in her early thirties, was left a widow with two small daughters and in dire circumstances. She lost their parsonage home as well as her husband's two sources of income. Even worse, she inherited his debts, incurred from his expensive research.

She applied to Maskelyne to ask if she could continue her husband's computing duties. Maskelyne, who knew her position, agreed. It was an easy matter in practice to substitute the name Mary for John in the official accounts. Mary further undertook to double her workload, calculating annually a full 12 month's worth of tabulations for the Almanac instead of six, and was able to pay off her debts in six or seven years. This new arrangement meant that she was responsible for one complete set of computations, or one half of the entire Almanac work.

Twice daily predictions of the Moon's co-ordinates, for noon and midnight, were needed. These consist of day and date, place in the zodiac; ecliptic longitude and latitude in degrees, minutes and seconds. In the case of the planets which move more slowly, data were given at less frequent intervals – three or six days. Each calculation of celestial coordinates involved several steps of trigonometry and arithmetic, carried out by following exactly a series of mathematical procedures laid out by Maskelyne, performed with the use of certain basic tables and of seven-figure logarithm tables. Mary Croarken estimates that for each tabular entry in the *Nautical Almanac*, the computer had to perform about 12 look-ups of tables and 14 seven- or eight-figure arithmetical calculations. William Herschel discovered the new planet Uranus (then called the Georgian star) in 1781. It was added to others in the Almanac and the computers, including Mary, were given a small rise in fees for the extra work.

After Maskelyne's death in 1811, Mary's fortunes went into reverse. She suffered a cut in income which she could ill afford, as she was supporting her still unmarried daughters. She died in 1815, aged 64. Her daughter Eliza who used to help her mother in her lifetime, took over the computing, and continued until 1832, when she lost her job with the creation of the Nautical Almanac Office in London. Under the new formalised regime, the work was centralised and only men were employed.[viii]

Mary Edwards had been employed officially as a computer for 29 years. If one adds the earlier years when she took on her husband's work, and her daughter Eliza's years of employment after her death, one reaches the remarkable result that the Edwards women, mother and daughter, performed a vital service to navigators and astronomers for a total of 55 years.

A century after its foundation, the Royal Observatory at Greenwich was well established. It was efficiently performing its principal duty, providing mariners at sea with accurate co-ordinates of astronomical bodies. The Astronomer Royal of the day, Nevil Maskelyne, had effected a great practical improvement by instituting the *Nautical Almanac* which provided for the year ahead all the astronomical information required by navigators and included the positions of bright stars as celestial markers.

The limitation to the Almanac's usefulness was that only stars which were observable from Greenwich could be included. The Navy, meanwhile, was roaming the oceans of the world. Under the directorship of Maskelyne's successor, John Pond, it was suggested that steps be taken to provide similar nautical information on stars visible in the southern hemisphere. From that sprung the decision in 1820 to set up a new observatory at the Cape of Good Hope, South Africa, which would be a counterpart of the Royal Observatory at Greenwich. The Cape, at the southern tip of the African continent, was at latitude 34° south. A plan for an observatory building was drawn up, and orders placed for a set of identical astronomical instruments with those at Greenwich. The person in charge of the new observatory, to be known as His Majesty's Astronomer at the Cape, was to receive the same salary as the Astronomer Royal, and was to have the services of a trained assistant. It was a historic mission. The Royal Observatory at the Cape would be the first permanent observatory in the southern hemisphere. Knowledge of the southern skies until then came from two temporary expeditions. One was Edmund Halley's to the Island of Ascension in the South Atlantic in 1677 where he spent a year and compiled a catalogue of some 350 stars,[ix] the other was an expedition to the Cape of Good Hope by the French astronomer Nicolas-Louis de Lacaille who was sent by the French Academy of Sciences to observe star positions and to measure an arc of meridian.[x]

Fearon Fallows, appointed to the newly-created post at the age of 31, probably on the recommendation of John Herschel, was a remarkable person, of a kind often found among astronomers of that era; to quote the formal tribute paid to him after his death, he was "an example, and, in this country, happily, not a solitary example, of the influence which talents and character may have on the fortunes of an individual under circumstances apparently untoward".[xi] He was the son of a hand loom weaver and was himself apprenticed to that trade; but, encouraged by his father he spent every spare moment on study, especially of mathematics. He was taken under the wing of the local Vicar who found sponsorship that eventually enabled him to become a student at Cambridge University. There he was a contemporary of John Herschel, being placed in third place in the mathematical Tripos (degree) examinations in which Herschel was first. He was a Fellow of his College, and was elected a Fellow of the Royal Society. He also took Holy Orders in the Church of England. Fallows was a practical observer as well as a theoretical astronomer, and on being chosen for the Cape post he spent some time at the Royal Observatory at Greenwich, familiarising himself with his future duties.

The task ahead was formidable. A site for a substantial observatory building had to be found in an unfamiliar country; a building had to be erected; and while awaiting the instruments which were to arrive in due course, he was instructed to observe a given list of southern stars and to determine their coordinates using portable instruments.

On 4 May 1821, Fearon Fallows, with his bride Mary Anne (née Hervey) (c. 1750–1815), eldest daughter of the clergyman who had encouraged him in his

Fig. 4.1 Cape Observatory c. 1830. (Brian Warner, *Astronomers at the Royal Observatory, Cape*)

youth, embarked for South Africa, ready to face the challenge. They reached the port of Simonstown fully expecting that there would be advance warning of their arrival, and Government help to unload the instruments and carry them to the Cape. There was no welcome; they were left to their own resources. The story of the next several years, recounted in detail by Brian Warner in his History of The Cape Observatory (Fig. 4.1) is one of almost unrelieved disaster and disappointment.[xii] Not only were there the huge practical problems of operating in an inhospitable environment – Fallows' successor Thomas Henderson always referred to the place as "Dismal Swamp"[xiii] – but also some unfortunate experiences with wayward staff. There were endless delays in communicating with and receiving goods from England (the voyage by ship took three months each way), and apparent neglect by the Colonial Government officials and by officials at home. There was also shortage of money. In the first year, Fallows, writing to John Herschel, described how "After a long conversation with my wife, we finally agreed to live as sparingly as possible and whatever we could possibly save out of our income, to apply it to the use of the temporary observatory which I wish to get ready without delay." Here, it would seem, Mary Anne was prepared – in the interest of astronomy as well as for love of her husband – to make sacrifices in the cause of duty.

It took seven years for the observatory building to be completed. The major instruments, supplied from England, were eventually installed and made ready for work. These were a large Transit instrument with which to time the passage of stars across the meridian, and a large Mural Circle with which to record the altitude of each star at the same instant. Two observers were needed for these basic operations. Fallows, now in poor health, was interrupted by sporadic illness, and in the end had to be carried in a blanket to observe and regulate his clock. To add to his misery, his sole assistant left and returned to England in 1830. From this time onwards, his only assistant was his wife, who learned to handle the Mural Circle. "He was relieved from this difficulty", wrote the Astronomer Royal, "by the intelligence and affection of Mrs. Fallows, who offered to undertake the circle observations while he was engaged with the transit. A very little instruction sufficed to render her perfectly competent for this task; and the Cape astronomer had, like Hevelius, the pleasure of finding his best assistant in the partner of his affections." (Johan Hevelius of Gdansk, the great seventeenth century astronomer was, as mentioned in Chapter 1, assisted by his intelligent and erudite wife Elisabeth[xiv] who published the last of his results after his death). Mary Anne Fallows, who was not herself without a wider knowledge of astronomy, discovered a comet in that same difficult year, 1830.

Fallows died in July 1831 in his 43rd year and was buried in front of the observatory. His wife had his grave marked by a slab of Robben Island stone. Such devotion to science and to duty as that shown by this husband and wife would seem to us today to be utterly excessive and foolish: a dying man, willingly aided and abetted by a wife who had already suffered more than her share of tragedy.

Mary Anne was left a lonely widow, her only surviving child having died before her husband. She returned to England bringing with her husband's (and her own) abundant observations, which were reduced (i.e., the necessary calculations applied) and published many years later by the next Astronomer Royal, Sir George Airy.[xv] They extended over the period from 1829–1830, in one third of which she herself had been involved.

Back in England she erected another memorial stone in the churchyard at Cockermouth, near her husband's birthplace. She kept in touch with her husband's scientific friends, including Sir John and Lady Herschel, Admiral Francis Beaufort and Thomas Maclear, who later held the Cape post under less daunting conditions. She remained anxious that her late husband's good work should be published. In 1835 she wrote to Beaufort about the matter. "I hope they do go [to the press] if they are not already gone, [and that] they will be put into the hands of some honourable man that will do them justice, as I believe a more faithful Catalogue of Stars never appeared before the public as far as they go". Her wish was sadly not granted in her lifetime; but, as already mentioned, the observations were eventually reduced and published by Airy in 1851.

On first returning to England, Mary Anne lived with her clergyman father. A pension, surely well deserved, for which she applied with support from astronomers and officials in Britain and South Africa, does not appear to have materialised. However, she seems to have been fairly well provided for by her husband, and a few years later, in 1835, she married an old friend, William Hall, a wealthy shopkeeper in London. Her new-found happiness did not last long. She died tragically in somewhat freak circumstances in 1838 from the effect of leeches which she used herself as medication. The wounds opened and she bled to death.[xvi]

Fearon Fallows' early death was hastened by the unprecedented difficulties which he encountered in setting up what was to become the most important astronomical institution in the southern hemisphere. Brian Warner, his biographer and the historian of the Cape Observatory, believes that he has not received sufficient recognition. "He deserves a more important niche in history as a Martyr to Science." Mary Anne, "the partner of his affections," deserves no less.

While astronomers such as the Fallows laboured to provide accurate positions of stars for the benefit of navigators at sea, the task of teaching the navigators how to make use of them was in the hands of specialist teachers on land. As ships sailed over the wide oceans and as commerce expanded, dozens of schools of navigation came into existence in London and at major ports.

The art of navigation by the Sun, Moon and stars was by no means simple. It involved, first of all, making appropriate observations of these bodies; and then, performing the calculations needed to derive the ship's geographical co-ordinates from these observations. The computational part required the use of spherical trigonometry, and it was this element that was taught in the navigation schools. The pupils learned to apply the relevant formulae by rote, but many of the teachers were mathematicians of talent who improved on the methods of calculation and even wrote books on the subject. These teachers were often associated with the navigational instrument-making business. Some had their own workshops, others were retailers of sextants, compasses, chronometers and other instruments to the multitudes of ships at sea.

One such prominent teacher was the "celebrated lady" Mrs. Janet Taylor[xvii] who had a Nautical Academy and Warehouse in the Minories in London, a district favoured by opticians and instrument-makers.[xviii] Janet Taylor (née Ionn) (1804–1870) was born in the north of England, in Wolsingham, Co Durham, one of the large family of a clergyman and schoolmaster from whom she received her education, including her knowledge of navigation. While still in her twenties, in 1830, she married George Taylor (1792–1853), another navigation teacher, a widower in his late thirties, with three children.[xix] They are listed in 1833 as having separate premises in the same street in London.

In 1835 Janet, aged 31, opened her Academy and Warehouse in the Minories. Her establishment was unique in that it was run by a woman who was a competent practitioner in her own right, and not a widow carrying on a late husband's business.[xx] According to the London directories, Janet was active between 1833 and 1859, whereas George's dates are 1833–1845. In that latter year Janet took over her husband's business, and was in fact the family's principal breadwinner. This is a remarkably energetic record for a woman who also brought up six surviving children.

Janet Taylor's Academy was recommended by the Admiralty, Trinity House (the establishment responsible for the nation's lighthouses) and the East India Company. Her head teacher there was James Griffin, himself the author of books on navigation and an examiner for Trinity House. (One wonders whether it was considered inappropriate for a woman to teach male classes in person.) Classes were held in the daytime and in the evening. Courses on Navigation, and on Longitude by Lunars and Chronometers were offered separately. The fee for the full course was four guineas (£4.20 in modern currency).

Janet Taylor's published treatises on navigational astronomy are proof of her knowledge and talent. The geometry of the celestial sphere was well known to mathematicians for centuries. The challenge for practical navigation was to convert observations of celestial bodies quickly and efficiently into useful information, specifically the geographical co-ordinates of the ship in mid-ocean. The raw observations taken at sea were the altitudes of celestial bodies, chiefly the noon Sun, measured with a sextant. These in theory supplied the ship's latitude. As described earlier, longitude was for a long time a much more difficult proposition, as it required a knowledge of Greenwich time as well as of astronomically determined local time. This problem was eventually solved at the end of the eighteenth century by the invention of the marine chronometer – a transportable clock, unaffected by the motion of the ship. Not all ships, however, were equipped with accurate chronometers which were still expensive, or were prepared to rely entirely on them; and the old method of finding the time by observations of the Moon remained an essential part of a seaman's skill. The instrument in this case was the quadrant. The datum required was the angular distance between the Moon and either the Sun or a suitable beacon star, known as the "lunar distance."

All computations required solutions of spherical triangles by numerical methods. The only practical aid available were logarithm tables, including trigonometrical logarithms. To make use of such tables, formulae had to be presented in suitable forms; procedures had to be made as simple as possible, and the number of steps in the computations reduced to a minimum. Speed was also a consideration, as the work was being done on a moving ship. Prizes for better and faster methods of reduction were often offered by the Admiralty and by shipping companies. Numerous navigational handbooks laying down different procedures both observational and computational were published.[xxi]

Janet Taylor's first publication, at the very beginning of her career, was *Lunar Tables*, published in 1833 (Fig. 4.2), which claimed that "the usual tedious preparations [are] avoided".[xxii] The book included a *Treatise on the Chronometer, under Patronage of the Lords Commissioners of the Admiralty* and carried a dedication to King William IV, successor of George III: "The talent from which it has sprung was fostered by the benevolence and under the immediate patronage of your illustrious mother" (i.e., Queen Charlotte). (Queen Charlotte died in 1818, but as young girl Janet had attended a school in London under her royal patronage). The book which explained an abridged method which the author had devised of computing lunar distances, brought her fame as well as "handsome rewards" in 1835 from the Admiralty, Trinity House and the East India Company. She was also awarded medals from the kings of Holland and Prussia. Her other books were *Principles of Navigation Simplified* (1834, third edition 1837), *Epitome of Navigation and Nautical Astronomy* (1842 and editions up to the eleventh in 1858, a fact that speaks for itself), and *Directions for using a Planisphere of the Fixed Stars* (1846). Her abbreviated rule for calculating latitude from altitudes by a manipulation of the geometrical formulae in the third edition of *Principles of Navigation* was described as "ingenious". Her books were well regarded and found places among the principal textbooks of the time.[xxiii]

In addition to her teaching, Janet Taylor engaged in every aspect of the navigational business. Her warehouse sold charts and offered instrument repairs and compass checking. Instruments signed with her name including barometers and a telescope, are on record. She therefore either employed her own mechanics, or

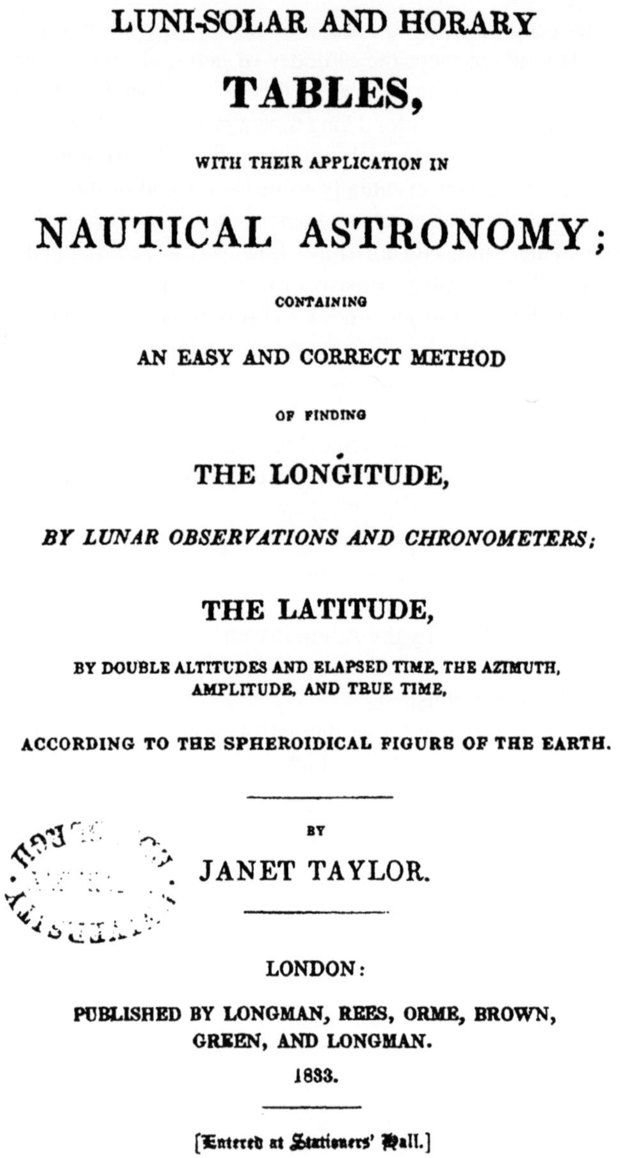

LUNI-SOLAR AND HORARY

TABLES,

WITH THEIR APPLICATION IN

NAUTICAL ASTRONOMY;

CONTAINING

AN EASY AND CORRECT METHOD

OF FINDING

THE LONGITUDE,

BY LUNAR OBSERVATIONS AND CHRONOMETERS;

THE LATITUDE,

BY DOUBLE ALTITUDES AND ELAPSED TIME, THE AZIMUTH, AMPLITUDE, AND TRUE TIME,

ACCORDING TO THE SPHEROIDICAL FIGURE OF THE EARTH.

BY

JANET TAYLOR.

LONDON:

PUBLISHED BY LONGMAN, REES, ORME, BROWN, GREEN, AND LONGMAN.

1833.

[Entered at Stationers' Hall.]

Fig. 4.2 Title page of one of Janet Taylor's books on Nautical Astronomy. (University of Edinburgh library)

ordered and purchased instruments from other instrument-makers. She was the sole agent for Dent's chronometers, a considerable business advantage: Edward John Dent (1790–1853) was the noted clock and watch maker, the designer of the clock on the Palace of Westminster, "Big Ben". Janet is also recorded as a patentee of a certain type of clock spring, and of improved instruments for measuring angles and lunar distances[xxiv] which were on display at the Great Exhibition of 1851.

Janet Taylor's husband, who supported his wife's career though his own was less effective, died in 1853. Janet herself continued working until 1868 and died two years later at the age of 66 in her sister's home in Co. Durham.

Clearly Janet Taylor was not only an accomplished mathematician and geometer but – what was exceptional among intellectual women of the early nineteenth century – a successful businesswoman in a man's world.

These were women whose exertions were of essential practical usefulness. Others, wielding their pens, contributed in the less arduous environment of the literary salons (next chapter).

Notes

[i] Derek Howse. 1989. *Nevil Maskelyne, the Seaman's Astronomer*. Cambridge Univerity Press: Derek Howse. *Oxford DNB.*

[ii] Eric Forbes. 1975. *Greenwich Observatory* Vol. 1, Chapters 6 and 7. London: Taylor and Francis.

[iii] Forbes. op. cit. p 121.

[iv] Forbes. op. cit. p 154.

[v] Mary Croarken. 2003. Mary Edwards: Computing for a living in 18th-century England. *IEEE Annals of the History of Computing* **25**, October–December issue, 9–15.

[vi] All references in Mary Croarken's paper. op. cit.

[vii] Edwards' work is recorded in Henry C. King. 1979. *The History of the Telescope.* p 89–90. New York (Dover edition).

[viii] Women were not appointed to Nautical Almanac computer posts again until 1928. Croarken, op. cit.

[ix] Michael Hoskin in Michael Hoskin (ed.). 1997. *The Cambridge Illustrated History of Astronomy.* p 251. Cambridge : University Press; (also in Michael Hoskin (ed.). 1999. *The Cambridge Concise History of Astronomy.* p 214)

[x] Brian Warner. 2002. Lacaille 250 years on. *Astronomy and Geophysics* **43**, 2.25–26.

[xi] G.B. Airy. 1851. Results of Observations by the Reverend Fearon Fallows at the Royal Observatory, Cape of Good Hope 1829. 1830, 1831. *Memoirs of the Royal Astronomical Society.*

[xii] Brian Warner. 1979. *Astronomers at the Royal Observatory Cape of Good Hope*, Cape Town and Rotterdam: A.A.Balkema, Chapter 1; Brian Warner. 1995. *Royal Observatory, Cape of Good Hope 1820–1831. The founding of a colonial observatory, incorporating a biography of Fearon Fallows.* Dortrecht/Boston/London: Kluwer Academic Publishers.

[xiii] Brian Warner, op. cit.1995. p 34.

[xiv] Michael Hoskin, op. cit. Chapter 6; Patricia Fara. 2004. *Pandora's Breeches. Women, Science and Power in the Enlightenment*, Chapter 7, London: Pimlico. and many other histories of astronomy.

[xv] G.B.Airy, op. cit.

[xvi] Brian Warner. 1995. op. cit.

[xvii] Charles Cotter. 1868. *A History of Nautical Astronomy.* p 250. London.

[xviii] E.G.R. Taylor. 1966. *The Mathematical Practitioners of Hanover Ian England 1714–1840*, 461–2 Cambridge University Press.

[xix] Susanne Fisher. Oxford DNB.

[xx] E.G.R. Taylor. op. cit. p 101.

[xxi] Charles Cotter. op. cit.
[xxii] Janet Taylor. 1833. *Luni-Solar and Horary Tables with Their Appication in Nautical Astronomy*. London: Longman Rees Orme Brown Green and Longman.
[xxiii] *Epitome* and the *Lunar Tables* was stocked in Edinburgh University library where a course of university lectures in Astronomy and Navigation was offered in the mid century.
[xxiv] E.G.R. Taylor. op. cit. p 462.

Chapter 5
Celebrities

Maria Edgeworth (1768–1849), famous "literary lady" and inventor of that description, is acknowledged as an early promoter of women's worth in all intellectual pursuits, that of science included. Her talents ranged widely. She was the most highly acclaimed novelist of the early decades of the nineteenth century, a precursor of her friend and admirer Sir Walter Scott. She was an influential educationalist, an advocate of women's education, and of the teaching of science and technology to children. In her long and sociable life she cultivated the friendship of many scientists including astronomers and left valuable accounts – and gossip – of the men and women among whom she moved. She was also a prolific letter writer.

The Edgeworths were an Anglo-Irish family with an estate at Edgeworthstown, County Longford. The father was the polymath Richard Lovell Edgeworth, landowner, agriculturalist, politician, educationalist and inventor.[i] He lived in England as a young man and moved in scientific circles that included Sir Joseph Banks, President of the Royal Society, Captain James Cook the explorer and Nevil Maskelyne, the Astronomer Royal. He belonged to the famous Lunar Society of Birmingham, a group of brilliant innovative scientists and engineers who included the renowned chemist Joseph Priestley and Erasmus Darwin, physician and grandfather of the biologist Charles Darwin. Darwin was a special personal friend of Edgeworth's with whom he corresponded throughout their lives: Darwin's last letter before his death in 1802 was to Edgeworth, as Maria herself recorded in the published Life of her father.[ii]

Back in Ireland, Edgeworth was a founder member of the Royal Irish Academy. His Irish social circle included John Brinkley, eminent professor of astronomy at Trinity College Dublin, whose successor, the young mathematical genius William Rowan Hamilton would become one of Maria's closest friends and confidants in later life. Within the immediate family, a daughter Anna married Thomas Beddoes, the physician and chemist renowned for his "pneumatic" treatment of patients with oxygen and less orthodox gases (nitrous oxide, or "laughing gas", had a great vogue). Another daughter married Beddoes' medical assistant John King. Beddoes was the master under whom the young self-educated Humphry Davy began his illustrious career;[iii] thus Maria was a personal friend of Davy and his flamboyant socialite wife when he was the Royal Institution's famous experimenter and lecturer who kept her supplied with his scientific papers, and whose discourses she attended

M. Brück, *Women in Early British and Irish Astronomy*,
DOI: 10.1007/978-90-481-2473-2_5,

in Dublin. Maria's stepmother Frances was a sister of Francis (later Admiral Sir Francis) Beaufort, Hydrographer of the Royal Navy and originator of the Beaufort scale of wind force. Frances and Maria, both aged about 30 at the time of that marriage, became deeply attached to each other, an extraordinarily close relationship that lasted until the end of Maria's long life. One of Frances' daughters became the second wife of the influential astronomer Thomas Romney Robinson, Director of Armagh Observatory. One of Maria's brothers studied chemistry under the renowned Joseph Black in Edinburgh where Maria, on a visit to him with her father, met Black and the mathematician John Playfair. Another brother, assisted by Maria's favourite sister Fanny, ran an observatory and weather station at their family home in Edgeworthstown.

Richard Edgeworth had several children for whom he devised an educational scheme in which the grown-up Maria, a child of his first marriage, collaborated actively. They published their ideas and their results in a treatise, *Practical Education*,[iv] in 1798 and were regarded as experts on educational methods. Maria also put them into practice in her popular children's stories[v] (which formed part of the educational diet of, among many others, the young Maria Mitchell (Chapter 8), the first woman astronomer in the United States of America[vi]). A novel component was the introduction of science to even quite young children by means of examples from everyday life. There should also be no distinction between girls and boys as regards what and how they are taught, a principle already expressed by Maria in her first published work, *Letters for Literary Ladies* (1795)[vii] . In that famous book, much revered by later feminists, written when she was 27, she had stated her views on the female intellect and argued the case for women's education. By "literary ladies" she meant "women who have cultivated their understandings not for the purpose of parade but to make themselves useful and agreeable" – agreeable, that is, as companions to their menfolk – or, if unmarried, to "store their minds with knowledge," so that they can "amuse or be amused in the company of well-informed people". Science, for a lady, ought to be cultivated as part of the general powers of the mind and the love of knowledge, not to make a her "merely a botanist, a mathematician or a chemist". Botany was already popular with women; chemistry was deemed suited to their situation, demanding no bodily strength, pursuable in the domestic scene, and in "no danger of inflaming the imagination". Hardly the modern image of a working scientist, but an advanced idea at the end of the eighteenth century.

Maria Edgeworth was acquainted with her father's old colleagues while he lived, and in the 32 years which she survived him relished the company of the younger generation with unabated enthusiasm. As a best-selling novelist and brilliant conversationalist, she was always received as a celebrity and had no difficulty in obtaining introductions to leading men and women of learning. Her letters home to her stepmother and sisters from her trips to London provide a wonderful record of social intellectual life in a metropolis in which the literary world overlapped with the scientific community.[viii] In this milieu, Maria met John Herschel, friend of her relative Francis Beaufort, Charles Babbage of computer fame, the geophysicist and surveyor Edward Kater, and Henry Hyde Wollaston, famous chemist and polymath.

Wollaston and she were particularly fond of each other. He left a platinum pen to be given to her after his death (in 1828) as a token of his friendship (he had made his fortune as a chemist in platinum). He was, she said, "in truth consistently great and good, living and dying".[ix] Maria was taken to Greenwich to meet the Astronomer Royal; John Herschel and the mathematician Charles Babbage were among the many guests at her Irish home in Edgeworthstown. She struck up a special friendship with two scientific couples of famed wives, the Marcets and the Somervilles (next chapter). These influential scientists were more than social trophies: she had a genuine interest in their work and studied their publications. Maria Edgeworth's life demonstrates very clearly the manner in which scientists and literary people, men and women, associated freely and naturally in early nineteenth century Britain.

Maria Edgeworth's association with the Herschel family is a happy example of this. On an early tour of Britain with her father and stepmother, she visited the great Sir William Herschel at Slough and was shown the famous 40-ft reflector, then in use. On a second occasion the 84 year-old Sir William was too ill to receive visitors, but young John devoted a whole day to Maria and her companions, showing them the various instruments and conversing about astronomy.[x] "The great telescope ... is there but the supports are decaying, as Lady Herschel observed with tears in her eyes. It is never used now", wrote Maria in a letter to a sister at home. The visit took place in early June, 1822. William Herschel died on August 25 of that year, so the Edgeworths were among the last visitors at Slough in that great astronomer's lifetime.

When Maria visited Slough again in 1831, she met John Herschel's young wife for the first time and was charmed by her. "Of all the people I have seen and of all the society I think the Herschels are the best worth cultivating ... in their ways of living, too, so comfortable and well regulated...", she reported.[xi] She read his weighty *Preliminary Discourse on the Study of Natural Philosophy*,[xii] and made notes in her copy of the book in order to ask for explanations "especially on the polarisation of light and the difference between prismatic and periodical colours".[xiii] Herschel next day brought out his apparatus and demonstrated various optical phenomena. He "then explained all that is known and all that has been imagined about them – alternating theory and experiment most beautifully and accurately – and so patient and kind and clear! How my father would have admired him." This extraordinary lesson in optics proves Maria Edgeworth's genuine interest in science – all the more remarkable at 63 years of age.

In the evening Herschel showed her the Moon and Saturn, its rings and satellites, through the telescope. The telescope was the one which she remembered from her previous visit – the 20 ft reflector in the corner of the garden. She ascended 18 feet to view from "a little stage of about 8 ft by 3 with a slight iron rod rail on three sides but quite open to fall in front". The great instrument was not in use, but the 4-ft mirror was there in a good state of preservation.

Herschel spent the years 1834–1838 observing the southern skies from the Cape, South Africa, as already described in connection with his aunt Caroline (Chapter 3), after which he set up home at Collingwood, Kent. There, in 1843, Maria Edgeworth, aged 75, was again the guest of the hospitable Herschels. She found

Herschel (now Sir John) hard at work on his Cape observations. He showed her drawings of sunspots "and their changes uncountable from day to day" and a large number of his daguerreotypes including one of the great telescope at Slough before it was dismantled.[xiv] Maria visited Collingwood again the following year, on her last ever journey out of Ireland. During this trip she also called on Michael Faraday, now a world-renowned scientist, to give him her supreme gift – a copy of her father's *Memoirs* which she had completed and published after his death.[xv]

At home in Ireland, Maria Edgeworth's most celebrated scientific friendship was that with the great mathematician and astronomer, William Rowan Hamilton.[xvi] It began in 1824 when she was already in her fifties and Hamilton a student of only nineteen. They first met in Edgeworthstown where Hamilton was brought by Richard Butler, a clergyman from Hamilton's home town of Trim, Co Meath, who married one of Maria's half sisters. Hamilton was already famed for his extraordinary gifts as a linguist and a mathematician, "a real prodigy of talents", wrote Maria, "who Dr. Brinkley [Dublin's astronomer] says may be a second Newton."[xvii] Hamilton for his part (though at first disappointed by her personal appearance, which, however, "seemed to improve, as if that of her mind cast reflected graces upon her person") was overwhelmed by Ms. Edgeworth's brilliant conversation.[xviii] They became close friends, as did their families. When the Chair of Astronomy in Dublin became vacant Maria Edgeworth was one of those who urged Hamilton, though still an undergraduate student, to be a candidate.[xix] He was appointed in 1827 at the age of only 21.

Edgeworthstown, already a famous calling point for visitors to Ireland, became and remained almost a second home to Hamilton, even after Maria's death. His own official residence at Dunsink Observatory, near Dublin, was also a gathering place for scientists and poets, where guests might be shown the Moon or Jupiter through the telescope, and spend the rest of the evening poetry reading. Rowan Hamilton had ardent literary aspirations and saw poetry as akin to astronomy as an intellectual and almost spiritual pursuit. While still a very young man he gained the friendship of the celebrated poets William Wordsworth and Samuel Coleridge who both belonged to Maria's literary circle. On his first visit to England Hamilton called on Coleridge in London before calling on John Herschel, and paid his first visit to Wordsworth at Rydal, his home in the Lake District. Wordsworth returned the visit to the Dublin Observatory. Such an association of poets and scientists was not unique at that time: Humphry Davy was a talented poet and a friend of Wordsworth, Coleridge and Robert Southey.[xx] Maria Edgeworth shone in that literary-scientific world.

Another of Maria Edgeworth's valued scientist friends was David Brewster, the Scottish physicist, best known to the public as the inventor of the kaleidoscope. In 1823 she paid her celebrated visit to her fellow writer Sir Walter Scott at Abbotsford, his romantic mansion on the river Tweed in the Scottish borders. It was one of the highlights of both their lives. There she met Brewster who was a friend and neighbour of Scott. They "commenced a most cordial friendship"[xxi] and carried on a lively correspondence for many years. Brewster visited Edgeworthstown, and when she came to write her last and most ambitious book of children's science lessons, he was her principal adviser.

The book, *Harry and Lucy Concluded*,[xxii] was published in 1825. It was the third and last in a series instituted by R.L. Edgeworth about two fictitious children who are happily learning science by the Edgeworth method. The first two books of the series, *Harry and Lucy* (1801 and 1813) were written jointly by himself and Maria. They were, said Maria, "the very first attempt to give any correct elementary knowledge or taste for science in a narrative suited to the comprehension of children, and calculated to amuse and interest, as well as to instruct."[xxiii] As the series progresses the children get older and the material gets more advanced; in the final number Harry is 14 and Lucy somewhat younger.

The story begins with Lucy expressing a wish to catch up with her brother's superior knowledge of science (like Eudosia and Leander in Ferguson's book), and culminates in the father taking the children on an educational tour of industrial England. They visit a cotton mill, a foundry, a coal mine, a pottery, a printing press, and other wondrous sites. The topics were those in which Maria's late father had been expert. They also visit a gentleman's laboratory (she may have had John Herschel's in mind) where they learned about optics and such delights as the camera obscura. Maria appealed to Brewster, her "Scottish gentleman-philosopher", to read over the manuscript, sent in relays to Edinburgh. Brewster complied, and tested the book on his own four boys. "The intense interest it excited was a true presage of the popularity which it continued to meet from all intelligent youthful readers", wrote Brewster's daughter.[xxiv] It also presumably met the approval of the Herschels, who ordered Maria Edgeworth's books for their own children.[xxv] Maria had a lifetime of experience in teaching children – her own many young siblings, and then their offspring (though her fictitious children seem incredibly priggish by modern standards).

While Maria Edgeworth cannot be said to have contributed directly to the advance of astronomy, her interest greatly helped the general cause of science in education (Ferguson's *Astronomy* was recommended in her Early Lessons series), placing it on a par with literature as a branch of culture. She maintained girls to be of equal mental aptitude with boys, and deplored superficial "female accomplishments" such as drawing, dancing and music, generally considered suitable but in reality a waste of time for the untalented.[xxvi] Her opinion of the value of scientific education for women emerges from the parable of *Harry and Lucy Concluded*. There, in the character of Lucy's mother, is Maria's ideal scientific-literary lady. "By acquiring knowledge", she assures Lucy who had overheard a gentleman say that "scientific ladies were his *abhorrence*", "women not only increase their power of being agreeable companions to their fathers, brothers, husbands or friends, if they are so happy as to be connected with sensible men, but they increase their own pleasure in reading and hearing of scientific discoveries; they acquire a greater variety of means of employing themselves independently, and at home".[xxvii] Women like herself and her friends Mrs. Marcet and Mrs. Somerville (next chapter) who could combine their careers with exemplary home life represented that ideal.

In Maria Edgeworth's social world, the cultivation of the mind – for women – was seen as the preserve of the upper classes, and for women in those ranks to have careers outside the home was unthinkable. (By contrast, it seemed perfectly natural that women and children should be employed, as if they were part of the machinery, in the Lancashire cotton mills that Harry and Lucy were shown in the course of their fictitious educational tour. She also had cautionary advice for mothers about the dangerous moral influence of servants below-stairs). Maria's vision for "Lucy" of an educated lady gracing a male dominated home was no advance on that offered by James Ferguson to "Eudosia" half a century earlier; indeed it was a step backwards. Nevertheless, Maria Edgeworth has to be recognised for proclaiming in her influential writings higher aims for women's upbringing, and a firm belief in their capacity to understand the sciences.

Among the books approved of by the Edgeworths for the education of children was Mrs. Jane Marcet's *Conversations on Chemistry*.[xxviii] Maria first met Mrs. Marcet (1769–1858),[xxix] who was about the same age, through her father's old friends, and got to know her intimately on her later visits to London. Jane was the daughter of an English father, a wealthy merchant in London, and a Swiss mother. Her Swiss husband, Alexander Marcet, was a physician, a graduate of Edinburgh University, who practised in London for twenty years and also did research in chemistry, becoming a Fellow of the Royal Society. The family inherited private means from her father, and Dr. Marcet could afford eventually to devote himself entirely to science. It was a very happy marriage. The couple had their home in London, but used to return to Geneva where they were in touch with the thriving circle of scientists there. They were, said Maria Edgeworth, at the centre of "the most agreeable as well as scientific society in London"[xxx] Men of science and literature on their way to and from meetings of the Royal Society would drop in at Mrs. Marcet's house for a half-hour's "pleasant tea-drinking", where they would enjoy her "good sense and amiable character and manners".[xxxi] Maria had high praise for Jane's informed conversation – something of which she was surely a good judge. "I never knew any woman – except Mrs. E [Maria's stepmother, née Beaufort] – who had so much accurate information and who can give it out in narration so clearly, so much for the pleasure and benefit of others without the least ostentation or mock humility. What she knows, she knows without fear or hesitation and stops and tells you she knows no more whenever she is not certain."[xxxii] She "never made any false pretensions", wrote the famous journalist Harriet Martineau. "She sought information from learned persons.... She simply desired to be useful; and she was eminently so."[xxxiii]

Jane acquired her scientific knowledge from listening to lectures at the newly founded Royal Institution in London. Two eminent men were appointed as professors there in 1801 – Thomas Young as Professor in Natural Philosophy (Physics) and Humphry Davy as Professor of Chemistry. The lectures were open to the public, and it was Young's particular ambition that they should be accessible to women who would make up a large proportion of the audience. "The many leisure hours, which are at the command of females in the superior orders of society, may surely be appropriated, with greater satisfaction, to the improvement of the mind and to the acquisition of knowledge, than to such amusements as are only designed for facilitating the insipid consumption of superfluous time". He saw the Royal Institution as "in some degree a subordinate University, to those whose sex or situation in life has denied them the advantage of an academical education in the national seminaries of learning".[xxxiv] So indeed it turned out, though of the two professors, Davy (later Sir Humphry), with his eloquent delivery and spectacular demonstrations, was by far the more popular. His lectures became social occasions among the fashionable leisured classes.

Young was a scientist of exceptional talents. A physician by profession, he was also a brilliant physicist. It was he who in 1800 demonstrated the wave nature of light in one of the fundamental experiments in physics – a revelation that appeared to contradict Newton's theory that light was transmitted in corpuscles and for that reason was not immediately accepted in England. Young also gave his name to a famous quantity in the properties of matter, Young's Modulus.

As a lecturer, however, he was less successful. He overestimated the intellectual capacity of his audience and was not a good communicator. For all his efforts and brilliance, his discourses were above the heads of the majority in the audience at the Royal Institution; the numbers declined and he gave up lecturing after two years.

Jane attended both courses. Her husband encouraged her to write popular accounts of what she learned that would be suitable for children – they had young children of their own – and as teaching aids. She planned two books, one on each course. The first to be published, *Conversations on Chemistry*, with the subtitle *intended more especially for the Female sex*, which appeared in 1806, was an overwhelming success; it went into numerous editions, the last when the author was 84 years old, and sold by the thousand. Mrs. Marcet published her work anonymously at first (believing perhaps that a woman writing on science would not be taken seriously). Some readers surmised that the real author was Mrs. Margaret Bryan – a surprising idea, as the styles of writing and presentation of the two women were very different.[xxxv]

The book's most famous reader was the young genius Michael Faraday when, as a struggling bookbinder's apprentice, he was eagerly educating himself. *Conversations on Chemistry* was one of the books that came into the workshop for binding; it instantly fired his imagination. When he, in his turn, succeeded Davy as the Royal Institution's chief attraction, he never forgot his debt to his "first instructress".[xxxvi] It has been suggested that his founding of the Institution's famous Christmas lectures for children in 1827 (still popular in the twenty-first century) was inspired by Mrs. Marcet's book.[xxxvii] "At the height of his fame", Mary Somerville recorded, "he always mentioned Mrs. Marcet with deep reverence".[xxxviii] He used to send her his publications "as a thank offering", and made sure that there was always a place for her at his lectures. All she had to do, he wrote to her, was to mention his name at the door.[xxxix]

Other *Conversations* followed – in Political Economy, Physiology, Botany, Zoology and, finally, in 1819, Natural Philosophy or Physics. Like its predecessors, *Conversations on Natural Philosophy* employed the popular convention of question and answer sessions between a teacher and pupils. The teacher was a woman, addressed as Mrs. B; the pupils were sisters, Emily and Caroline, the elder prim and serious, the younger uninformed and sceptical. They all spoke in a stilted and earnest manner. In the Preface to this book, Jane Marcet confesses her "ignorance of mathematics" and "imperfect knowledge of natural philosophy". The book was therefore offered "with great diffidence" for the use of very young pupils – though without the qualification "more especially for the female sex". Mrs. Marcet need not have been so apologetic. The book was deservedly successful, and remained in print throughout the author's lifetime: the updated twelfth edition came out in 1851.[xl]

Though it was the last of the series to be published, the text of *Conversations on Natural Philosophy* had been Mrs. Marcet's very first effort in scientific writing, planned to precede the *Conversations on Chemistry*.[xli] It had evolved from Thomas Young's lectures at the Royal Institution in 1801–1803. Though he abandoned his spoken addresses, he committed an expanded version of his course of 31 lectures to print a few years later.[xlii] Jane Marcet's *Conversations* clearly made use of these published lectures. They discuss Properties of Matter, Gravity, Laws of Motion and Mechanics; optics and optical instruments, the camera obscura, the wave theory of light, the spectrum, theories of colour. The chapters on Astronomy cover the geometry and motions within the solar system, the physical appearance of planets, satellites, and stars. Double stars are described, as are William Herschel's observations of faint stars in huge numbers. The 12th edition of 1851 also includes the discovery of the planet Neptune in 1845. What is remarkable about the book is the accuracy with which all these matters are conveyed in simple language, without the aid of mathematics or even a single formula.

Jane Marcet, however, did not venture to publish the book until 1819. One may ask what induced her to change her mind 15 years after the manuscript was first penned and shelved. Might it have been connected with Mary Somerville's visit to Geneva in 1818, the first lengthy meeting between these two scientifically-minded women, who were to remain ever after "on terms of affectionate friendship"?[xliii] Did Mary, with her superior competence in mathematics and astronomy and her admiration for Young, reassure and encourage her? "So many books have now been published for young people", Mrs. Somerville wrote 50 years later, "that no one at this time can duly estimate the importance of Mrs. Marcet's scientific works. To them is partly owing that higher intellectual education now beginning to prevail among the better classes in Britain". Her *Conversations on Chemistry*, said Harriet Martineau, "opened an entirely fresh region of ideas to the mind of the rising generation of that day, to whom the very nature of chemical science was a revelation".[xliv]

Conversations on Natural Philosophy, we may assume, received the seal of approval of the distinguished experimental physicist David Brewster whose theory of colour is discussed in the book: soon after it was published, we find the delightful Scottish writer Elizabeth Grant of Rothiemurchus entertaining "the clever authoress" and Sir David and Lady Brewster at her Highland home.[xlv]

Conversations in Natural Philosophy has received less attention from historians than *Conversations in Chemistry*, because of the latter's association with the great Michael Faraday. But, in fact, *Conversations in Natural Philosophy* was every bit as admirable a contribution to the teaching of science in its day: that it remained in demand even after the author's death testifies to this. As an elementary mathematics-free exposition it replaced Margaret Bryan's book on the same subject which by then belonged to another generation.

Alexander Marcet died suddenly in London in 1822. Jane, who survived him by over 30 years, continued writing extensively on a variety of topics, but her role as an interpreter of physics and astronomy went no further. In her later years she wrote mainly for little children. She outlived most of her scientific contemporaries – Maria Edgeworth died in 1849 and the Somervilles had gone to live in Italy – and in old age "she inhabited a world of her own, untouched by events in the outside world."[xlvi] She died in her daughter's home in London at the age of 90.

Notes

[i] Desmond Clarke. 1965. *The Ingenious Mr Edgeworth.* London: Oldbourne.

[ii] R.L. Edgeworth, 1969. *Memoirs of Richard Lovell Edgeworth Esq.*, begun by himself and concluded by his daughter Maria Edgeworth. Reprinted with an Introduction by Desmond Clarke. Shannon, Ireland: Irish University Press.

[iii] Sir Harold Hartley. 1972. *Humphry Davy*. Wakefield: EP Publishing.

[iv] Maria and R.L. Edgeworth. 1898. *Practical Education* (3 vols). London (reprint of second revised edition 1801).

[v] Maria Edgeworth. 1890. *Stories for Children, or the Parents' Assistant.* London: George Bell and Sons.

[vi] Helen Wright. 1969. *Sweeper in the Sky*. p 4. Nantucket: Maria Mitchell Association.

[vii] Claire Connolly (ed.) 1993. *Maria Edgeworth. Letters for Literary Ladies.* London: Everyman (original edition 1795).

[viii] Christina Colvin(ed.) 1971. *Maria Edgeworth. Letters from England 813–1844.* Oxford: Clarendon Press.

[ix] Maria Edgeworth, Letter to W.R. Hamilton. In Robert Percival Graves.1882. *Life of William Rowan Hamilton* vol 1. p 382. Dublin: Hodges Figgis and Co.

[x] Colvin, op. cit. p 411.

[xi] Colvin, op. cit. p 499.

[xii] J.F.W. Herschel. 1830. *A Preliminary Discourse on the Study of Natural Philosophy.* London.

[xiii] Periodical colours were those produced by diffraction.

[xiv] Günther Buttmann.1974. *The Shadow of the Telescope, a Biography of John Herschel* (tr B. Pagel), Guildford and London: Lutterworth Press. It was Francis Beaufort, Maria's relative, who first introduced Herschel to photography.

[xv] James Hamilton. 2003, *Faraday: The Life.* p 309 London: HarperCollins Publisher.

[xvi] Robert Percival Graves. 1882. *Life of Sir William Rowan Hamilton*, 3 vols. Dublin: Hodges Figgis and Co; London: Longmans Green and Co. This wonderful biography includes selections from his poems and correspondence.

[xvii] ibid., quoting a letter from Maria to a sister. p 161.

[xviii] ibid. Hamilton, letter to his sister. p 162.

[xix] Brinkley was made Bishop of Cloyne but continued his interest in astronomy.

[xx] Raymond Lamont-Brown.2004. *Humphrey Davy, Life beyond the Lamp.* Stroud: Sutton Publishing. This biography contains several examples of Davy's poetry.

[xxi] Margaret Maria Gordon. 1879. *The Home Life of David Brewster*, Chapter 8. Edinburgh.

[xxii] Maria Edgeworth. 1825. *Harry and Lucy Concluded*, (4 vols). London: R. Hunter.

[xxiii] Introduction to *Harry and Lucy Concluded.*

[xxiv] Gordon op. cit.

[xxv] D.S. Evans, T.J. Deeming, Betty Hall Evans, S. Goldfarb. 1989. *Herschel at the Cape, Diaries and Correspondence of Sir John Herschel 1834–1838*. p 126. Cape Town: A.A. Balkema.

[xxvi] Maria and R.L. Edgeworth. op. cit. *Practical Education*, vol 3, Chapter 20: On Female Accomplishments, Masters and Governesses.

[xxvii] *Harry and Lucy Concluded*, vol 1, p 10. op. cit.

[xxviii] Mrs Jane Marcet. 1896. *Conversations in Chemistry*, London: Longman.

[xxix] Elizabeth M. Morse. Oxford DNB.

[xxx] Christina Colvin (ed.) 1971. *Maria Edgeworth. Letters from England 813–1844*. p 383 Oxford: Clarendon Press.

[xxxi] Maria Edgeworth. letter to Hamilton, February 1838. in Robert Percival Graves, *Life of Sir William Rowan Hamilton*, vol 2. 1885. p 242. Dublin: Hodges Figgis and Co.

[xxxii] Letter to her sister, 1822, quoted in Colvin, op. cit. p 308.

[xxxiii] Harriet Martineau. 1869. *Biographical Sketches*, 386–392. London.

[xxxiv] Alexander Wood, completed by Frank Oldham. 1954. *Thomas Young Natural Philosopher 1773–1829*. p 126. Cambridge University Press. His own wife was one such intelligent woman; she is mentioned in the next chapter.

[xxxv] Margaret Alic. 1986. *Hypatia's Heritage*. p 177. London: The Women's Press. Another suggestion was that it was written by her husband.

[xxxvi] quoted from his correspondence by James Hamilton. 2003. *Faraday: The Life*. p 220. London: HarperCollins Publisher.

[xxxvii] James Hamilton op. cit. p 208.

[xxxviii] Mary Somerville (next chapter). *Queen of Science*. p 92.

[xxxix] Quoted in Hamilton, op. cit. p 207.

[xl] Mrs Marcet.1851 (twelfth edition). *Conversations on Natural Philosophy: in which the Elements of that Science are familiarly explained and adapted to the comprehension of young persons*. London: Longman, Brown, Green and Longmans.

[xli] Preface to Mrs Marcet's *Conversations on Natural Philosophy*.

[xlii] *A course of lectures on Natural Philosophy and the Mechanical Arts*, 1807.

[xliii] Mary Somerville. *Queen of Science*. p 92.

[xliv] Harriet Martineau, op. cit.

[xlv] Elizabeth Grant. 1898. *Memoirs of a Highland Lady* (ed. Lady Strachey). London: John Murray (Reprint 1928). p 349.

[xlvi] Oxford DNB.

Chapter 6
Queen of Science

Caroline Herschel was in her eighties when she was made an honorary member of the all-male Royal Astronomical Society in 1835. The same honour was awarded on the same occasion to Mary Somerville (1780–1872),[i] a woman 30 years her junior (Figs. 6.1 and 6.2). "To be associated with so distinguished an astronomer was in itself an honour", wrote Mary Somerville in her *Recollections*.[ii] By a quirk of fate the two great women never met. When Mary Somerville and her husband called at Slough on their way to London from Scotland in 1816, Caroline for once was not at home.

Mary Somerville was described by her eminent friend and contemporary David Brewster as "certainly the most extraordinary woman in Europe – a mathematician of the very first rank, with the gentleness of a woman, and all the simplicity of a child"[iii] and was deemed after her death "the most remarkable woman of her generation". "Her endowments were enhanced by rare charm and geniality of manner, while the fair hair, delicate complexion and small proportions which had obtained for her in her girlhood the sobriquet of the Rose of Jedburgh formed a piquant contrast to her masculine breadth of intellect",[iv] a comment that, incidentally, demonstrates how the notion persisted that physical beauty in women was incompatible with mental prowess.

Like Caroline Herschel, Mary Somerville in her old age wrote her captivating memoirs. She was born Mary Fairfax in 1780, the daughter of Vice Admiral Sir William Fairfax (1739–1813) of the Royal Navy. Serving under Admiral (Viscount) Duncan, Fairfax had distinguished himself at the sea battle of Camperdown against the French-Dutch alliance in 1797, for which he was knighted. Mary described him as "very good looking, of a brave and noble nature, and a perfect gentleman both in appearance and character". Mary's mother, Lady Fairfax, the daughter of a Scottish law officer, belonged to a cultivated family with connections in academic and literary circles.

The Fairfax family, for all that, lived modestly. Their home was at Burntisland in Fife (the house is still preserved), a small town on the north shore of the Firth of Forth, directly opposite the city of Edinburgh which could be reached only by ferry, a distance of several miles across often rough waters. She had a happy and carefree childhood roaming with her brother on the seashore, watching the fish and the birds, collecting seashells and fossils. The brother attended the famous High School in

DOI: 10.1007/978-90-481-2473-2_6,

Fig. 6.1 Mary Somerville, portrait by Thomas Phillips 1834. It was described as "one of the most intellectual of his female portraits". (Scottish National Portrait Gallery)

Edinburgh, the Alma Mater of many illustrious citizens of that city, which obliged him to live during term time with grandparents in town. There was no equivalent establishment for girls, and like most girls of her day, Mary received little formal education (apart from an unhappy year at a boarding school where she learned little and felt like a caged animal). Her mother taught her to read and write. The village schoolmaster who came to the house in the evenings for a time showed her how to "use the globes" (elementary geography and astronomy), perhaps from James Ferguson's book, but it was not considered appropriate for her to attend the lessons in Latin and navigation that he gave to local boys.

When Mary was somewhat older, her mother used to rent an apartment in Edinburgh for the winter, as was usual among Scottish families of distinction who lived outside the city.[v] Mary could then attend classes in reading and arithmetic. Of greater interest to her were the art classes she took with the renowned landscape painter Alexander Nasmyth. Nasmyth was the father of a remarkably talented family. Eight of his eleven children became artists, including six daughters who were encouraged to earn their own living.[vi] His youngest son James, who also had artistic talent, had a successful career as an engineer and made a fortune as the inventor of the steam-hammer, after which he became an equally successful astronomer, one of the "Grand Amateurs" of the nineteenth century.[vii] Painting, mainly watercolours, became one of Mary's lifelong relaxations. Her other abiding pleasure was music.

Fig. 6.2 Mary Somerville, portrait (chalk drawing) by J.R. Swinton, 1848. This is the portrait for which Mary's daughter Martha carved a beautiful frame, now lost. The original portrait, property of the National Portrait Gallery, hangs in Bodelwyddan Castle, Wales. (Mary Somerville, *Physical Geography*)

An uncle gave her a piano (a considerable gift in those days). She had lessons in Edinburgh from an Italian master, and practised hard: her favourite composer was Beethoven.

Accounts of Mary Somerville's early life tend to emphasise her lack of educational opportunity and hint at unsympathetic parents. Mary certainly complained about the handicap of girls in the matter of education; but this was the general situation for all young ladies in her circumstances. In fact, she was fortunate to have contacts from childhood with Edinburgh, the cradle of the Scottish Enlightenment, a lively city which an eighteenth century traveller had called a "hotbed of genius". Its university was famous throughout Europe for philosophy and medicine. There was a flourishing social life, theatres, a Musical Society, the Assembly Rooms in the New Town (the architecturally magnificent extension to the city built at the end of the eighteenth century) where evening dances were regular features of life among the young, as Mary Somerville describes in her *Recollections*.

Mary, however, had her own more serious ambitions – to study, to read books, to learn. There was one person who fully appreciated her "ardent thirst for knowledge". This was her uncle by marriage the Reverend Thomas Somerville, historian and minister of the Church of Scotland in Jedburgh in the Scottish Borders. His neighbour and friend, the renowned writer Sir Walter Scott, regarded him a "venerable member the literary brotherhood" and introduced him to the poet William Wordsworth and his sister on their tour of Scotland in 1803.[viii] Dr. Somerville recalled in his memoirs:[ix] "[Mary] often resided in my family, was occasionally my scholar and was looked upon by my wife and me as if she had been one of our own children". He assured the ambitious young girl that there was no reason why women should not become scholars, citing examples from ancient times. Each morning before breakfast she came to his study for an hour or two's reading of Virgil.

Mary first encountered the world of pure mathematics at Nasmyth's drawing class. The master was explaining the art of perspective: "You should study Euclid's Elements of Geometry, the foundation not only of perspective but of astronomy and all mechanical sciences". "Here in the most unexpected manner", wrote Mary, "I got all the information I wanted, for I at once saw that it would help me to understand some parts of Robertson's *Navigation*" – a standard text presumably belonging to her naval officer father. With the help of her brother's tutor, Mary acquired a copy of Euclid and began to teach herself mathematics.

In 1804, at the age of 24, Mary Fairfax married a second cousin on her mother's side, Samuel Greig, Russian Consul in London. Samuel was the son of the distinguished Sir Samuel Greig (1735–1788), Admiral of the Russian Navy who rose to the highest rank in that service, was responsible for modernising the Navy and decorated by the Empress Catherine.[x] Sir Samuel's elder son Alexis became also an Admiral in the Russian Navy, and in Mary's time commanded the Black Sea fleet.[xi] Mary's husband Samuel, too, was an officer in the Russian navy, and it was during a visit to the Forth on a frigate that Mary met him. Not trusting the political situation in Russia, Mary's father consented to the marriage only when Samuel was appointed to a safe post as Russian Consul in London where the young couple made their home.

The marriage does not appear to have been particularly happy. The couple lived in a cramped apartment with no garden, a contrast to the open spaces on the seaboard at Burntisland and the gracious squares and gardens of Edinburgh's New Town. Mary continued her studies "under great disadvantages. For although my husband did not prevent me from studying, I met with no sympathy whatever from him, as he had a very low opinion of the capacity of my sex, and had neither knowledge of nor interest in science". Mary's daughter (by her second marriage) was at pains to emphasise, and perhaps exaggerate, this point when she published her mother's *Recollections*: "Nothing could be more erroneous than the statement, repeated in several obituary notices of my mother, that Mr. Greig, her first husband, aided her in her mathematical and other pursuits. Nearly the contrary was the case. Mr. Greig took no interest in science or literature, and possessed in full the prejudice against learned women which was common at that time".[xii] These unflattering views of Samuel Greig's character are somewhat at odds with what is known of his brother Alexis who took a lively interest in science and was an early member of the Royal Astronomical Society. According to one report, "there is no question that the successful building and endowment of Pulkowa [the great Russian observatory founded by the Emperor

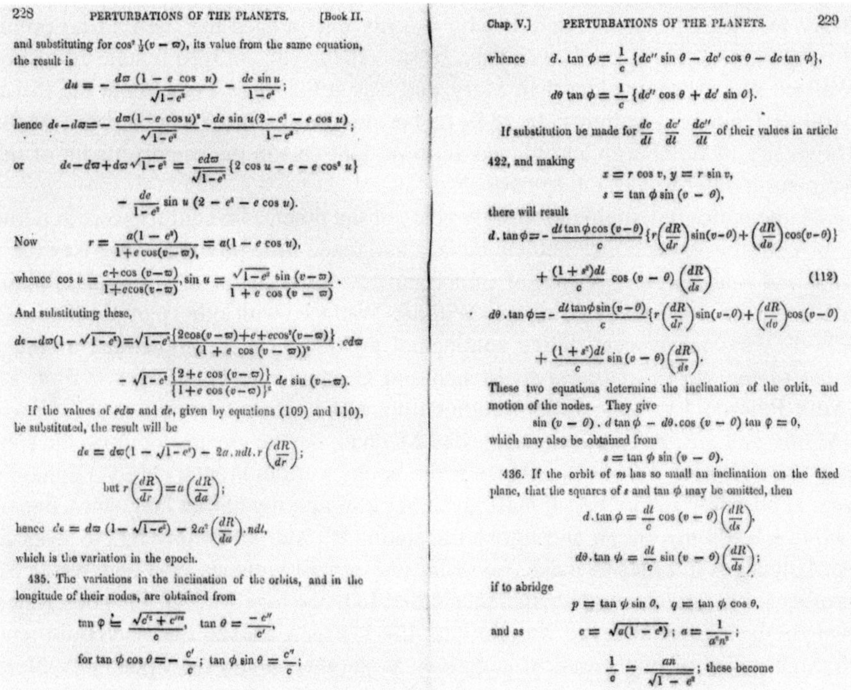

228 PERTURBATIONS OF THE PLANETS. [Book II.

and substituting for $\cos^2 \frac{1}{2}(v - \varpi)$, its value from the same equation, the result is

$$du = -\frac{d\varpi\,(1 - e\cos u)}{\sqrt{1-e^2}} - \frac{de \sin u}{1-e^2};$$

hence $d\epsilon - d\varpi = -\frac{d\varpi(1-e\cos u)^2}{\sqrt{1-e^2}} - \frac{de\sin u(2-e^2-e\cos u)}{1-e^2}$;

or $$d\epsilon - d\varpi + d\varpi\sqrt{1-e^2} = \frac{ed\varpi}{\sqrt{1-e^2}}\{2\cos u - e - e\cos^2 u\}$$
$$- \frac{de}{1-e^2}\sin u\,(2 - e^2 - e\cos u).$$

Now $$r = \frac{a(1-e^2)}{1+e\cos(v-\varpi)}, = a(1 - e\cos u),$$

whence $\cos u = \frac{e+\cos(v-\varpi)}{1+e\cos(v-\varpi)}$, $\sin u = \frac{\sqrt{1-e^2}\sin(v-\varpi)}{1+e\cos(v-\varpi)}$.

And substituting these,

$$d\epsilon - d\varpi(1-\sqrt{1-e^2}) = \sqrt{1-e^2}\frac{\{2\cos(v-\varpi)+e+e\cos^2(v-\varpi)\}}{(1+e\cos(v-\varpi))^2}.ed\varpi$$
$$-\sqrt{1-e^2}\frac{\{2+e\cos(v-\varpi)\}}{\{1+e\cos(v-\varpi)\}^2}de\sin(v-\varpi).$$

If the values of $ed\varpi$ and de, given by equations (109) and 110), be substituted, the result will be

$$d\epsilon = d\varpi(1-\sqrt{1-e^2}) - 2a.ndt.r\left(\frac{dR}{dr}\right);$$
$$\text{but } r\left(\frac{dR}{dr}\right) = a\left(\frac{dR}{da}\right);$$

hence $d\epsilon = d\varpi\,(1-\sqrt{1-e^2}) - 2a^2\left(\frac{dR}{da}\right).ndt$,

which is the variation in the epoch.

435. The variations in the inclination of the orbits, and in the longitude of their nodes, are obtained from

$$\tan\varphi = \frac{\sqrt{c'^2 + c''^2}}{c}, \quad \tan\theta = \frac{-c''}{c'},$$
$$\text{for } \tan\phi\cos\theta = -\frac{c'}{c};\ \tan\phi\sin\theta = \frac{c''}{c};$$

Chap. V.] PERTURBATIONS OF THE PLANETS. 229

whence $d.\tan\phi = \frac{1}{c}\{dc''\sin\theta - dc'\cos\theta - dc\tan\phi\}$,

$$d\theta\tan\phi = \frac{1}{c}\{dc''\cos\theta + dc'\sin\theta\}.$$

If substitution be made for $\frac{dc}{dt}$ $\frac{dc'}{dt}$ $\frac{dc''}{dt}$ of their values in article 422, and making

$$x = r\cos v,\ y = r\sin v,$$
$$s = \tan\phi\sin(v-\theta),$$

there will result

$$d.\tan\phi = -\frac{dt\tan\phi\cos(v-\theta)}{c}\left\{r\left(\frac{dR}{dr}\right)\sin(v-\theta) + \left(\frac{dR}{dv}\right)\cos(v-\theta)\right\}$$
$$+\frac{(1+s^2)dt}{c}\cos(v-\theta)\left(\frac{dR}{ds}\right) \qquad (112)$$
$$d\theta.\tan\phi = -\frac{dt\tan\phi\sin(v-\theta)}{c}\left\{r\left(\frac{dR}{dr}\right)\sin(v-\theta) + \left(\frac{dR}{dv}\right)\cos(v-\theta)\right.$$
$$+\frac{(1+s^2)dt}{c}\sin(v-\theta)\left(\frac{dR}{ds}\right).$$

These two equations determine the inclination of the orbit, and motion of the nodes. They give

$$\sin(v-\theta).d\tan\phi - d\theta.\cos(v-\theta)\tan\varphi = 0,$$

which may also be obtained from

$$s = \tan\phi\sin(v-\theta).$$

436. If the orbit of m has so small an inclination on the fixed plane, that the squares of s and $\tan\phi$ may be omitted, then

$$d.\tan\phi = \frac{dt}{c}\cos(v-\theta)\left(\frac{dR}{ds}\right),$$
$$d\theta.\tan\phi = \frac{dt}{c}\sin(v-\theta)\left(\frac{dR}{ds}\right);$$

if to abridge

$$p = \tan\phi\sin\theta, \quad q = \tan\phi\cos\theta,$$

and as $$c = \sqrt{a(1-e^2)};\ a = \frac{1}{a^2n^2};$$
$$\frac{1}{c} = \frac{an}{\sqrt{1-e^2}};\ \text{these become}$$

Fig. 6.3 A page from Mary Somerville's *Mechanism of the Heavens*. (Royal Observatory, Edinburgh)

Nicholas I in 1835] are mainly owing to his care and intelligent guidance".[xiii] James Nasmyth, the engineer-astronomer who met Grieg in St Petersburg, confirmed this assessment.[xiv] Mary kept in touch with her brother-in-law in later life and was able to arrange to have him provide Russian observations of the tides to British scientists. When she published her *Mechanism of the Heavens* (Fig. 6.3) she offered (and presumably sent) copies to Tsar Nicholas reminding him of her family's association with Russia.[xv]

After only 3 years of marriage Samuel died, aged only 29 (Mary does not record the cause of his death in her *Recollections*: it has been suggested that it was cholera[xvi]), and Mary, left with two small sons, returned to live with her parents in Burntisland. She had by now studied plane and spherical trigonometry, conic sections and Ferguson's *Astronomy*.[xvii] She resumed her studies, and during her quiet five years of widowhood, in between looking after her children, got to grips with the higher branches of mathematics.

It was Mary's good fortune at this stage in her life to make the acquaintance of a sympathetic adviser in the person of the mathematician William Wallace. Wallace (1768–1843), like herself a native of Fife and largely self-educated, had been a mere printer's apprentice who, through hard work and application, succeeded in reaching Edinburgh University to become a pupil of the great mathematician John Playfair.

Mary was later to know Playfair who warmly encouraged her: he had no prejudice against women – on the contrary, so she tells us, he enjoyed female company. Wallace was 10 years older than Mary, and was at that time a teacher at the Royal Military College, Sandhurst. In 1819 he became Professor of Mathematics at the University of Edinburgh and played an important part in the establishment of the Edinburgh (later Royal) Observatory.[xviii]

Mathematics in Britain in the early years of the nineteenth century were at a low ebb. With few exceptions, mathematicians had made little progress since Newton's time and many were ignorant of important developments in differential calculus made on the continent of Europe. William Wallace, with others of the Scottish school, was an advocate of the continental methods. He recommended to Mary a list of the works of the great French and German mathematicians – Biot, La Croix, Poisson, Lagrange, Euler – culminating with Laplace's monumental *Traité de Mécanique Celeste* (Treatise on Celestial Mechanics), the greatest opus on the subject since Newton's *Principia*. Most of the books were in French (Mary's unhappy year at boarding school had at least given her a smattering of that language), one or two in Latin, only one in an English translation.[xix] Mary therefore had to work at her languages in order to make use of these essential volumes. She and Wallace's brother John, an able mathematician, studied Laplace together, but she soon found that she was as quick as he was. In later life it was remarked that she could read through pages of mathematical equations as another would read poetry.[xx] Mary acquired her "treasure" of a library, now housed at Girton College Cambridge, at about the time of her second marriage.

Mary's second husband, whom she married in 1812 at the age of 32, was her first cousin, William Somerville, son of her beloved aunt and uncle in Jedburgh. William, an army surgeon, aged 41, already had an adventurous career in the colonies and was now Inspector General of Military Hospitals in Scotland. He had the opportunity of a similar post in Canada, but, to his father's great delight, he stayed instead, and married Mary.[xxi] William too had been married before and was a widower with one son.[xxii] Mary's younger son from her first marriage died at an early age, as did two children of the second marriage. Her eventual surviving family consisted of her son Woronzow Greig (named after his godmother, Countess Catherine Woronzow, daughter of the Russian Ambassador in London), her stepson James Somerville, and two surviving daughters of the second marriage, Martha and Mary. Woronzow eventually went to Cambridge and became a successful barrister as well as taking an interest in the sciences (he was a Fellow of the Royal Society) and being helpful to his mother in various practical ways. James studied medicine at Edinburgh University and became a doctor.

The Somervilles lived in Edinburgh for 4 years. One of William's duties during that period was an official visit to the field of Waterloo, in the company of the Professor of Military surgery at Edinburgh University.[xxiii] Edinburgh was still in its Golden Age of literati and philosophers among whom was the coterie of brilliant young writers behind the liberal *Edinburgh Review*. The Somervilles moved in these intellectual circles. William, a keen amateur of science, was elected a Fellow of the Royal Society of Edinburgh in 1813. Mary continued her mathematical studies, while together the couple took up the hobby of mineralogy – the beginning of Mary's serious interest in geology which she made use of in later life. Edinburgh was the birthplace of geology as a science, John Playfair being its leading exponent. The Somervilles could not have begun their married life in more congenial surroundings.

Dr. William Somerville was transferred to London in 1816 and a year or two later was appointed physician to the Chelsea Hospital for retired soldiers, a post he held for 20 years. It did not take long for the Somervilles to establish themselves in the metropolis. They took advantage of the Royal Institution's lectures, open to all, and through Edinburgh expatriates were introduced into the scientific scene. They were already acquainted with the young (24 year-old) John Herschel whom they had met on their honeymoon trip when William Wallace took them to call on his father, the venerable Sir William, at his observatory in Slough. "As a celestial explorer, Sir William Herschel had but one legitimate successor, and that successor was his son",[xxiv] who became the most revered and influential scientist in Britain, and probably in Europe. He was to be Mary Somerville's closest and most trusted friend and adviser for the rest of their lives. Through Herschel they met his former university class-mate and close friend Charles Babbage of calculating machine fame, and the astronomer Sir James South with whom Herschel was collaborating. South made the Somervilles welcome at his private observatory at Campden Hill in London, where he showed them how to observe double stars.[xxv]

They also formed a friendship with Edward Sabine, an army officer charged with magnetic surveys throughout the British empire, and with the geodesist Edward Kater, designer of a famous reversing pendulum for accurate measurement of gravity, and their talented wives. Mrs. Sabine assisted her husband and translated Alexander von Humboldt's famous work, *Cosmos*. Mrs. Kater, who was also "of great assistance to her husband" though she had a large family, was renowned at social gatherings in her home for her beautiful singing voice which Caroline Herschel had also admired. Another friend was Charles Lyell, the leading geologist of the younger school.

In the older generation they were privileged to be encouraged by the remarkable polymath William Hyde Wollaston, the first to discover dark lines in the solar spectrum (in 1802), Wollaston, some 20 years older than Mary Somerville, was by profession a medical doctor and also a gifted research chemist. In his simple spectroscope, sunlight entered through a slit between the window shutters into a prism. With this, he demonstrated the dark lines of the Sun's spectrum to Mary and gave her the prism to keep: it was one of the few scientific instruments she ever possessed. He also helped the Somervilles with their analysis of minerals, one of their hobbies. Mary Somerville had a special regard for Wollaston. She had a portrait of him painted for herself,[xxvi] and he bequeathed his cabinet of minerals to her in his Will (in which he also remembered Maria Edgeworth).

Another important friend was Thomas Young, the first to demonstrate the wave nature of light, whose lectures on physics Mrs. Marcet had attended at the Royal Institution. Mary Somerville found Young's published lectures "a mine of riches". Young, a scholar of extraordinarily wide knowledge, had already achieved fame in another specialist branch of learning, as the interpreter of Egyptian hieroglyphics.[xxvii] In her memoirs Mary describes how she and her husband, returning home late at night from observing the stars at the Katers' house, saw a light still burning in Young's window and rang his doorbell. He invited them in to show them a papyrus of astrological interest that he had just deciphered. Young was of "kindly disposition", and also musical (he played the flute). Mary was a friend of his wife Eliza, a clever woman who shared her husband's interest in mathematics.[xxviii] Wollaston, Young and Kater had all died before Mary herself rose to fame as an author.

Mary Somerville's wide circle in London overlapped the literary world, and included some famous women, all well established in London society. There was the revered aging Scottish poet and dramatist Joanna Baillie, known to Caroline Herschel;[xxix] the beautiful Mary Berry, writer and literary hostess, also Scottish, whom Mary already knew from earlier days; and, of course, the Irish literary magnet Maria Edgeworth.

Within only one year of their arrival in London William Somerville was elected a Fellow of the Royal Society which in those days readily accommodated supporters of science as well as active practitioners. Women, however, were ineligible. For Mary, her husband's Fellowship meant invaluable access by proxy to that society's scientific activities. In the same year, Mary met the French physicist and mathematician Jean Baptiste Biot, author of a work on analytical geometry and astronomy which she had studied, who was on a scientific mission to Britain. Learning of Mary's talents, he pressed her and her husband to visit Paris, which they did in the course of an extensive tour they made of the Continent in 1817.

In Paris they were received as distinguished guests by the city's scientific luminaries including the 68-year old Marquis de Laplace himself who had been informed that Mary had read his *Mécanique Celeste*. It was the couple's honour to be invited to a dinner at Laplace's country house at Arceuil, a village outside Paris, where the celebrated Club of Arcueil consisting of a dozen or so of the the country's leading men of science, used to meet.[xxx] The illustrious company included Biot, Dominique F. Arago, future Director of the Paris Observatory, the astronomer Alexis Bouvard, renowned compiler of astronomical tables, and their wives – with Mary Somerville seated next to Laplace who gave her an inscribed copy of his non-mathematical exposition of his work, *Systeme du Monde*, which pleased her "exceedingly".[xxxi] It was the confirmation of her status as a serious mathematician. Proceeding to Geneva, they met the Marcets who had their second home there, the occasion, perhaps, when Mrs. Marcet decided to publish her *Conversations in Natural Philosophy*.

Back in London, the Somervilles remained involved in scientific society. Charles Babbage was one of those whom they visited frequently when he was making his calculating machines.[xxxii] His first Difference Engine, completed in 1832 – a working machine which may be seen today in the Science Museum in London – would be on display at parties in his London home. Babbage's Saturday evening gatherings, "one of the great meeting places of liberal intellectual Europe",[xxxiii] were usually attended by two or three hundred guests, with the Somervilles among the regulars.[xxxiv] The tragic Ada (later Lady Lovelace), the daughter of Lord

Byron, attended Babbage's parties for the first time in 1833. There she met Mary Somerville, by now famous, who advised her to study mathematics, gave her instruction and helped with her difficulties. She became very attached to Mary, would often stay in her house, and kept up a correspondence with her until the end of her short life.[xxxv] Ada became actively involved in Babbage's work, and is nowadays acclaimed as the world's first computer programmer.[xxxvi] She died of cancer at the age of 37.

The recognition of her talents in high places did not go to Mary's head; to quote her father-in-law, "she never displayed any pretensions to superiority".[xxxvii] When the future eminent Edinburgh physician, Sir Robert Christison, a fellow student of her stepson James in Edinburgh, called on the family at Chelsea, he found "Mrs. Somerville the polished, highly accomplished unaffected lady of the household".[xxxviii] Maria Edgeworth on first meeting Mary in Wollaston's house described her as "timid, not disqualifyingly timid, but naturally modest." "While her head is among the stars, her feet are firm upon the earth".[xxxix]

For 10 years Mary Somerville kept up-to-date with the progress of science while attending to family duties. She even tried some experiments on solar radiation, using the apparatus Wollaston had given her, published in a paper communicated by her husband to the Royal Society in 1826.[xl] It was the first research paper by a woman to be published by that prestigious body. Entitled "On the magnetising properties of the more refrangible solar rays", it attempted to understand the nature of ultra-violet radiation (beyond the visible limit of the solar spectrum). This radiation was known to affect certain chemicals, specifically salts of silver (a fact that was important in the evolution of photography) and was thought perhaps to be in some way associated with magnetism, then a mysterious phenomenon. Mary Somerville repeated an experiment carried out originally in 1813 by Domenico Morichini, a scientist in Rome. Her apparatus, set up with advice from John Herschel,[xli] was in principle very simple, though delicate to use. It consisted in a light rod of steel (she used a sewing needle) exposed to the deep violet rays of the Sun to determine whether it became thereby magnetised. The Roman experimenter claimed to have obtained a positive result; other observers disagreed.[xlii] Mary Somerville's conclusion, that the effect was real, was accepted by the diplomatic Sir John Herschel as "feeble though unequivocal indication of magnetism", though "it was not surprising that it should have been regarded by many as insufficient to decide the issue".[xliii] Mary herself was later embarrassed about her paper, as her result proved to be erroneous,[xliv] and she "consigned all copies to the flames" (Mary's daughter, in editing her mother's *Recollections*, removed all references to this paper from the published version).

At the time, however, it made a deep impression and put Mary's name on the map. One of those who found the result inspiring was Joseph Mallord William Turner, the brilliant artist whom Mary Somerville knew well and whose work she greatly admired. Turner took a keen interest in science; he was a friend of Faraday, himself a good amateur artist, and knew men like Wollaston, Davy and Babbage with whom he could discuss matters of light and colour.[xlv] One of his famous canvases, *Ulysses deriding Polyphemus*, depicts Ulysses and his companions sailing north by their compass, in their return to Greece. "It can be no coincidence", writes the artist's recent biographer James Hamilton, "that Turner has floated a pale violet tone in the northern sky, the same colour that, in 1826, his friend Mary Somerville showed had the power to magnetise a needle, to make it point to the north."[xlvi]

The most important outcome of Mary Somerville's published paper was that she became "advantageously known to the philosophical world".[xlvii] "Suddenly and unexpectedly", she was invited by a Scots acquaintance, Henry Brougham (later Lord Brougham and Vaux), reformer, educationalist, lawyer and politician, to write an account in English of Laplace's *Mécanique Celeste*. Brougham began his career as a brilliant student of mathematics in Edinburgh and was in his youth a highly opinionated scientific controversialist.[xlviii] He was instrumental in the foundation of the University of London in 1827, and aimed at bringing important works of literature and science to a wider readership through his Society for the Diffusion of Useful Knowledge. Among them was Laplace's *Mécanique Celeste* which he invited Mrs. Somerville to tackle. He suggested two treatises, one to give the popular view, the other the analytical.

Mécanique Celeste, published in successive volumes between 1799 and 1827, the year of Laplace's death, was an intellectual masterpiece, "a divine work", Brougham called it. It contained solutions to formerly almost intractable problems in celestial mechanics – the deviations from perfect ellipses of the orbits of the various planets caused by their mutual gravitational influences, the motions of Jupiter's satellites, the Moon's complex motion around the non-spherical Earth and the theory of the tides. The treatise also included theoretical discoveries about the dynamics of the solar system. It was hardly a suitable choice for popularisation, for, as Mary pointed out to Brougham, it could not be expounded without the use of calculus and some preparatory mathematical theorems. Nevertheless the project went ahead, though not surprisingly it outgrew its original purpose.

Mary Somerville spent four years on her exposition of Laplace's opus, working secretly for fear of failure. She dreaded visits from well meaning friends, who would call to chat and pass the time with her – visits such that of Maria Edgeworth and her two sisters who "breakfasted at Mrs. Somerville's and sat in her painting room. Left her at one o'clock",[xlix] probably the same occasion when Maria "put on for her a blue crape turban to show her how one of Fanny's was put on, with which she had fallen in love".[l]

Mary Somerville's version of *Méchanique Celeste* was not a direct translation; the material was presented in the way she felt to be most lucid, using sometimes her own rather than the original methods of proof. At Mary's wish, Sir John Herschel read the manuscript and wholeheartedly endorsed it. Brougham had by now lost interest in it. The book was brought out by the famous publisher John Murray at his own risk. The result was *The Mechanism of the Heavens*, published in 1831 when Mary Somerville was 50.

The Mechanism of the Heavens was noticed in the serious British literary journals of the time, the *Quarterly* and *Edinburgh Reviews*, and in France. The reviewer in the *Quarterly Review*[li] was John Herschel himself who declared "we know not the geometer in this country who might not reasonably congratulate himself on such a work". On the question of the author being a woman, Herschel commented on "the simplicity of character and conduct, the entire absence of anything like female vanity or affectation.... In the pursuit of her object it yet never appears to have suggested itself to her mind that the acquisition of such knowledge, or the possession of such knowledge, by a person of her sex, is in itself anything extraordinary or remarkable".

The writer in the *Edinburgh Review*,[lii] Thomas Galloway, mathematician son-in-law of William Wallace, quoted a sonnet in her honour composed by William Whewell (later Master of Trinity College Cambridge) which ended:

> "An honoured name
> Be yours; and peace of heart grow with your growing fame."

Mary reproduced the poem in her *Recollections*. Whewell recommended the book to the students of mathematics in Cambridge where, as a result of the efforts of John Herschel and his like-minded friends, continental mathematics were at last being introduced. Whewell's recommendation guaranteed sales of several hundred copies of a book which was distinctly not in the Useful Knowledge category.

As a way of doing her honour, since they could not elect her to their number, Mary Somerville's friends in the Royal Society commissioned a bust from the famous and fashionable sculptor Sir Francis Chantrey, himself among the Somervilles' social acquaintances. This bust was placed in the Society's rooms where it still stands.

In April 1832, the Somervilles were delighted to receive an invitation from Whewell to visit Cambridge. They were to be the guests of the Airys (Professor George Airy who became Astronomer Royal in 1835) at the Observatory, but as this was some distance from the town it was decided that the couple (and the lady's maid) should reside at Trinity College in the rooms of the astronomer Richard Sheepshanks who happened to be away – a very special arrangement in a bachelor establishment. Adam Sedgwick, the geologist, planned a round of engagements for them.[liii] They received "every honour from the heads of the University"; it was an experience of which Mary bore "a proud and grateful remembrance".

The Irish mathematician William Rowan Hamilton has left an amusing account of that memorable week. He and his student Viscount Adare were on a packed scientific tour (dutifully described by Hamilton to Maria Edgeworth[liv] who liked to keep up to date with all the gossip) which began in London and ended in Cambridge where Hamilton was astonished to find his normally staid colleagues in a social whirl. "Since I came to Cambridge with my pupil", he wrote to his friend the poet William Wordsworth, "we have been nominally at Professor Airy's Observatory, but really in a continual round of breakfasts, dinners and evening parties at the university, especially Trinity College. At these we met Mrs. Somerville, a lady who has lately distinguished herself by publishing a commentary on Laplace and who happens to be now visiting Cambridge." Hamilton had first heard of Mrs. Somerville from Maria Edgeworth when he was still a student. He was clearly entranced by her. "Her visit to Cambridge exactly fell in with ours, for she spent there the same week that we did. The consequence was that we lived for that week in a continual round of engagements, and found Cambridge so gay that Airy, who hates ladies' parties, complains that we shall have gone away with quite a false and unjust notion of the University". He repeated his account to Samuel Taylor Coleridge. "I spent a week in Cambridge", he wrote, "where I met many eminent men, and one distinguished woman, Mrs. Somerville, who has lately published a kind of commentary on Laplace, which shows high mathematical attainments".[lv]

After Cambridge, as she mentions briefly in her *Recollections*, Mary Somerville and her husband spent a week in Oxford, at the invitation of the establishment geologist William Buckland and his wife, where the company included Roderick Murchison, a rising star of the younger generation who promoted the doctrine of the long-age universe. Perhaps it was not such fun as the week in Cambridge. Buckland tried hard to reconcile the Biblical account of the origin of the Earth with the findings of modern geology. Mary Somerville took the modern view and was criticised for it in some ecclesiastical quarters; but, as she remarked, "Dr. Buckland was obliged to join the geologists at last".[lvi]

Mary Somerville well deserved the honours she now received from learned institutions. The first was membership of the Royal Irish Academy, instigated by Rowan Hamilton. She was unanimously elected an Honorary Member "without the form of the ordinary ballot" at the meeting of 26 May 1834,[lvii] becoming the Academy's very first woman member. Caroline Herschel was elected in 1838 at the great age of 85, and Maria Edgeworth, the third women member, in 1842.

The Royal Astronomical Society which by its statutes could not elect women to the normal Fellowship devised a new category of "honorary member". Mary Somerville and Caroline Herschel were both elected to this rank in 1836. The inspiration to honour them owed more perhaps to Mary Somerville's winning personality than to Caroline's distinguished connections. Neither ever attended either of the learned societies.

The Societé de Physique et d'Histoire Naturel of Geneva also made Mary an honorary member (in April 1834). Her friend Jane Marcet was given the pleasant task of announcing the award, and added her own sentiments: "You received great honours, my dear friend, but that which you confer on our sex is still greater, for with talents and acquirements of masculine magnitude you unite the most sensitive and retiring modesty of the female sex".[lviii]

Mechanism of the Heavens was too specialised for the ordinary reader, but there was a 70-page introduction where the gist of the science was outlined in words. Separate copies of this much praised *Preliminary Dissertation* were reprinted and given to literary friends such as Maria Edgeworth and Joanna Baillie. Maria Edgeworth responded: "There is one satisfaction at least in giving knowledge to the ignorant, to those who know their ignorance at least, that they are grateful and humble … I can only assure you that you have given me a great deal of pleasure; that you have enlarged my conception of the sublimity of the universe beyond any ideas I had ever before been enabled to form."[lix] The *Preliminary Dissertation* was the basis of Mary Somerville's second book, a non-mathematical treatise incorporating allied branches of physics. Its preparation, however, required a considerable amount of extra research.

The new book, *The Connexion of the Physical Sciences* (where *connexion* is given its unusual spelling) covered properties of matter, optics, heat, electricity and magnetism, in addition to astronomy (the longest section); and showed that all these sciences were connected through the universal laws of physics. The author's dedication, with gracious permission, to the Queen (Adelaide, wife of King William IV) expressed her "endeavour to make the laws by which the material world is governed more familiar to my countrywomen". But though nominally for women (i.e., popular) it was in fact rather like a student's textbook, markedly more advanced than interpretations such as Mrs. Marcet's *Conversations on Natural Philosophy*. Explanations of such topics as the three-body problem, or the secular inequalities of the Moon's motion, or the polarisation of light in terms of the wave-theory, required considerable concentration on the part of the reader. At the same time, being pleasantly written, the words could communicate the thrill of discovery to the non-scientific reader. The difficult bits of geometry and the diagrams (which she drew herself) were consigned to an appendix.

The idea that all laws of nature might be interconnected. was surely inspired by John Playfair's lectures to students in the University of Edinburgh,[lx] published in her youth, which ended with the prediction that "a principle more general than all of these [laws of nature] and connecting all of them with that of gravitation, appears highly probable". Mary Somerville's opening words in her new book were something similar: "The progress of modern science has been remarkable for a tendency to simplify the laws of nature, and to unite detached branches by general principles." It was the aspired "Theory of Everything" of the day.

The book was written at a momentous moment in the history of physics, that which saw the discovery in 1831 of electromagnetic induction by Michael Faraday. Faraday, the successor of Humphry Davy at the Royal Institution, demonstrated his experiments in a series of evening discourses in that year and enunciated its laws in several papers published by the Royal Society. Mary Somerville was already a friend and correspondent of Faraday. She attended his lectures and included an account of the phenomena of electricity and magnetism leading up to these discoveries in her book. She sent him the proofs of these chapters for his comments. He admired her, advised her, and sent her copies of his papers throughout his life. She valued his "approbation and friendship" and regarded him as "the greatest experimental philosopher and discoverer next to Newton".[lxi]

Other chapters in the book were devoted to acoustics, the science of sound and vibrations. These were also entirely up to date. They made use of the work of Charles (later Sir Charles) Wheatstone (subsequently famous for his pioneering work in telegraphy) who performed fascinating experiments with musical instruments and on the patterns produced on vibrating surfaces, and wrote an important memoir on the subject, published in 1833.

The second edition of *On the Connexion of the Physical Sciences* came out only one year after the first. It included a Supplement on the most recent observations of the physicist Macedonio Melloni of the University of Naples, on infra-red (i.e., heat) radiation. Melloni visited Faraday at the Royal Institution and gave an important lecture there on the subject in January 1835. Mary Somerville succeeded in writing her additional chapter in time for the new edition which was in print within a few months.

The book was a triumph, acclaimed alike by scientists and general readers. Mary Somerville immediately became a celebrity. She was received by royalty and met the little Princess Victoria, the future queen. In 1835 the Government awarded her a Civil list pension of £200 a year (later raised to £300) which gave her an income for life. Indeed, between her pension and her royalties, she was, in her mid-50s, a very successful career woman. The book sold thousands of copies and was translated into German and Italian. It remained in demand throughout Mary Somerville's lifetime and beyond: the ninth edition appeared in 1858, 25 years after the first, and a tenth, revised by a respected woman populariser of science, Arabella Buckley, was published after her death.[lxii]

Before the epoch-making researches of Laplace and his contemporaries, it had been far from obvious that the solar system could remain stable under the maze of complex motions caused by the mutual action of the various bodies on each other. No longer. It was now demonstrated mathematically that the system is stable, and that the observed "perturbations" or disturbances, however complicated, are all cyclical and consequently predictable. Mary Somerville described the position as follows in the early editions of her popular book: "descending from the principle of gravitation, every motion in the solar system has been so completely explained, that the laws of any astronomical phenomena that may hereafter occur, are already determined."[lxiii]

But even as she wrote, unexplained discrepancies between the positions of Uranus, the outermost of the known planets discovered by William Herschel in 1781, and those predicted from Tables were beginning to emerge. In the next edition of her book (1836) Mary Somerville had to modify that statement: "The tables of Jupiter agree almost perfectly with modern observation; those of Uranus, however, are already defective, probably because the discovery of the planet in 1781 is too recent to admit of much precision in the determination of its motions, or that possibly it may be subject to disturbances from some unseen planet revolving about the sun beyond the present boundaries of our system."

Some years later, when the Somervilles were spending Christmas 1847 with Sir John Herschel and his family, John Couch Adams told William Somerville that it was this passage about "some unseen planet" which he read in the book's sixth edition (1842), that "put it into his head to calculate the orbit of Neptune",[lxiv] as that duly discovered planet was named. William, the adoring husband, may have over-interpreted Adams' kind remarks. Adams had resolved already in 1841 while still a student to investigate the irregularities in the motion of Uranus "in order to find whether they may be attributable to the action of an undiscovered planet beyond it".[lxv] It is possible, however, that the speculation in Mary's book was an additional spur. The story of how Adams at Cambridge and Urbain Leverrier in Paris simultaneously and independently computed the orbit of the new planet is well known. The actual discovery of the planet was made in Berlin in 1846 on the basis of Leverrier's predictions. It is a pity that Mary did not have the confidence to try the calculations for herself.

In 1835 Halley's comet was due to return (it went through perihelion on 15 November). Mary Somerville contributed an excellent essay on the subject to the *Quarterly Review*,[lxvi] a learned journal owned by Mary's publisher John Murray and edited by John Gibson Lockhart, son-in-law of Sir Walter Scott, whom Mary knew personally. It covered the theory of periodic orbits and their perturbations, the history of Halley's and other individual comets, the physical nature of comets, and a discussion on the possibility and outcome of a comet striking the earth. One must regret that this now little known essay was not circulated as a separate pamphlet, following the example of Sir John Herschel whose essays in the *Edinburgh* and *Quarterly Reviews* were published in book form.

These were very busy years. Apart from her writing, Mary Somerville, inspired by Melloni's lecture on infra-red radiation, resumed some spectroscopic experiments of her own with interesting results. Melloni, who began his pioneering researches in 1831 established the fact that infra-red or heat rays, like light, come in wavelengths, and that this invisible radiation is simply a continuation of the normal visible spectrum or rainbow colours. At the other end of the spectrum, radiation in the invisible ultra violet was known as "chemical rays" because of its blackening effect on certain silver salts – a fact that was the foundation of photography which began in 1834. Michael Faraday supplied Mary Somerville with some silver nitrate solution for experiments similar to Melloni's on the transparency of various media in terms of their blackening effect. Her mentor John Herschel being absent in South Africa, she communicated her results to her friends in Paris. Her paper, an early experiment in the science of photography, was published in French in 1836.[lxvii] the English version appeared a year later. Circumstances prevented her from pursuing these experiments any further at this stage. She resumed them 10 years later when living in Rome.

As regards mathematics and astronomy, her two brilliantly successful books, *Mechanism of the Heavens* and *The Connexion of the Physical Sciences*, and the essay on comets, all published between 1831–1835, represent Mary Somerville's short but intensely productive period. Apart from revisions, she was never to publish more on these favourite subjects. In 1832, at the suggestion of the French mathematical physicist Simeon Poisson, pupil of Laplace and Professor at the Sorbonne, she began, and apparently completed, an extensive work on the rotation of the Earth and on spheroids in general. It was never published: the writing *of The Connexion of the Physical Sciences* came in the way. On completing the latter, she began a purely mathematical work on analysis (calculus), but this, too, was shelved when a new edition of her popular book was demanded. The success of *The Connexion of the Physical Sciences*, and of her later book on Geography, meant the doom of Mary Somerville's career in the higher branches of science. The rest of her life was one litany of updating and re-editing – necessitated perhaps to some extent by financial necessity. One wonders whether her devoted husband fully appreciated where his wife's true talent lay.

In 1838 William Somerville, now in his late sixties, retired from his post at the Chelsea Hospital. The family went to live in Italy for the sake of his health. They were never to return to Britain except on visits. It was in Italy that Mary

Somerville wrote her second highly successful book, *Physical Geography*,[lxviii] published in 1848. She had always had a lively interest in the Earth sciences. From the mineralogy of her Edinburgh days she had progressed to the broader aspects of geology. Through her knowledge of astronomy she was familiar with the allied field of geodesy and the sciences of surveying and mapmaking. The friends of her own generation, the geophysicist Edward Sabine and the geodesist Captain Henry Kater, were both now dead, but the leading geologists of the younger generation, Sir Charles Lyell and Sir Robert Murchison, were close friends and fellow Scots.

Mary also had a romantic interest from childhood, through her father's tales, in travel and exploration. In that age of Arctic voyages, she followed closely the exploits of Parry and Franklin; a tiny island off the coast of British Columbia (Somerville Island, 54.3 N, 131 W) was named after her by Sir William Edward Parry on his third voyage in 1829. Before setting out on that expedition Parry had shown her round his ships when she presented him with a supply of marmalade, personally prepared, for the voyage.

On her first visit to Paris in 1817 and again in Germany in 1824 Mary had met the celebrated explorer and naturalist Baron Alexander von Humboldt whose record of his perilous five year-long travels through South America in 1799–1804, published over a period of years, were available in an English translation.[lxix] Another friend was Joseph Pentland, former British Consul in Bolivia who had explored the Bolivian Andes and had made some important geological observations. Pentland, now settled in Rome, provided Mary Somerville with geographical information and helped to see her book through the press.

Physical Geography, written intermittently, was the result of many years' accumulation of material. When it was finally ready for printing Mary learned to her dismay that the 80-year-old Humboldt's massive tome, *Cosmos*, the compendium of his scientific travels, was about to appear. Her first reaction was to destroy her manuscript, but Sir John Herschel encouraged her to publish. It was good advice. Humboldt himself, to whom she sent a copy of the book, wrote expressing his "*respecteuese admiration*" for her fine work which, he said, "had charmed and instructed him." He pointed out that there was room for both her book and his. Indeed the books were very different, Humboldt's being a first hand account of his adventurous explorations, while *Physical Geography* aimed at giving a scientific account of planet Earth. Mary Creese classes it as a pioneering work, which broke away from the usual pattern of describing the world country by country within political boundaries.[lxx] It opened with an excellent up-to-date outline of geology and the evolution of the Earth. Other sections dealt with meteorology, climatology, the distribution of flora and fauna, and even ethnology. Mary Somerville was a highly conscientious researcher. The book, a veritable encyclopaedia of quantitative data, including a list of the heights of hundreds of mountains peaks, proved every bit as popular as *Connexions*. It went into several editions and was on the book list in the universities of Oxford and Manchester until the 1890s.[lxxi] The revision of the sixth and final edition, which came out when Mary Somerville was aged 88, was reprinted in 1877, five years after her death.

Mary Somerville's contribution to geography was acclaimed in Britain and Italy. She was awarded the Victoria Medal of the Royal Geographical Society in 1869, and was made an honorary member of the Rome and Florence Geographical societies.[lxxii]

Mary Somerville's fame rests on her books. Her efforts in experimental physics are also worth remembering as an example of how a woman, working alone, felt the urge to undertake research in the manner of her male compatriots. Mary never parted with the rudimentary apparatus she used early in her career in London. Years later (in 1845), now living in Rome, she investigated the reaction of various coloured liquids (actually juices extracted from vegetables and flowers, what Herschel called floral dyes) to light of different wavelengths. She used paper moistened with the liquid under investigation and exposed it to the spectrum of the Sun. The illuminated strip showed a different pattern for each liquid, becoming bleached by some of the colours and not by others. Mary made drawings of the patterns produced by 18 different juices and sent her results to Herschel, who. communicated them to the Royal Society where they were published in a substantial paper.[lxxiii] The patterns she found are in effect low-resolution spectra of the liquids, though at that time such a scientific explanation was not available. Her work received little attention at the time, but it is satisfying to find this early work recognised in a modern history of spectroscopic techniques.[lxxiv]

It is moving and sad to contemplate Mary Somerville at the age of 65 in her Italian exile (for such it was) doing experiments with the only apparatus she possessed – Wollaston's little prism and lens, the same that she first used 20 years earlier in London. Herschel urged her to continue with her "delightful experiments".[lxxv] But no more is heard of Mary Somerville's efforts in experimental physics. A seventh edition of *Physical Sciences* was coming out and a new edition of *Physical Geography* was in preparation. In truth, her life in Italy was scientifically dull compared with the glorious 22 years in London among the colourful geniuses of an exceptional age. Even her acclaimed *Physical Geography*, though a huge commercial success, was to a great extent a waste of her talents: the chapters she felt obliged to include on human statistics and the races of man are a far cry from the rigorous mathematics of Laplace, and were a denial of the natural workings of her mind.

In 1860 Mary Somerville's husband William died at their home Casa Caponi, in Florence, aged 89; Mary herself was 80. He had been her greatest admirer and indefatigable amanuensis throughout 48 years of marriage, copying manuscripts, searching out material and attending conferences on her behalf. He was also her promoter and business manager. Mary said of him: "The warmth with which Somerville entered into my success deeply affected me; for not one in ten thousand would have rejoiced at it as he did; but he was of a generous nature, far above jealousy, and he continued through life to take the kindest interest in all I did."

William's devotion to his wife is charmingly evoked in an account by a visitor to the Somerville home in Florence in 1858, two years before his death. The visitor was Maria Mitchell, America's first woman astronomer, who was making an educational Grand Tour of Europe (Chapter 8). Armed with an introduction from the Herschels, Maria called on the Somervilles and was shown into their parlour. Dr. Somerville, "an exceedingly tall and very old man, in the singular head-dress of a red bandanna" entertained her while they awaited Mrs. Somerville. Presently she came "tripping into the room", looking 20 years younger than her 77 years and speaking "with the vivacity of a young person". Another visitor present, a voluble English woman, tried to bring the conversation round to the subject of art. Dr. Somerville, who was seated by the fire busily toasting a slice of bread on a fork, "grew fidgety, moved the slice of bread backwards and forwards as if the fire were at fault, and when, at length the English lady had fairly conquered the ground, and was started on a long sentence, he could bear the eclipse of his idol no longer, but coming to the sofa on which we sat, he testily said "Mrs. Somerville would rather talk on science than on art" ".[lxxvi]

Feeling the loss of her husband, Mary looked for something to do. She considered revising some chapters of *Physical Sciences*, but her daughters, foolishly, encouraged her to attempt something new. The result was *Molecular and Microscopic Science*,[lxxvii] a work concerned with recent discoveries in biology. Indeed, one wonders if the daughters had any real understanding of their mother's gifts: how, if they knew her genius for mathematics, could they have suggested a book about biology and zoology as a suitable occupation to divert her in her widowed old age? She spent five years on this book, published in 1869 when she was almost ninety. It was a formidable achievement for someone of her age; but biology was not her strength and she regretted having written it. "Mathematics are the natural bent of my mind", she wrote, and once the book was published (it had just the one edition) she returned to her favourite subject.

The last years of Mary Somerville's life were spent in Naples, with her two unmarried daughters. Her memoirs, written in those years, tell how she devoted some hours of every day to mathematics, her brain as sharp as ever. She also remained amazingly well informed of developments in astronomy and geophysics. She took an interest in recent observations in solar spectroscopy and the solar corona, and in the question of whether matter in the interplanetary medium might affect the movement of comets. Several astronomers passing through Naples on their way to observe the total eclipse of the Sun in Sicily in 1870 called on her. One of them, Professor Benjamin Pierce of the US Coast Survey, sent her on his return home some recent mathematical volumes. Other books she ordered from England, among them Peter Guthrie Tait's treatise on Quaternions and Rowan Hamilton's new system of algebra.

The sunspot maximum of 1871 provided many bright aurorae (northern lights), visible as far south as Italy. Mary Somerville commented in her *Recollections* on their influence on telegraphic communications, and recorded her own observation of the brilliant display of 4 February 1871 which lasted for several hours. "The whole sky from east to west was of the most brilliant flickering white light, from which streamers of red darted up to the zenith". She also left a vivid account of the powerful volcanic eruption of Vesuvius on 26 April 1872, having already witnessed the earlier outburst of 1868. It was visible from her house, but to get a better view she installed herself in a hotel opposite the mountain where she watched all day from a window. Such was the remarkable power of observation of a person of 92, only months before her death.

Mary Somerville died, in perfect peace, on 29 November 1872, one month short of 92 years of age. She is buried in the English Cemetery in Naples, her grave marked by a handsome marble monument.

After her death her elder daughter Martha edited her mother's *Recollections*, dividing the material into chapters, adding some interesting letters and words of explanation where needed. They were published soon afterwards, and are the source of most of our knowledge of Mary Somerville's long and interesting life.[lxxviii] The early part, with its vivid pictures of a Scottish childhood and of life in Edinburgh in the early nineteenth century, forms part of her native country's intellectual inheritance. Her experiences afterwards in London and Italy are a fascinating record of social life in that era: being an international celebrity she was received, without loss of her natural simplicity, in the company of scientists, artists and literary people.

In 2001, the *Recollections* were reprinted, with a scholarly commentary by Dorothy McMillan, an academic and expert on Scottish literature.[lxxix] This new edition gives Mary Somerville's original text (preserved among her papers in the Bodleian Library, Oxford) as well as the published version, which shows up some changes made with the best intentions by the daughter. They appear to have been aimed at giving Mary a more lady-like image, concentrating on her domestic virtues, and omitting any element that might detract from her status as a model of perfection – harmless references to personal vanity such as "I was still very good looking and was aware of it.", or "I must confess that I was fond of dress". Suggestions of weakness in her mother's formidable intellect were also purged: like a remark once made that she "regretted having spent so much time on science and [on] the dead instead of the living languages". More serious was the censoring of a reference to the experiment on sunlight which produced the erroneous result; in her intended manuscript Mary, an honest scientist and an honest autobiographer, had freely recorded the truth. Martha, however, eliminated it.

Mary Somerville was without doubt a person possessed of brilliant talents. Chief among these was her innate gift for mathematics, though she also had a keen talent for observation and a taste for experimentation. But did she make the most of them? Was she, as has been argued, "an astronomer and mathematician of underused talents"?[lxxx] Though she inspired immense admiration in the highest scientific circles, no-one, apart from John Herschel, encouraged her to turn her mind to original research. Her work on interpreting Laplace ought to have been the start of a full life in the academic world. In a different age one can well imagine that she would have joined the active Scottish school of mathematicians and become a worthy successor to William Wallace in the Chair of Mathematics at Edinburgh in 1838. Or she might have been appointed to the Chair of Astronomy in the same university, which was filled in 1836. Richard Proctor, astronomer and popular communicator of a later generation, put this view very strongly: "We shall never know certainly, though it may be that hereafter we shall be able to guess, what science lost through the all but utter neglect of the unusual powers of Mary Fairfax's mind."[lxxxi] However, her most recent biographer Allan Chapman takes a different view, preferring to consider

Mary Somerville in the context of the age in which she lived, in which exposition rather than doing of science was the best outlet for her talents. She was, he says, the Jane Austen of science. She belonged to "the world of the Grand Amateurs, and it was in this Grand Amateur world that Mary found her scientific voice and established her reputation".[lxxxii]

Mary Somerville herself deplored the lack of opportunities for women in her own lifetime. "Age has not abated my zeal for the emancipation of my sex from the unreasonable prejudice too prevalent in Great Britain against a literary and scientific education for women", she wrote when she was already over 90. She signed John Stuart Mill's petition to Parliament on women's suffrage in 1868, corresponded with him and thoroughly approved of his book *Subjection of Women* (1869). She rejoiced to see, in that same year, the foundation of Girton College for women at Cambridge.[lxxxiii]

Yet there was some ambiguity in her attitude. At the height of her fame she confesses to feeling "much less elated than might have been expected". "I was conscious that I never made a discovery myself, that I had no originality. I have perseverance and intelligence but no genius, that spark from heaven is not granted to the sex." (this passage was purged by the daughter from the published text). Dorothy McMillan believes that it was a common problem with women achievers to wonder if there is a difference between men's and women's brains as there is between their bodies.[lxxxiv] Mary Somerville was also influenced by the death in 1823 at the age of 10 of her eldest daughter, "a child of intelligence and acquirements far beyond her tender age" and feared that she had "strained her young mind too much."[lxxxv] She obviously did not put much strain on the remaining two daughters, who grew up cultivated and polished, devoted to their mother but without her high intellectual ambitions, without careers, and apparently supported by her in comfort at home throughout their lives.[lxxxvi]

Mary Somerville was an icon in her time, and is admired to this day for both her talents and her personality. Dorothy McMillan, taking the overall view of her full varied life, finds the key in the word "connection" – not just the connection she sought in the physical sciences, but in the domestic and spiritual aspects of living. "For her, the desirable life is the connected life and it is a life that does not categorise lest it stigmatise life's more homely areas."[lxxxvii]

Notes

[i] Elizabeth Chambers Patterson. 1983. *Mary Somerville and the Cultivation of Science, 1815–1840*. The Hague: Nijhoff; E.C.Patterson. 1975. *Dictionary of Scientific Biography* **12**, 521. New York: Charles Scribner's Sons; Mary Creese. 2004. Oxford DNB. Patterson's book is the authority on the life and work of Mary Somerville, particularly on her life in Scotland and London, and on her work in mathematics, astronomy and natural philosophy.

[ii] Mary Somerville. 1873. *Personal Recollections from early life to old age (with selections from her correspondence by her daughter Martha Somerville)*. London: John Murray. For a modern edition of *Personal Recollections* see *Queen of Science, Personal Recollections of*

Mary Somerville (edited and introduced by by Dorothy McMillan) Edinburgh: Canongate 2001. References given hereafter refer to this edition, as *Queen of Science*, and unattributed reerences to events i Mary Somerville's life quoted in this chapter are taken from this edition.

[iii] David Brewster. 1829. Quoted in James Hamilton. 2002. *Faraday, the Life*, p 274. London: Harper Collins.

[iv] Ellen Mary Clerke. Oxford DNB Archive.

[v] Jane McKinlay. 1987.*Mary Somerville 1780–1872*. Edinburgh: University of Edinburgh.

[vi] *Alexander Nasmyth and his Family: an Exhibition to mark the 150th anniversary of the death of "The Father of Scottish Landscape Painting"*. 1990. Exhibition Catalogue, Malcolm Innes Gallery, Edinburgh and London.

[vii] Samuel Smiles (ed.). 1883. *James Nasmyth, an Autobiography*. London: John Murray. Alan Chapman. James Nasmyth: Astronomer of Fire, in Patrick Moore (ed.) 1996. *The Yearbook of Astronomy*, 143–167. London: Macmillan; David Gavine. 1996. James Nasmyth (The Lorimer Lecture). *Astronomical Society of Edinburgh Journal* **35**.

[viii] Dorothy Wordsworth. 1981 (Reprint edition). *A Tour of Scotland in 1803*. Edinburgh: Mercat Press.

[ix] Thomas Somerville D.D. 1861. *My Own Life and Times 1741–1814*. Edinburgh.

[x] Oxford DNB. Admiral Samuel Greig's father was the first cousin of Mary's mother.

[xi] Oxford DNB.

[xii] *Queen of Science* p 72.

[xiii] R.A.Sampson in J.L.E. Dreyer and H.H. Turner (eds.). 1923. *History of the Royal Astronomical Society* Volume 1 Chapter 3. London: Royal Astronomical Society. (reprinted for the Society by Blackwell, Oxford, 1987).

[xiv] Nasmyth (ed. Samuel Smiles).1883. op. cit. p 291.

[xv] John H. Appleby. Entry on Worontsow Greig (Mary's son). Oxford DNB.

[xvi] Allan Chapman. 2004. *Mary Somerville*. Bristol: Canopus Publishing.

[xvii] James Ferguson. 1756. *Astronomy Explained upon Sir Isaac Newton's Principles*. London: Strahan (11th edition, 1803).

[xviii] H.A. Brück 1983. *The Story of Astronomy in Edinburgh*. Edinburgh.

[xix] This was Euler's *Algebra*, originally written in Latin in 1780, translated into English in 1810. The first introduction of the works of the French mathematicians to Cambridge was due to John Herschel, then a young man, who translated La Croix's *Calculus* into English in 1816, some years after Mary began her studies.

[xx] Patterson. op. cit. The comment was made by Somerville's publisher, John Murray.

[xxi] Thomas Somerville. op. cit.

[xxii] Patterson. op. cit.

[xxiii] James Simpson. 1816. *A Visit to Flanders in July 1815 Being Chiefly an Account of the Field of Waterloo*. Edinburgh: William Blackwood.

[xxiv] The descrip[tion is Agnes Clerke's.(*Popular History of Astronomy during the Nineteenth Century*).

[xxv] All references from *Queen of Science*.

[xxvi] Oxford DNB entry on William Hyde Wollaston.

[xxvii] Andrew Robinson. 2006. *The Last Man who knew Everything*, Oxford: One World.

[xxviii] Alexander Wood, completed by Frank Oldham.1954. *Thomas Young Natural Philosopher 1773–1829*, 190–91. Cambridge University Press.

[xxix] Caroline Herschel. *Memoirs* p 122. Her plays were among the books given to Caroline as a farewell gift by Lady (William) Herschel when she left England.

[xxx] Sir Edmund Whitaker. 1949. Laplace. *The Mathematical Gazette* **33**, no 303. 1–13.

[xxxi] The volume is now in the archives of Girton College, Cambridge.

[xxxii] *Queen of Science*, p 115.

[xxxiii] Anthony Hyman. 1982. *Charles Babbage Pioneer of the Computer*, p 175. Princeton University Press.

[xxxiv] ibid. p 174.

[xxxv] *Queen of Science*, p 125.

[xxxvi] Dorothy Stein. 1985. *Lady Lovelace, a Life and a Legacy*. MIT Press.
[xxxvii] Thomas Somerville. op. cit. p 389.
[xxxviii] *Life of Sir Robert Christison by His Sons*. Vol 1, 1885, p 120. London: Blackwood.
[xxxix] *Queen of Science*, p 127. Quotation from a letter in 1822.
[xl] Mrs M. Somerville. 1826. On the magnetising properties of the more refrangible solar rays. *Phil. Mag*. **68**, 168–173.
[xli] *Queen of Science*, p 109.
[xlii] James Hamilton. 2003, *Faraday: The Life*. p 99. London:HarperCollins. When Sir Humphry Davy, accompanied by his novice helper Michael Faraday, made his great scientific tour of Europe in 1814, the pair called on Professor Morichini to witness a demonstration of the alleged magnetising rays. Faraday was convinced: the experienced Davy was sceptical.
[xliii] Quoted in R.A.P. [roctor]. 1873. Obituary notice of Mary Somerville. *Mon. Not. Roy. Astron. Soc*. **33**, 190–197.
[xliv] Mary R.S. Creese. 1998. *Ladies in the Laboratory*, p 211. Lanham and London: Scarecrow Press. Mary Somerville's result had some support at first.
[xlv] Hamilton, op. cit., p 223.
[xlvi] James Hamilton. 1997. *Turner a Life*, p 243 London: Hodder and Stoughton, Sceptre paperback.
[xlvii] John Herschel. 1833. Mechanism of the Heavens (book review). *Quarterly Review* No 99. Reprinted in Sir John Herschel 1857. *Essays*, p 41. London: Longman, Brown, Green, Longman & Roberts.
[xlviii] Brougham was one of the founders of the *Edinburgh Review* where he wrote notoriously scathing criticisms of Thomas Young's research papers on the wave theory of light (1801–1803). (Andrew Robinson, op. cit. Chapter 8). Mary Somerville deplored the *Edinburgh Review* attack in her memoirs. (*Queen of Science*, p 221). B.
[xlix] Maria Edgeworth, Letter to her aunt Mrs Ruxton, 4 March 1822, quoted in Augustus Hare. 1894. *Life and Letters of Maria Edgeworth*. Volume 1, p 75. London.
[l] Maria Edgeworth, quoted in Christina Colvin. 1971. *Maria Edgeworth, letters from England 1813–1844*. p 386. Oxford:Clarendon Press 1971. To be fair, this particular visit occurred before Mary Somerville began writing.
[li] John Herschel, *Quarterly Review* **47**, 537, 1832; reprinted in his *Essays*. op. cit.
[lii] Thomas Galloway. 1832. *Edinburgh Review* **60**, 1.
[liii] Mary Somerville's Recollections gives the date on Sedgwick's letter as 1834, but from Hamilton's well-documented correspondence it was unmistakably 1832.
[liv] R.P. Graves, *Life of Sir William Rowan Hamilton*, Volume 1. p 551. Dublin: Hodges Figgis. Letters to Maria Edgeworth, 1832 March 29 from London and April 3 from Slough.
[lv] ibid. p 559. Letter, June 14, 1832.
[lvi] *Queen of Science*, p 106.
[lvii] Academy Minutes and Council Minutes, Royal Irish Academy Dublin, 1834.
[lviii] Queen of Science, p 168. Letter from Jane Marcet, 6 April 1834.
[lix] ibid., p 163. Letter dated May 31, 1832.
[lx] John Playfair. 1816. *Outlines of Natural Philosophy, being heads of lectures delivered in the University of Edinburgh*, vol 2 (2nd edition) Edinburgh.
[lxi] *Queen of Science*, p 238.
[lxii] 10th edition, corrected and revised by A.B. Buckley, London 1877.
[lxiii] *On the Connexion of the Physical Sciences*, p 4, second edition, 1835.
[lxiv] *Queen of Science*, p 235.
[lxv] H.M. Harrison. 1994. *Voyager in Time and Space, the life of John Couch Adams, Cambridge Astronomer*, p 24. Lewes: The Book Guild. Adams' note is dated 3 July 1841.
[lxvi] Mary Somerville. 1835. *Quarterly Review* **55**, 195–233, 1835.
[lxvii] Mary Somerville. 1836. *Comptes Rendus* **3**, 473–76.
[lxviii] Mary Somerville. 1858. *Physical Geography*. London: John Murray.
[lxix] McGillivray's translation, 1832.

[lxx] Mary R.S. Creese. 1998. *Ladies in the Laboratory?* p 203. Lanham Md and London: Scarecrow Press.

[lxxi] ibid.

[lxxii] In all, according to Elizabeth Patterson, she was elected an honorary member of 11 Italian and two American learned societies during the last three decades of her life.

[lxxiii] Mrs M. Somerville. 1846. On the action of rays of the spectrum on vegetable juices, extract of a letter to Sir J.F.W.Herschel, Bart. dated Rome September 20, 1845. *Phil. Trans. 1846.* p 111–21.

[lxxiv] Klaus Hentschel. 2002. *Mapping the Spectrum*, p 202. Oxford University Press.

[lxxv] *Queen of Science*, p 227, a letter from John Herschel 1845.

[lxxvi] Phebe Mitchell Kendall. 1896. *Life, Letters and Journals of Maria Mitchell.* Boston: Lothrop, Lee and Shepard Co. Inc.

[lxxvii] Mary Somerville. 1869. *Molecular and Microscopic Science*. London: John Murray.

[lxxviii] Mary Somerville. 1873. *Personal Recollections from early life to old age*, edited by Martha Somerville. London: John Murray.

[lxxix] Dorothy McMillan (ed). 2001. *Queen of Science. Personal Recollections of Mary Somerville*, edited and introduced by Dorothy McMillan. Edinburgh: Canongate Classics.

[lxxx] Mary T. Brück. 1996. Mary Somerville, mathematician and astronomer of underused talents. *Journal of the British Astronomical Association* **106**(4), 201–206.

[lxxxi] R.A. Proctor, Obituary. *Monthly Notices of the Royal Astronomical Society*, op. cit.; Mary T. Brück, op. cit.

[lxxxii] Allan Chapman. 2004. *Mary Somerville*, p 122. Bristol: Canopus.

[lxxxiii] Elizabeth Patterson, op.cit.

[lxxxiv] Dorothy McMillan. Introduction to *Queen of Science*, p xxiv.

[lxxxv] Dorothy McMillan. *Queen of Science*, p 124.

[lxxxvi] Martha practised wood-carving: a beautiful ornate frame which she made for a portrait of her mother, now in Bodelwyddan Castle in Wales, is unfortunately lost. *A frame by Martha Somerville, a Victorian carver in Italy*. National Portrait Gallery website.

[lxxxvii] Dorothy Mcmillan. op. cit. Introduction p xl.

Chapter 7
In the Shadow of Giant Mirrors

Mary Somerville lived until 1872. She had emigrated to Italy in 1838; her last scientific paper was published in 1845, and her last revision of *Connexion of the Physical Sciences* was done back in 1855. What was going on in the world she had left behind in Britain?

In the realm of science, and of astronomy in particular, it was the great age of the independent amateur. Unlike the situation in France and Germany where the sciences were well on their way to being state sponsored professions, innovative science continued to be the province of the private individual who could afford to do research at his own expense and according to his own ideas. Pioneers like the biologist Charles Darwin and the geologist Charles Lyell were men of independent means; in astronomy, the most illustrious amateur (by this definition) was Sir John Herschel, whose inherited wealth allowed him to decline a university career and to organise his life exactly as he wished.[i] The British amateur tradition not only survived but flourished until the last quarter of the nineteenth century.

According to *Encyclopaedia Britannica*, 25 private observatories operated at one time or other in the British Isles in the nineteenth century, half of them still functioning in 1884.[ii] Their owners were men of inherited wealth, or men who having made their fortunes in business chose to acquire first class telescopes, perhaps hiring educated young assistants to do the hard work. Others had ambitions that went further: they constructed their own large mirror telescopes, taking up the quest for more light begun by William Herschel.

The largest mirror ever constructed before the 1840s, the 40-ft long reflector with its mirror of 48 in. diameter, had been the great achievement of William Herschel; but for many reasons the giant unwieldy instrument was a disappointment, and after his death was allowed to fall into disuse. (The great survey of the sky completed by John Herschel in South Africa in the 1830s, was carried out with his father's smaller 20 ft instrument with its 18 in. diameter mirror.) Now, decades later, with improved technology, it was time to try again. But, as Allan Chapman points out, building and utilising large reflecting telescopes was as costly in money and manpower as maintaining racehorses,[iii] and those who took up the challenge had to be persons of wealth and leisure. Most spectacular of these who made the attempt were two men from very different backgrounds – Lord Rosse, an Irish landowner, and William Lassell, "a brewer by profession but an astronomer by inclination".[iv] A third was

M. Brück, *Women in Early British and Irish Astronomy*,
DOI: 10.1007/978-90-481-2473-2_7,

Lassell's associate and adviser James Nasmyth, son of Mary Somerville's former drawing master in Edinburgh, who built up such a lucrative business as an engineer and inventor that he could retire before the age of fifty and throw himself into enjoying his beloved hobby.

Lord Rosse (William Parsons) (1800–1867),[v] third Earl of Rosse, a title he inherited on his father's death in 1841, studied at Trinity College Dublin and at Oxford where he graduated with first class honours in mathematics (Fig. 7.1). He had a talent and taste for mechanics and chose to become an entrepreneurial astronomer according to his own inclination. He began by experimenting with alloys from which to cast mirrors, and went on to constructing machines for grinding and polishing. His first success was a telescope with a metal mirror of 36 in. in diameter completed in 1839. The images were of the highest quality which led Rosse to aspire to construct another of double the diameter. The story of how he achieved his ambition, employing only local labour and machinery in his own workshops and forges in the grounds of Birr Castle in Ireland, is one of astronomy's great legends. The giant mirror, of 72 in. diameter and 54 ft in focal length, was cast in 1842, and the actual telescope, dubbed the Leviathan, mounted between two gigantic pillars, was ready for use in 1845 (Fig. 7.2). It was the wonder of the astronomical world, and would remain unsurpassed in size until the 100-in. telescope on Mount Wilson in California was erected in 1917. Hardly was the Leviathan turned on the sky when it made its famous discovery: the spiral structure of certain nebulae, objects that would eventually prove to be external galaxies. The first spiral to be revealed, early in 1845, was the famous Whirlpool Nebula (M51 of Messier's Catalogue) which had shown little if any structure in the best existing drawings of the Herschels. Altogether, fifteen spiral nebulae were confirmed with the great Leviathan and many more were suspected.

It is no surprise that intelligent women who witnessed such exciting activities should have been drawn into participating in them. In Lord Rosse's case, the women involved were his wife and his young cousin Mary Ward.

Lord Rosse's wife Mary (1813–1885), Countess of Rosse, whom he married in 1836, played an important role in the rise of the great observatory and may be regarded as in many ways her husband's close partner.[vi] Crucially, it was she, an English heiress, who financed the major project which, at a cost of £12,000, was beyond the means even of an extensive landowner.[vii] This fact was not publicised at the time, perhaps by her own wish.[viii] For this indispensable contribution to astronomy, Mary Rosse could claim a place in the female pantheon alongside Anne Sheepshanks, sister of the Cambridge astronomer Richard Sheepshanks, who was made an Honorary Member of the Royal Astronomical Society in 1862 in recognition of her benefactions to the Society and the University. It is also known that the planning and implementation of operations in the Birr venture owed a great deal to Mary Rosse's organisational skill and to her energetic commitment to discipline and hard work. She ruled her large family and her staff with a rod of iron, every member having their prescribed duties, which, including the children's lessons, commenced each morning at six o'clock.

Fig. 7.1 The Earl of Rosse showing his drawings of spiral nebulae to the Countess, May 1850. He showed the drawings at the Royal Society the following month. Watercolour by Charles Piazzi Smyth. (Royal Observatory, Edinburgh, Piazzi Smyth Collection)

The energetic Mary Rosse also established an independent place of her own as a distinguished photographer in the early days of that new art. This separate career began in 1853, prompted by Lord Rosse's wish to photograph the Moon. There is no record that he persevered with this ambition; but it led to his wife's own serious interest in photography. She became a skilful and artistic photographer whose work was admired by the great pioneer William Fox Talbot and exhibited in London and Dublin. She was a founder member of the Irish Photographic Society, and one of its prizewinners. Her historic photographs of the giant telescope and buildings, still preserved in Birr, remain a valuable legacy to the history of astronomy.

Also associated with Lord Rosse's observatory was his young cousin Mary King (1827–1869) (afterwards Mrs. Ward),[ix] daughter of a neighbouring country clergyman, who witnessed the Birr activities from 1840 when she was the only 13 until the Leviathan came into operation in 1845 and afterwards. Like most young ladies of her time and class, her accomplishments under the tutelage of a governess included drawing, in which she became extremely skilful. In numerous sketches and watercolours which are still preserved, she recorded the construction of the giant telescope in all its stages.

At Birr Castle and in her own scholarly home she came to know leading astronomers including Lord Rosse's particular friend and mentor Thomas Romney Robinson of Armagh Observatory, Sir James South, collaborator of John Herschel, and the physicist David Brewster for whom she illustrated scientific papers. These scientists also had ties of friendship with Maria Edgeworth and Mary Somerville (previous chapters): Robinson married Maria Edgeworth's half-sister Lucy as his second wife in 1843, adding to the complex ramifications of the Edgeworth family;

Fig. 7.2 A photograph by Lawrence, 4th Earl of Rosse of the Leviathan of Birr, with the observer's gantry, where Mary Ward described being suspended. (Sir Patrick Moore, from his *Astronomy of Birr Castle*)

the Rosses knew the old but still indomitable Maria herself, while Mary Somerville was a friend of South, Hamilton and Brewster, and a correspondent of Lord Rosse.

Robinson and South participated closely in the progress of the great reflector and were in Lord Rosse's company when it was directed on the sky for the first time in February 1845. They viewed Jupiter, some double stars and the Orion Nebula; the spiral structure of the Whirlpool Nebula was discovered by Lord Rosse in April. The 18 year-old Mary King was among the first to have the privilege of looking through that wondrous instrument. We are not told what object she viewed on that occasion; but she has the distinction of being the only woman to have made the perilous ascent of the galleries of the gigantic edifice, where the observer was dangerously "suspended over a chasm 60 ft deep". There, as she told William Rowan Hamilton, "she had more than once stood in bitter cold long after midnight".[x]

Birr now became a magnet for visitors, and Mary is sure to have met other notable astronomers there, such as William Lassell and the group including George Airy the Astronomer Royal, Colonel Sabine and Charles Piazzi Smyth (next chapter) that convened there in 1852 to plan the mapping of the Moon. She was one of a number of women who took part in the meeting of the British Association in Belfast that same year. She married in 1854, after which she lived in Dublin and elsewhere in Ireland, but kept in contact with Birr and her astronomer friends.

William Rowan Hamilton was impressed by Mary Ward's knowledge of astronomy. He found her "so well informed and so much in earnest as a student of the stars" that he gladly assisted her in her enquiries about Donati's comet which appeared spectacularly in 1858, and obtained copies of useful publications for her. Hamilton was an encourager of women in all intellectual pursuits. He had three devoted sisters who lived with him in his young bachelor days at the observatory whom he hoped would become his scientific partners. "You know how desirous I have been that you should learn astronomy, both for your own sake and for mine", he wrote to one of them, Eliza, who, like himself, was something of a poet, "for yours, because I consider the study of science useful to all minds, but especially to the female, and still more especially to a poetic, mind". He hoped that at least one of his sisters would emulate Caroline Herschel, but they (fortunately for them) did not have Caroline's dog-like admiration for their brother. They did, however, perform calculations for him from time to time, and the eldest, Grace, grew to be "quite a diligent observer" who on one occasion located a newly announced comet in her hand-held telescope: Hamilton himself who had little taste for observing had not troubled to look for it.

Astronomy was not, however, Mary Ward's only interest, or even her chief one. Her favourite pursuit was entomology. She became an expert naturalist and collector of botanical and biological specimens. She possessed a microscope (acquired on Brewster's recommendation) and made her own beautiful slides and delicate drawings from nature. While bringing up a family of several children, she found time to publish books on biological and entomological subjects, culminating in her acclaimed *Microscope Teachings* (1864) (renamed *The Microscope*) which sold by the thousand. A similar volume dedicated to astronomy, *The Telescope* (originally published in 1859 as *Telescope Teachings*) was equally successful and went into several editions.[xi]

The Telescope, subtitled *A familiar sketch combining a Special Notice of Objects coming within the range of a small telescope*, was aimed at the ordinary amateur, plainly written and beautifully illustrated in delicate colour, what she called "light literature of the sublime science". Based on the author's own experience with a good quality 2-in. instrument, acquired at Lord Rosse's suggestion, it describes the appearance of these objects – members of the solar system, certain double stars, a few star clusters and the Orion Nebula – and gives straightforward information from trusted sources such as Herschel's *Treatise on Astronomy*, Humboldt's *Cosmos*, Gall's *Sky Atlas* and *Monthly Notices of the Royal Astronomical Society*.

The Telescope was published soon after the appearance of Donati's comet in 1858, of which Mary Ward included in her book a particularly charming coloured sketch in a country landscape, as she herself had observed it on the evening of 11 October (Fig. 7.3). It is a depiction that more than matches the many famous representations made at that time. A complete set of her drawings of the

Fig. 7.3 Donati's comet, 11 October, 1858, as seen near Dublin, observed and sketched by Mary Ward. (Mary Ward, *The Telescope*, 1859)

comet, observed over a period of 10 days, showing its changing place against the background of the stars, was published in the first edition.[xii]

Another recent event included in her book was a near-total eclipse of the Sun in 1858, visible from Britain and Ireland, shown illustrated in an authentic cloudy sky. The year was close to sunspot maximum, and a fine series of drawings of spots was also included.[xiii] All in all, her charming and instructive book bears the stamp of the naturalist who records in plain words and illustrations what she sees with her own eyes, without speculation or theorising. For this reason, Mary Ward's *The Telescope* is timeless as a beginner's guide to astronomy with a small telescope. Its value in its day was in stimulating an interest in astronomy among the general public through a combination of clear writing, and excellent illustration.

Mary Ward's talent as an observer was put to good use in her naked eye account of the great meteor display of 13–14 November 1866. This annual meteor shower which arises from a dust stream in a 33 year orbit round the sun, recurs with special intensity every 33 years, and was well anticipated. It did not disappoint: "God sent enough fireworks tonight", said an Irish railway porter, quoted by Mary Ward in a vivid account in the *Irish Times* newspaper two days later.[xiv] Her observations were made from her home near Dublin and reported in a long paper published in 1867[xv] where it appeared together with the account of Alexander Herschel, John Herschel's son, a specialist on meteors, who observed the same display from Glasgow. Her paper, besides describing her own observations in careful detail, included an explanation of the origin and occurrence of meteors, drawing chiefly on John Herschel's exposition in *Outlines of Astronomy*. Had Mary Ward lived, she would undoubtedly have continued to make observations and would have earned a place as an original researcher as well as a populariser of astronomy.

This, however, was not to be. Only two years later, in 1869, Mary Ward died at the age of only 42, in an accidental fall from a mechanical road locomotive in the town of Birr. It was doubly tragic, in that the machine was an invention of her cousin, the fourth Earl of Rosse, son of the great telescope builder who had died at the relatively early age of 67; and that Mary was visiting Birr to pay tribute to his memory. Her popular book, which went into numerous editions and stimulated a genuinely intelligent interest in the subject among the general public, was her legacy to astronomy. Her wide range of talents, her courage in adversity and her entrepreneurship, are deservedly recorded by her present-day admirer, Susan McKenna-Lawlor.[xvi]

Lady Mary Rosse retired to England after her husband's death. The observatory at Birr continued to operate under the fourth Earl of Rosse (1840–1908) who inherited his father's passion for astronomy and his flair for mechanics.[xvii] Assisted by a succession of distinguished assistants, he used both the Leviathan and the more manageable 36-in. reflector to observe nebulae for some years; but with the next step in technology – the construction of large silvered glass mirrors – the giant and clumsy telescope had lost its edge and, around 1878, ceased to be used. The smaller instrument, however, was employed for delicate observations of heat rays from the Moon, results which were of the greatest significance, because they proved that the only heat which the Moon possesses is due to sunlight on its surface.

When the fourth Earl of Rosse died in 1908, the Leviathan was dismantled and the huge metal mirror sent to the Science Museum in London where it is still preserved. The observatory ceased to function in 1916. Its history, fortunately, has not been lost. The seventh Earl of Rosse, descendent of the great telescope makers, has refurbished the giant instrument and has established a splendid museum in the original, almost unchanged, setting of the great observatory which is now open to the public. There one can see preserved many relics of the tools and instruments used in the construction of the historic telescopes; and also, the well-preserved contents of Mary Rosse's darkroom, believed to be the oldest extant darkroom in the world.

Mary Ward's archives, including her drawings, are preserved in Castle Ward in Co. Down, the home of her husband's family where she was a regular visitor during her married life, and which, after her death, was inherited by her widower who succeeded his brother as Viscount Bangor. It is now the property of the National Trust of Northern Ireland.

Lord Rosse's fellow-master in the art of constructing large astronomical mirrors was the brewer turned astronomer, William Lassell (1799–1880) of Liverpool. Unlike Lord Rosse, who had the advantage of a university education and leisure to spend as he chose, Lassell had to earn his living and his later wealth. He began his astronomical career in a simple way, sharing his hobby with young friends of similar interests. He experimented with producing alloys for mirrors, and with making mirror telescopes. His business prospered, and by the time he was 40 years of age – half way through a long life – he could afford to devote himself entirely to astronomy. He was "an astronomical phenomenon", being as talented as a telescope user as he was a telescope designer, and went on to become one of Europe's most admired astronomers, on a par with the greatest of the academics.[xviii]

Lassell's great achievement at this time was a beautiful reflecting telescope with what he could truthfully claim was an "almost perfect" mirror of 24 in. in diameter. The telescope was equatorially mounted, that is, movable, so as to follow the rotation of the sky, the first large instrument ever to be so arranged, which was in itself a major feat of engineering. Lassell erected this telescope at his combined observatory and home, Starfield, in the outskirts of Liverpool. The instrument's very first success, and proof of its superb quality, was to reveal a moon (later named Triton) of the planet Neptune within 17 days of that planet's historic discovery in 1846. Lassell had experienced no difficulty in seeing the planet's image distinctly as a disk as soon as its position in the sky had been announced, while inferior instruments were unable to distinguish its image from that of a star. Had he known where to look, it is possible that the famous discovery of Neptune would have been made in Liverpool, rather than in Berlin.[xix] Lassell also discovered Saturn's satellite Hyperion, and the two satellites of Uranus. In 1852 he transported his telescope to the better climate of Malta, in the Mediterranean, for a short period, where observing conditions were much better than in smoky Liverpool.

Like Lord Rosse in Birr, Lassell built his own workshops and polishing machines on his own premises, and went on to construct an even larger mirror of 48 in. diameter and 40 ft focal length. He moved the telescope incorporating this mirror, and all its auxiliary machinery, to Malta in 1861 for a four-year sojourn dedicated chiefly to the observation of nebulae. So massive were this telescope and its driving mechanism that two men were needed to operate it. Lassell also employed a professional astronomer, Albert Marth, a talented German university graduate, to make observations. This was an operation that must have involved a great deal of organisation and expense – a specially erected large observatory building, a huge telescope with all its works, manned by a substantial staff including local helpers. There was also the domestic side – a large house and servants, the home of Lassell, his wife and three daughters.

William Lassell's wife Maria, whom he married in 1827 while still in his twenties, was the daughter of a teacher of navigation in Liverpool[xx] whom he met through her brothers, keen amateur astronomers and companions of Lassell's early days. Of their five children (a son and four daughters of whom the two eldest were married before the Malta sojourn of 1861), two are mentioned in the annals of astronomy. They are the second and third daughters Jane (1831–1920) and Caroline (1833–1918). (The youngest daughter Charlotte is reputed to have shared an interest in her father's work[xxi] but she has left no tangible legacy). These two talented women may have owed some of their understanding of astronomy to their mother as well as to their father. No obituaries were published after their deaths, and no record exists of how they were educated: but one assumes that they were taught, as was usual in their circumstances, by governesses who would have provided, for example, the command of German which was one of their accomplishments.

The Lassell daughters were still children (the eldest was 15) at the time of the exciting discovery of Neptune's satellite. The first – and very rare – glimpse we have of them as adults is when the pioneering American woman astronomer and comet-finder Maria Mitchell (next chapter) on her tour of Europe in 1857 came to visit. Lassell invited Maria to dine, showed her his two telescopes in their domes, and was "very genial and pleasant". Maria was impressed by the Lassells' four "accomplished daughters", then in their twenties, and was pleased to hear that one of them had observed "her" comet of 1847, though the young woman was under the impression that it had been discovered by their father's friend Mr. Dawes[xxii] (William Rutter Dawes, another leading figure from the great age of amateur astronomers). "They take photographs of each other which are very beautiful, make their own picture frames and work in the same workshop as their father", she reported. This small vignette throws valuable light on the young women's lives. It shows them as seriously interested in experimentation, and in photography which they were probably pursuing as an artistic hobby.

Charles Piazzi Smyth, who with his wife Jessie visited Malta in 1864, left an account of Lassell's magnificent observatory there, and also of the family home – "really a splendid house, for size of halls, rooms and staircases, paved with stone and [rooms] 20 ft high".[xxiii] Mrs. Lassell and the daughters entertained them to a very good lunch, but unfortunately the account gives no hint of the nature of the daughters' other activities. Their ages were respectively 33, 31 and 30. They continued to live with their parents, and moved with them to Maidenhead in Berkshire when Lassell had completed his observations in Malta (resulting in a catalogue of 600 nebulae) and the great telescope had finished its work. Lassell erected his smaller 24 in. telescope at his new home, a large house called Ray Lodge, and continued to observe there. In 1883, three years after his death, the daughters presented the telescope to the Royal Observatory at Greenwich where it was re-erected and used for some time.[xxiv]

Jane and Caroline never married; Charlotte did not marry until the age of 51, some years after the deaths of both their parents. One is reminded of Mrs. Somerville's two daughters who similarly devoted themselves to their parents to the extent of forgoing independent lives of their own. They no doubt provided support – but it is not easy to find out how much they actually assisted their father with his astronomical work, or whether they made their own independent observations. It is most unlikely that they would have used the large 48-in. in Malta, as this elaborate and massive instrument required an observer and an operator, and was in the hands of a professional astronomer.

It is, however, quite possible that Jane and Caroline made use of the more manageable 24 in. at Ray Lodge, Maidenhead. According to Gerard Gilligan, expert on the Lassell family, they continued their father's astronomical observations after his death. The telescope remained in place for three years after that event; and in his later years, when no longer able to observe owing to failing sight, Lassell may have delegated the work to his daughters. Jane and Caroline, therefore, may well have employed the telescope during and after their father's lifetime.[xxv]

Whatever their involvement in the reflector, what is known for certain is that Jane and Caroline took up the study of spectroscopy after the family moved to Maidenhead. They appealed to their father's friend William Huggins for advice on a suitable textbook from which they might study this new subject. William Huggins was the pioneer of astronomical spectroscopy in Britain and one of its founders on the world stage (Chapter 11). He recommended a book on *Spectrum Analysis* in German which the sisters not only studied but translated into English.[xxvi] The author, Heinrich Schellen, headmaster of a college in Cologne, was in communication with leading astronomical spectroscopists in Europe and the United States. His book was based on a series of lectures delivered in 1869 to a Scientific Society in his city. In the Preface to the translation, Huggins explained that the daughters of his friend Mr. Lassell (then President of the Royal Astronomical Society) had asked him to edit the translation. He found no need to make any changes. He merely declared himself "not responsible for the views of the author" nor "for the relative importance which he has given to the work of different investigators". Huggins was always quick to maintain his priority over his rivals in the matter of discoveries and observational matters, but in this instance his minor comments were kept in footnotes.

Schellen's book could hardly be called elementary. It covered the entire subject of laboratory and stellar spectroscopy as known at that time.[xxvii] It was well illustrated, with a number of colour plates of laboratory spectra and of recent solar coronae and prominences, to which the "translatresses" added some recently published lists of spectral wavelengths. They also included a note of their own observations of the great aurora of 1870 which they had seen from their home at Maidenhead. In this work, the Lassell sisters performed a valuable service to astronomical spectroscopists, and demonstrated their confident knowledge both of science and of the German language. A long review in the *Quarterly Journal of Science* (a most useful publication edited by Sir William Crookes, the chemist) declared that "those students of spectroscopic analysis in England and America who do not read German owe thanks to the Misses Lassell for undertaking the labour of translation, and to Dr. Huggins [referred to as "the Herschel of the Spectroscope"] for editing and annotating his excellent treatise". "It is a treatise without which no scientific library can be regarded as complete".

This was not the Lassells' only work of translation from the German. In 1873 they also translated a life of Alexander von Humboldt, author of *Cosmos*. That biography, by Karl Bruhns which had appeared only the previous year, was "excellently translated by the Misses Lassell"[xxviii] and recommended by Agnes Clerke as the best available.

As regarded practical spectroscopy, it would appear that the Lassell sisters continued to make astronomical observations of some kind which were not published. In 1886, Piazzi Smyth sent them a copy of his recent volume of observations of the solar spectrum, reproduced in colour. Jane and Caroline were now the only two remaining members of the family. In Jane's letter of thanks on behalf of them both, she wrote: "We are having I fear a disastrous spectroscopic season – I should think very little good work has been accomplished in this climate for many months. Here at least it has been exceptionally cloudy both night and day until within the last week or so."

It is hoped that more will be discovered about the Lassell sisters' activities in astronomy in order to give due recognition to these very intelligent and highly educated women whose entire lives were spent in the shadow of large telescopes.

Women's customary education at home, certainly in the British tradition, concentrated on the feminine accomplishments of art and languages. Translating foreign publications, as the Lassells did, was one way in which these skills could be put to use. Humboldt's *Cosmos* was translated by Mrs. Sabine (as will be described), but being such a popular work, it attracted other (women) translators. One was Elise Otté (1818–1901),[xxix] a Danish-born translator of books of various kinds, whose version was published in 1858. Humboldt and Bonpland's *Travels in Equatorial regions* of America was translated from the French by Thomasina Ross,[xxx] to replace (according to the publisher) a previous version by Helen Maria Williams.[xxxi]

The women did not always get full credit, however. On Mary Somerville's first visit to Paris in 1817, her hostess at dinner was the charming wife of Jean-Baptiste Biot, the Somervilles' original contact with France. Madame Biot, wrote Mary, was "a well-educated woman, and had made a translation from the German of a work which was published under the name of her husband".[xxxii] This was an example of a practice that may have been more widespread than can be discovered, of female family members helping their scientific menfolk anonymously behind the scenes.

A glaring example of this transference of attribution from wife to husband is that of the Sabine translations of the works of Alexander von Humboldt. Elizabeth Juliana Sabine (née Leeves) (1807–1879) married the famous geodetist and explorer

Sir Edward Sabine (1788–1883) in 1826 when she was only 19. Sabine,[xxxiii] an army officer (eventually a General), began his distinguished scientific career as an astronomer on the Arctic expedition to seek the north-west passage, led by Sir James Ross in 1818. His research led him later to geodesy – the study of the shape of the Earth – and geomagnetism – the pattern of the Earth's magnetism and its variation from place to place on the Earth's surface.

The compass needle was from very early times a navigational instrument, and it was not necessary to understand the workings of the Earth's magnet to put it to that use. The direction of the magnetic needle is made up of two angles or elements – its declination or direction with respect to the meridian (the north-south line), and its dip, or direction with respect to the vertical – and these were the useful pieces of information noted by standard magnetic instruments. These quantities varied slightly with time, both daily and on longer scales. However, they did not tell the whole story, because the overall strength of the Earth's magnetism also varied, as was realised quite particularly by the greatest scientific traveller of the age, Alexander von Humboldt, during his voyages in South America. In 1828 Humboldt, home in Germany, proposed an international effort to tackle the problem of terrestrial magnetism from a scientific angle, and his idea was taken up by the great Carl Friedrich Gauss in Göttingen, who set up the country's first magnetic observatory there. Humboldt followed this with an appeal to the Royal Society in London in 1833 to join the "magnetic crusade", with the result that a number of magnetic stations was set up throughout the British Empire under Sabine's direction. Sabine, analysing the data from stations as far apart as Toronto and Tasmania noted spasmodic large variations in the Earth's magnetic elements, what he called "magnetic storms", recurring with a period of about 11 years. He announced this unexpected discovery in 1852, shortly after the Scottish born astronomer Johann von Lamont (John Lamont) (1805–1879) director of the observatory at Munich, had found a similar periodicity in the German magnetic records. An additional extraordinary result, uncovered by Sabine, was the similarity between this period of magnetic activity, and the period of an apparently unrelated phenomenon: the number of spots on the Sun. The discovery of the sunspot cycle was the work of an assiduous German amateur astronomer, Felix Schwabe, who had made sunspot observation his life's work. Sabine's correlation was confirmed by other observers; it also correlated with the frequency of polar aurorae, those spectacular illuminations in the north and south regions of the Earth. That the Sun, a heavenly body millions of miles away, should influence the Earth in such a material manner was an astounding revelation that gave rise to a new important branch of enquiry (pursued later in the century by the Maunder husband and wife team: Chapter 14) which remains active and exciting today. It was Sabine's great legacy to astronomy.

Sabine went on the become President of the Royal Society and a highly influential figure in the administration of British science, as well as an authority on all aspects of geophysics. It was well known that his wife, a popular figure among his scientific contemporaries, was his constant helper in his researches which continued even in their retirement. Mary Somerville, who knew both Sabines as close friends, described her as "a lady of talent and scientific acquirements. She translated *Cosmos* and assisted and calculated for her husband in his laborious work".[xxxiv]

In 1843 the Sabines were at Lord Rosse's observatory at Birr where the fabulous 6-ft diameter Leviathan was in operation, as members of a distinguished group that spent 2 weeks there observing and "telescope viewing". William Rowan Hamilton was also of the number. He wrote from there to a friend: "I have known Sabine for many years, and his wife Mrs. Sabine is another old friend of mine. She is rather a learned lady, and has translated many foreign, especially German, papers for Taylor's *Memoirs*, having no children to occupy her otherwise; and I remember that with her husband she attended a course of lectures that I gave at Trinity College Dublin." (It was the custom at Trinity College for women to be allowed to listen to the inaugural lectures of the professors at the beginning of the academic year). He went on to describe sitting with her in the little building near the telescope where "ladies could warm themselves at intervals", and having "a cosy fireside chat" discussing mutual friends.

Humboldt's inspirationally titled *Cosmos* in four volumes (1849–1858) was a popular but also erudite account of the Universe, incorporating geography, astronomy, meteorology and natural science generally, and owed its attraction to the fact that the author, famous explorer as well as scientist, had first hand experience of the phenomena he recorded. It was one of the best-loved scientific books of all time. The Sabine translation was authorised by Humboldt himself and carried out by Mrs. Sabine. Her name, however, did not appear on the title page, which was described as "Translated under the Superintendence of Lieut-Col. Edward Sabine, R.A., For. Sec. R.S."[xxxv] . An "Editor's Preface" stated that the editor was indebted to the earlier writings of the author for awakening in him a taste for his present interests: "long cherished feelings of gratitude for this obligation, combined with those of personal regard, have been motives with himself and with Mrs. Sabine – by whom the Translation has been made – to surmount the hesitation, which they might otherwise have felt in venturing on a task embracing so extensive a range of subjects." The seven words between dashes in this Preface to Volume 1 represents all the credit given to the real translator; her name is not mentioned at all in the other volumes. However, John Herschel took care to give credit where credit was due. In a combined review of *Cosmos* in the original German and of the Sabine translation, he wrote:[xxxvi] "The author has been especially fortunate in his translator (*translatress* we should say, since in the style of its execution, we have no difficulty in recognising the same admirable hand that gave an English garb to Baron Wrangel's *Expedition to the Polar Sea*.) So perfect a transfusion of the spirit and force of a very difficult original into another language, with so little the air of a translation, it has rarely been our fortune to meet."

Ferdinand von Wrangel, author of the book mentioned by Herschel, was a Russian statesman and explorer whose book was published in German in 1839 and in an English translation by Mrs. Sabine in 1844. She also translated, from the French, Francois Arago's *The Aspects of Nature* (2 vols, London 1849–1851) and his *Meteorological Essays* (London 1855). All carried the same formula: "translated under the superintendence of Col Sabine". Arago, physicist, astronomer and Director of the Paris Observatory was another of Mary Somerville's friends who figures in her *Recollections*.

Lady Sabine's (as she became on her husband's knighthood in 1869) role as her husband's closest helper continued throughout their lives. In 1871, Lord Kelvin (Sir William Thomson), the country's most senior and most influential scientist, in his Presidential address to the British Association for the Advancement of Science, recalled Sabine's mighty work on terrestrial magnetism since 1838 which was based on data from magnetic stations all over the globe. "Silently, day after day, night after night, for a quarter of a century, he has toiled with one constant assistant by his side to reduce these observations and prepare for the great work. At this moment, while we are here assembled, I believe that in their quiet summer retirement in Wales, Sir Edward and Lady Sabine are at work on the magnetic chart of the world. If two years of life and health are granted to them, science will be provided with a key which must powerfully conduce to the ultimate opening up of one of the most refractory enigmas of cosmical physics, the cause of terrestrial magnetism."[xxxvii] The vast challenge of the Earth's magnetism could not be solved once for all in such a short span. Sabine (with his wife, one assumes) continued to work until 1876, but his health began to fail and he retired from the army a year later. Elizabeth, his eclipsed satellite, predeceased him by four years. She died in 1879 at the age of 72.

Notes

[i] The only salaried post undertaken by John Herschel was a spell as Master of the Mint in later life.

[ii] J.L.E. Dreyer. 1884. Article 'Observatory' in *Encyclopaedia Britannica*, Ninth edition, volume 17. Edinburgh: Charles and Adam Black.

[iii] Allan Chapman. 1998. *The Victorian Amateur Astronomer*. p 94. Chichester: Praxis Publishing Ltd. The same point was made by William Rowan Hamilton in 1830 in a letter to the poet William Wordsworth. "You see, a telescope may be as expensive as a racehorse". R.P. Graves. *Life of Sir William Rowan Hamilton*, Volume 1, p 399. Dublin: Hodges Figgis and Co.

[iv] Agnes Clerke's description. *History of Astronomy*, p 83.

[v] Patrick Moore. 1971. *The Astronomy of Birr Castle*. Birr: The Telescope Trust.

[vi] Susan Barry. 1997. Photographs from the Birr darkroom, in *Stars Shells and Bluebells, Women Scientists and Pioneers*. Dublin: WITS 1997, 48–55.; Susan M.P. McKenna-Lawlor. 1998. *Whatever Shines should be observed*, Chapter 1. Dublin: Samton Press.

[vii] Allan Chapman, op. cit., p 96.

[viii] Susan M.P. McKenna-Lawlor. op. cit.

[ix] ibid., Chapter 2 and references therein; Ita Kavanagh 1997. In *Stars Shells and Bluebells*. Dublin: WITS.

[x] R.P. Graves. *Life of William Rown Hamilton* Volume 1, p 104. Dublin: Hodges Figgis.

[xi] The Hon. Mary Ward. [n.d. 1858]. *The Telescope, illustrated by the author's original drawings*. Sixth Edition. London: Groombridge and Sons. *The Telescope* was an abridged version of *Telescope Teachings*.

[xii] The set of drawings is reproduced in McKenna-Lawlor, op. cit, Chapter 2.

[xiii] The drawing which is not entirely accurate, was probably made after the event and mistakenly showed the outline of the invisible moon, but it served its purpose of illustrating the geometry of the phenomenon. I thank Terry Moseley who made a computer reconstruction of the eclipse for these comments.

[xiv] *Irish Times*. November 15, 1866, in the Correspondence column.

[xv] The Hon. Mrs. Ward. The November Meteors. *The Intellectual Observer*, January 1867, 449–58.
[xvi] Susan McKenna -Lawlor. op. cit.
[xvii] Patrick Moore, op. cit; last chapters.
[xviii] Allan Chapman. 1994. *Yearbook of Astronomy*; Robert W. Smith, ODNB; Margaret Lindsay Huggins. 1880. The late Mr William Lassell LL.D., F.R.S. *The Observatory* **3**, 587–90.
[xix] This interesting possibility has been the topic of some intriguing discussion. (Richard Baum. 1996. William Lassell and 'the accident of the maidservant's carelessness', or Why Neptune was not searched for at Starfield. *Journal of the British Astronomical Association* **106**).
[xx] Allan Chapman. *The Victorian Amateur Astronomer*, p 100.
[xxi] Gerard Gilligan. Private communication 2005, from his extensive research into the Lassell family.
[xxii] Phebe Mitchell Kendall. 1896. *Maria Mitchell. Life, letters and journals*. p 87–88. Boston: Lothrop, Lee and Shepard.
[xxiii] Mary T Brück. 2003. An Astronomer calls: extracts from the diaries of Charles Piazzi Smyth. *Journal of Astronomical History and Heritage* **6**(1), 37–45.
[xxiv] Derek Howse. 1975. *Greenwich Observatory*, Vol. 3, p 118. London: Taylor and Francis.
[xxv] Agnes Clerke (archive DNB) records that a large drawing of the Orion Nebula, observed with the 24-in. telescope, made by Caroline under her father's supervision in 1868, was presented to the Royal Society.
[xxvi] H. Schellen. 1872. *Spectrum Analysis*, 2nd enlarged edition. tr. Jane and Caroline Lassell. London: Longman Green and Co.
[xxvii] Hentschel. 2002. *Mapping the Spectrum*. Oxford University Press.
[xxviii] Karl Bruhns. 1873. *Life of Alexander von Humboldt*. Translated by Jane and Caroline Lassell. London.
[xxix] Oxford *DNB*.
[xxx] Alexander von Humboldt and Aimé Bonpland. 1851. *Travels in Equatorial regions of America 1799–1804*, translated and re-edited by Thomasina Ross. London and New York: George Routledge and Sons. Thomasina Ross (d. 1875) was a contributor to Charles Dickens' magazine *Household Words*.
[xxxi] Alexander von Humboldt. 1814. *Researches concerning the institutions and monuments of the ancient inhablitants of America*. Translated by Helen Maria Williams. London: Longman, Hurst Rees etc. Helen Maria Williams (1761–1827) was a well-known writer, poet and traveller.
[xxxii] Mary Somerville. *Queen of Science*, p 89.
[xxxiii] Gregory Good. Entry on Sabine. *Oxford DNB*.
[xxxiv] Mary Somerville. *Queen of Science*, p 118.
[xxxv] Alexander von Humboldt. 1846. *Cosmos: Sketch of a Physical Description of the universe*. Translated under the superintendence of Lieut.-Col. Edward Sabine, R.A. For. Sec. R.S. London: Longman, Brown, Green, and Longmans; and John Murray.
[xxxvi] John Herschel. *Edinburgh Review* January 1848; reprinted in his *Essays*.1857, p 360. London: Longman, Brown, Green, Longmans and Roberts.
[xxxvii] Lord Kelvin. Presidential Address to the British Association for the Advancement of Science, Edinburgh, 1871.

Chapter 8
The Admiral's Circle

A leading figure in Britain's golden age of amateur astronomers was William Henry Smyth (1788–1865), a naval officer (he attained the rank of Admiral) who, after service in the Napoleonic wars followed by 10 years command of the Anglo-Sicilian fleet at Messina, decided to devote himself seriously to astronomy (Fig. 8.1). He had developed his ardent interest in the stars under the influence of his friend Giuseppe Piazzi, Director of the Observatory at Palermo and famous discoverer of the first minor planet Ceres on 1 January 1801. On retiring on half-pay from the Navy, Smyth set up his "Temple of Urania" in Bedford, about 50 miles from London, in 1830, considered the best-equipped private observatory in the country. It included a beautiful 6-in. refractor, equatorially mounted and driven by clockwork, one of the first in Britain to have that facility. There he embarked on a programme of measurements of double stars and of observations of nebulae and star clusters. The result was a famous catalogue of 850 objects, known as the Bedford Catalogue, published as part of a larger treatise, *The Cycle Celestial Objects*, in 1844. The Bedford Catalogue,[i] in which the astronomical data are interspersed with a charming mixture of useful information and classical lore, became – and remains (it was reprinted in 1986) – one of the most popular works on astronomy in the English language. It gained for Smyth the Gold Medal of the Royal Astronomical Society in 1845, its highest accolade. In that society, which he had joined in its early days and of which he was President in 1845–1847 and Council member for longer, Smyth belonged to the coterie of elite astronomers that included Sir John Herschel and Sir George Airy, the Astronomer Royal, and placed him in a position of considerable influence in the astronomical community.

The energetic and flamboyant Smyth had many interests as well as astronomy. He was a member of the Royal Geographical and the Antiquarian Societies, and was an honorary or corresponding member of "at least three quarters" of the literary and scientific societies of Europe.

Throughout all his pursuits, Smyth was supported by his wife Annarella (née Warrington) (1788–1883) whom he had met and married in Messina in 1815 while he was stationed in the Mediterranean. She was the only daughter of an English banker and merchant who was for 40 years British consul at Naples where Annarella was born and brought up. She was described as "a lady of great abilities and rare accomplishments, who through all his scientific labours was his devoted companion

DOI: 10.1007/978-90-481-2473-2_8,

Fig. 8.1 Admiral W. H. Smyth and his wife Annarella. The Admiral shows off the Royal Astronomical Society's Gold Medal, awarded to him in 1845. (National Portrair Gallery)

and assistant",[ii] a gifted artist and musician who is reported to have sung before the court of Naples in her youth.[iii] She passed on her artistic gifts to her children. especially to her son Charles Piazzi (next chapter) who was born in Naples and named after his astronomer godfather. Throughout their married life she "shared [her husband's] tasks and so constantly contributed to his development of them, that those who knew the Admiral during the latter half of his life will ever associate with their remembrance of the genial, sensible and humorous character of the man, the quiet seriousness, the keen intelligence, and gentle goodness of the wife".[iv] An example, no doubt typical, is Annarella's correspondence with the ill and aging astronomer Baron Franz Xavier von Zach who liked to write to her in French since her French was much better than her husband's. Zach who had lived for many years in Genoa, was a friend of Smyth's from his naval days.[v]

The equatorial telescope (Fig. 8.2), when its main purpose was completed, was re-erected at the historic Hartwell House[vi] near Aylesbury, the mansion of Smyth's wealthy friend and patron of the arts and sciences, Dr. John Lee (1783–1866). The Smyths lived within walking distance in a house called St John's Lodge and continued to have the use of their beloved instrument whenever they wished. Hartwell House under the hospitable Lee was a famous centre for social scientific gatherings where guests, including wives and daughters, were treated to "astronomical delights" viewed through the telescope.[vii] Maria Mitchell, America's first woman

Fig. 8.2 Smyth's famous equatorial telescope, now in the Science Museum. Drawing by Annarella Smyth in *Aedes Hartwellianae* a lavish book by Smyth extolling the observatory and mansion of his friend Dr. Lee. (Royal Observatory, Edinburgh)

astronomer and first comet-discoverer, who visited Britain at the start of her great European tour in 1857, left her own similar impression of the Smyth household at St John's Lodge where she experienced the exuberant admiral's amusing conversation and "the society of the cultivated members of his family". She was taken to see Hartwell House and its grounds where the famous telescope was now installed. Smyth explained to her that he and his wife had a private key to the observatory, and when their host was away from home, had the use of a room in the house, "the Admiral's Room", where breakfast would be prepared for them after a night's observing.[viii] Among the furnishings of Hartwell House were *two* busts of Mrs. Somerville "from which", Maria wrote to her sister, "I received the impression that she is handsome, but Mrs. Smyth tells me that she is not so." Maria may have uncovered a little female jealousy here: Annarella, somewhat younger than Mary Somerville and an old friend, was herself a handsome woman, to judge from her portrait. Maria was to acquire her own bust of Mary Somerville which was given pride of place on her desk at Vassar College.[ix]

Among Annarella's work are illustrations for Smyth's lavish account of the treasures of Lee's mansion and observatory, *Aedes Hartwelliana*, published in 1851.[x] Smyth explained in the Introduction that the plates in the book were engraved from drawings expressly made by members of his family – his wife, his daughter Ellen Philadelphia, his sons Charles Piazzi and Henry, and his son-in-law Baden Powell – "who, having all, at various times. enjoyed the hospitalities of Hartwell House, were much interested in the undertaking". Annarella's contributions were a beautiful engraving copied from an old painting of the mansion and grounds, and the scientific illustrations. The latter included a drawing of the famous equatorial telescope in its dome. In 1861, when Annarella was already in her seventies, her expertise in this field was called upon in an astronomical debate concerning the interpretation of the mottled appearance of the photosphere (the bright surface) of the Sun. It began when the amateur astronomer and telescope maker James Nasmyth produced drawings of the Sun that showed what he called a "willow leaf" pattern, interpreted by him as indicating a multitude of elongated intertwining filaments. Once the idea was floated, several astronomers, including the now elderly Sir John Herschel, convinced themselves that they could see such a pattern. The veteran observer William Dawes took a different view, maintaining (correctly) that the elongated shape of the random mottles was imagined. Warren de la Rue, a telescope maker of a younger generation and one of the pioneers of astronomical photography (his lunar work will be described later), though tending to support the willow leaves, devised an independent test. He asked an artist who had no knowledge of astronomy to make drawings of a projected image of the Sun which he then submitted to Mrs. Smyth for her opinion. After careful examination with a magnifier, Annarella professed to finding no particular character in the markings.[xi] She was right: but the controversy continued for a further four years. (The markings, usually called granulation, are due to blobs or cells of gas, small by comparison to the dimensions of the Sun and persisting for only a few minutes, that rise to the visible surface from a hotter layer below and show up bright against the slightly cooler regions between them.) This observation of Annarella's is historically interesting. In the same year, 1861, De la Rue obtained a photograph of sunspots which he used as a test of the quality and faithfulness of details on reproduced images.[xii] Photography, he concluded, was not yet capable of giving a definite answer to such problems.

Maria Mitchell had a special regard for Admiral Smyth who had played a part in launching her to fame. She was almost 30 years of age when in 1847 she made her great and exciting discovery of a comet. Her patrons in Boston supported her claim to it, and, with the intervention of the well-connected Admiral Smyth and the Astronomer Royal's official confirmation, Maria was deemed to be the first to have observed the object. Establishing priority was important, as it involved a valuable trophy – a medal and prize from the King of Denmark. The prize had been instituted in 1831 to encourage telescopic comet-seekers, and there were elaborate rules to decide the earliest claimant in the case of more than one observer. Maria was the America's first comet-discoverer, and the first woman to follow Caroline Herschel in this field – though another ran close. As John Herschel recorded it, 'Eight comets were discovered by Ms. C. Herschel, who, however, is not the only female observer of these bodies, the comet of 1847 having been independently detected by two ladies, Ms. Maria Mitchell of Nantucket, U.S., and Madame Rümker of Hamburg, the priority lying with the American astronomeress'.[xiii] Madame Rümker, née Mary Hannah Crockford (1809–1889), was the English wife of George Rümker, director of Hamburg Observatory, who had worked for some years in the observatory at Durham University. She, too, was a comet-huntress, but trophyless.[xiv]

Maria had therefore reason to be grateful to the kindly Admiral who had described her as "a young lady, industrious and vigilant, a good astronomer and mathematician."[xv] The comet set her on her career path. She was recruited to compute the ephemeris of the planet Venus for the American Nautical Almanac, that planet being allotted to her "in a spirit of gallantry", though she was not too pleased at the connotation, being strongly opposed to all feminine occupations. She was paid a modest salary for this work which she did at home, and retained that duty for 20 years, even after she became a college professor. It was the same type of work that Mary Edwards did in her day for the Greenwich-based Nautical Almanac.

Maria Mitchell left diaries and letters which were later published.[xvi] Written with her characteristic mixture of earnestness and humour, no better record exists of the scientific and social world of astronomers in mid-nineteenth century Britain. She was intrigued to learn from Smyth that in England a wealthy man would buy a telescope "as an ornament to his house", and hire an astronomer to work for him and make a name for him – perhaps gaining a knighthood for himself in the process. This, it seemed to Maria, was different from America, where the amateur was likely to be a poor schoolmaster instructing his bright pupils, or a watchmaker seeking to keep the time – in short, a hardworking man. She was mistaken, however, in imagining that poor hardworking amateurs did not exist in Britain: she simply did not come across any in her round of academic and well-heeled society. On the other hand, her observations on the wealthy amateur were pretty exact. Astronomy in Britain, in the fields both of instrument construction and research, was almost entirely in the hands of independent workers. The "hired young men of talent" were also a reality: Lee and William Lassell, just back from Malta, whose modern observatory Maria also visited, were among those who had highly competent paid assistants in their private observatories.[xvii]

On her packed journey through England and Scotland Maria, armed with introductions from the Harvard astronomers, was entertained in style in the most exalted centres, and found much to comment on. At the Royal Observatory in Greenwich she was a guest in the home of George Airy, the Astronomer Royal, and his friendly wife, Richarda. Airy, she found, though "naturally a despot" in the Observatory, was "cheery in the drawingroom" and fond of good conversation. Maria sat at dinner in the company of Mrs. Somerville's friends the Sabines (the husband was now a General), the Powells (Baden Powell, Oxford mathematician and theologian, and his wife, a daughter of William Henry Smyth), and the great Otto Struve, Director of Pulkovo Observatory who was visiting from Russia. Maria was given the place of honour next to her host, a special arrangement that Mrs. Airy had agreed with Mrs. Sabine who would normally have been placed there as the lady of highest rank: indeed, Maria learned, "properly all married ladies should precede me". Maria found the erudite Mrs. Sabine "very agreeable and not a bit of a blue stocking". Professor Powell was fat, and his wife, she thought, "overdressed".[xviii]

A visit of some days to Cambridge was planned by the Airys. Mrs. Airy and one of her young daughters accompanied the American visitor, with an introduction to no less a person than William Whewell, Master of Trinity College. Arriving at the Master's Lodge, Mrs. Airy informed Maria that "although we are invited to be the guests of Dr. Whewell, he is quite too mighty a man to come and meet us". They were met by the Airys' two student sons, and escorted to the august presence. Maria was not only surprised but displeased to be left standing for 15 minutes before being offered a chair – it would not have been so in an American gentleman's house – and even more annoyed at their host's arrogant manner and his sarcastic comments about America and Americans. "There was a tone of satire in Dr. Whewell's remarks which I think was not amiable", thought Maria, who boldly held her ground. Next day, however, she lunched again at the Lodge and attended an evening party where she met other distinguished scientists, including the kindly old geologist Adam Sedgwick, one of Mary Somerville's favourite friends. In company with Mrs. Airy, she visited the Observatory and was received by James Challis, Airy's successor as professor, and met John Couch Adams, "a merry little man [who] loves games and is a favourite with young ladies" (Maria does not disclose how she acquired the latter piece of information). Adams devoted an entire morning to her, and showed her the room where he had made his calculations of the position of the unknown planet Neptune in 1845, though he was unfortunately beaten to the discovery. The Cambridge academic fraternity made a great fuss about Maria, reminiscent of Mary Somerville's visit of 20 years earlier – apart from the attitude of Whewell who on the previous occasion had so admired the guest of honour as to compose a sonnet in her praise. The plain-featured Maria Mitchell, in her severe black gown, who found "some of the [ladies'] dresses in English society too low for [her] taste" and disapproved of the extent of wine-drinking at the Cambridge table, could not compare with the beautiful Mary Somerville in her elegant Paris lace. Maria went away with mixed feelings about Cambridge University and its arcane traditions.

She suffered no social embarrassment, however, when she visited Sir John and Lady Herschel at their home in Collingwood. The personal invitation had come from Lady Herschel through Mrs. Airy. Maria experienced the same unforgettably warm hospitality at Collingwood, "a house like no other" as Mary Somerville described it, that she and Maria Edgeworth and many others had enjoyed. Eight of the Herschels' children were there, including Mary Somerville's 9 year-old god-daughter, and baby Constance, future author of the *Herschel Chronicles* (Maria carefully made a note of all their names). Maria found Sir John "an old man [he was only in his sixties], slightly bent, with perfectly white hair sticking out every way" – the very image that Julia Margaret Cameron captured in her famous photograph. Sir John showed her some of his father's and his Aunt Caroline's manuscripts, and gave her a sample of Caroline's handwriting. He discussed his observations at the Cape, and showed her the telescope he had used there, now stored without its optics in a barn. Before she left, Lady Herschel gave her a letter of introduction to Mrs. Somerville, which led to their delightful meeting in Florence later in the year.

An English social custom new to Maria was the evening party or "rout". She was invited to such a function in the London home of the Smyths' daughter, Mrs. Powell, whom she had already met at the Airys. Henrietta had married Baden Powell, a widower almost 50 years old with four children, when she was only 22. She was now in her early thirties, and an experienced hostess. At a famous gathering of the British Association in Oxford in June 1847 she entertained a galaxy of eminent scientists including John Herschel, David Brewster, William Fox Talbot, George Airy, and, most notably, the famous astronomer Urban Leverrier of Paris, the discoverer of Neptune, who played the violin, accompanied by Henrietta on the piano.[xix] The large throng of guests included many distinguished persons, whose names were already familiar to Maria.

Maria's tour demonstrated how the wives of British astronomers made up a close circle of friends, women who were not only socially known to each other but took a lively and intelligent interest in their husbands' affairs.[xx] Wives and daughters, cultivated and educated, were happy, or appeared so, in the supportive domestic scene idealised by Maria Edgeworth. The Smyths' daughters embodied this continuing tradition. Besides Henrietta, Baden Powell's wife, there were Rosetta (romantically named after the Rosetta Stone: W.H. Smyth was also a keen antiquarian), who married the biologist Sir William Flower, Director of the Natural History Museum in London; and Ellen, whose husband Captain Henry Toynbee (1819–1881), a naval officer of scientific talent, fitted ideally into the family. He commanded several vessels, became an expert navigator, and collaborated with his brother-in-law Piazzi Smyth in designing navigational equipment for scientific purposes. He became Marine Superintendent of the Meteorological Office, was an authority on weather in the southern oceans and advised the Astronomer Royal on plans for the Transit of Venus expeditions of 1874. Ellen accompanied him on all his distant voyages and artistically illustrated his meteorological logs.

Within the wide Smyth web, the Admiral's eldest son Warrington (1817–1890) married Antonia Story-Maskelyne, granddaughter of Nevil Maskelyne, the Astronomer Royal of an earlier generation (and friend of the elder Herschels), thus uniting two well-known astronomical families. Antonia's sister-in-law Thereza (next section) was a skilful astronomer.

Yet it is possible that women of the younger generation would have wished to spread their wings beyond the country house. Henrietta Smyth, according to the biographer of her illustrious son Lord Baden Powell, Boer War hero and founder of the Boy Scout movement, felt overshadowed by her three successful brothers – (Sir) Warrington Wilkinson Smyth, educated at Cambridge, became a distinguished geologist and mineralogist; Charles Piazzi held the Chair of Astronomy in Edinburgh and (Sir) Henry Augustus was a General in the Army – and resented the lack of opportunities for their sisters who were no doubt equally intelligent. Her eagerness to marry a husband more than twice her age sprung, he suggests, from her determination to secure the social and intellectual lifestyle which she experienced at Hartwell House on visits with her admired father.[xxi] The Smyth family, far from disapproving of her choice, was proud of the superior Powell connection.[xxii] Marriage, after all, was the only career open to such women with little or no dowries. The eminent archaeologist Sir Flinders Petrie recounts in his autobiography that his father, William Petrie, who as a young engineer moved in the Smyth circle, was attracted to Henrietta, but that Mrs. Smyth, "who was a careful mother and knew her duties", whisked her daughter away for more promising prospects elsewhere.[xxiii]

Another daughter of the Smyths was assigned to be housekeeper to her (then) bachelor brother – a common enough fate of spare unmarried daughters – but died young; while one of the last of the family remained at home with her parents, and predeceased her loving father by one year. On the other hand, their Scottish sister-in-law Jessie, Charles Piazzi's wife, led a very active life in science. A visit to the Edinburgh Smyths was naturally on Maria Mitchell's itinerary (next chapter), when she travelled north to Scotland, calling also on Professor John Pringle Nichol, writer and astronomer, at Glasgow University, and stopping off at Abbotsfort to pay her respects to the memory of the great bard Sir Walter Scott.[xxiv]

Maria Mitchell continued her European tour with visits to many famous scientists and scholars on the Continent. They included the aloof Urbain Leverrier in Paris; the now old but welcoming and gracious Baron von Humboldt in Paris; the astronomical spectroscopist Angelo Secchi in Rome and – most memorable of all, as already described – her deeply admired Mary Somerville in Florence. She arrived home in Nantucket in June 1858 and hastened to her telescope, to be rewarded with spotting Donati's comet while it was still only a faint telescopic object. It had been discovered by Donati on June 2 (Maria was then still on board ship) but she found it independently before the end of that month and alerted Harvard Observatory where it was observed in the middle of July.[xxv] She continued patrolling the heavens for comets. (She never discovered another.) True recognition came when Vassar College, the brainchild of a wealthy philantropist, opened in Poughkeepsie, New York, in 1865, and Maria Mitchell, at the age of 47, was appointed Professor of Astronomy, one of eight professors of whom two were women. It was a development that would have been unheard of in Britain.

Maria's chief success in her academic life was as a teacher and inspirer. She was adored by her students in whom she inculcated a true appreciation of learning and a down-to-earth belief in women's capability for science. Her high-minded philosophy was perfectly summed up in an anecdote related by a later staff member.[xxvi] It concerned a young man who was being shown round the observatory where Maria and her students happened to be observing Venus. She kindly offered to show the planet to him. He looked, and remarked "in a foppish way" that he had seen prettier Venuses in the College. She flashed him an angry glance and said: "There are no Venuses here. We are all Minervas".

Further out in the Smyth orbit was a young Minerva who deserves greater recognition from posterity, Thereza Mary Llewelyn (later Story-Maskelyne) (1834–1923) (Fig. 8.3). Thereza, who on her marriage acquired a proud astronomical surname, was well-versed in astronomy and an expert observer well before that event. She was born into a scientifically minded family and became the wife of another scientist. She also belonged by marriage to the extended Smyth clan. Like the Lassell sisters of whom she was a close contemporary, she practised both astronomy and photography at home as a young woman, though it is as a photographer that she has

Fig. 8.3 Thereza Llewelyn, photographed by her father. (Swansea Museum)

left her most significant mark. Her marriage was an ideal one, based on romantic love and a shared enthusiasm for scientific pursuits with her husband, Nevil Story-Maskelyne.[xxvii]

Thereza's father, John Dillwyn Llewelyn (1810–1882),[xxviii] a wealthy landowner and independent scientist, lived in an elegant country mansion in beautiful landscaped grounds and gardens at Penllergaer near Swansea in Wales. His wife Emma (née Talbot) (1808–1881) was a cousin of the pioneer photographer William Henry Fox Talbot, whose activities inspired him to begin experimenting with photography as early as 1840. Emma, and in the course of time their daughter Thereza, collaborated in his photography and in his darkroom experiments. John Llewelyn holds an important place in the history of early photography. He was the first photographer in Wales; many hundred of his beautiful photographs of rural scenes, seascapes, botanical specimens and family portraits, are preserved at the Museum in Swansea and in the National Library of Wales. He was a founder member of the Photographic Society of London (now the Royal Photographic Society), and exhibited a set of photographs at the Paris Universal Exhibition of 1855, gaining a silver medal and placing Britain in first place. While his chief field of renown was photography, Llewelyn, like John Lee of Hartwell House, was active in other fields of science including botany, chemistry, meteorology and astronomy. He was a Fellow of the prestigious Royal (1836) and Linnean (1837) Societies, and in 1852 became a Fellow of the Royal Astronomical Society,[xxix] his proposer being none other than William Henry Smyth.[xxx] That last mentioned year, 1852, coincided with the building of an observatory on his estate for his 16-year old daughter Thereza, from which it may be inferred that his interest in astronomy was fuelled by his daughter's enthusiasm.

Thereza, the eldest of six children of whom four were girls, was a remarkable young woman who collaborated in her father's scientific activities. She helped him with his photographic experiments which were concerned with a process of his own devising for sensitising collodion plates,[xxxi] and was herself a talented photographer who prepared and processed her own material. A collection of her photographs is preserved with her father's in the Museum in Swansea, depicting charming country scenes of activities such as harvesting, though in fact the participants were cleverly posed for long exposures. Her enlightened father also encouraged his daughter's other interests, principally botany and astronomy. She had an excellent collection of botanical specimens, and received advice from the well-known botanist George Bentham, a frequent guest at the Llewelyn home, who deposited her botanical log-book with the British Association for the Advancement of Science at their meeting in 1856. She had her own laboratory, and – like Mary Ward with her entomology (Chapter 7) – a microscope for which she made her own slides.[xxxii] Yet another interest pursued by father and daughter was meteorology. The British Association had a scheme whereby volunteer watchers maintained weather stations and kept daily records on official forms which were then collected by the Association. Thereza looked after the meteorological records at her father's station, and hoped that she would be allowed to present her work in person at a meeting of the Association which – unlike other scientific societies in the land – was open to women.

Unfortunately her conventional father's support for his daughter's scientific ambitions did not extend to allowing her to attend the BAAS meeting which, in the year in question, 1856, was being held in Cheltenham. Thereza, a devoutly religious girl, was disappointed, but accepted the situation dutifully. "I am sorry for it", she wrote in her diary, "as I should have enjoyed the lectures etc. very much, but I ought not to think about it, as I know everything is ordered for the best".[xxxiii]

Thereza's fascination with astronomy went back to her early childhood and lasted all her life. In her memoirs, written (like Mary Somerville's) in old age,[xxxiv] she recollected seeing Venus through the window and "as a little nursery child ... sitting alone behind the nursery curtain to watch the great resplendent planet in the evening sky near sunset. The wonder and deep admiration I felt was surely something quite outside me, coming from another side of existence, of which I knew nothing, only that it was enchanting. That feeling has returned several times in my life and always with the sighting of the starry sky." When she and her sister Emma were moved out of the nursery they were given a room of their own. "That room looked north", recalled Thereza, "and as later on we had a series of winters when the aurora borealis was very prominent, I saw some most magnificent displays with no trouble from my own window which I probably should never have heard of had my window not looked north. That again was a phenomenon that appealed to me much as the sight of Venus had done before – quite from an outside sense – bringing in the unknown and mysterious outside certainty". So it was that the orientation of bedroom windows brought the wonders of the sky to the attention of the young girl. There was a very high sunspot maximum in 1847 when Thereza was 13, and the years around that date produced unusually frequent aurorae.

Thus inspired, Thereza tried to learn all she could about astronomy, learning German to enable her to read deeper, perhaps with the help of her governess, Ms. Deutschik, about whom one would like to know more: she was governess to other high class families who figure in Thereza's life, including that of her future husband;[xxxv] but, like so many loved and loyal tutors of that era, her own erudition remains unsung. Thereza possessed a small telescope, which appears in one of her father's photographs of her.[xxxvi]

Seeing how much his daughter enjoyed astronomy, her father – so Thereza recalls in her *Memoirs* – "did what he could to foster my love of that wonderful science, building an observatory on the estate and buying an equatorial telescope, a 4-in. achromatic lens."[xxxvii] The foundation stone was laid on 7 July 1851 in a family ceremony in which the grandparents also took part. Thereza (aged 17) laid the first stone, her sisters the second and the third.[xxxviii] The observatory was a substantial stone building topped by a revolving copper-clad drum some 19 ft in diameter, with a pillar to support the equatorially mounted telescope. The telescope, however, did not have a drive and had to be moved about by hand. Her father showed Thereza "how to set the telescope on a star or planet before opening the shutters,"[xxxix] i.e., to set by coordinates, which means that the observatory was also equipped with a sidereal clock. In the opinion of John Birks,[xl] who has a close knowledge of the building and its structure, the size of the dome suggests that it may have once housed a refractor of 6–8 in. diameter. The whereabouts of neither telescope are not now known.[xli] There is also a mysterious reference in Thereza's journal to a "speculum [i.e., metal mirror] telescope."[xlii]

The exact date when the telescope came into use is not recorded, but in an intermittent journal begun by Thereza in May 1856,[xliii] one finds her making observations of the Sun and Venus on 3 August 1856, and of more elusive objects on the following night. "We looked at γ Delphini, the nebula in Scγutum and a cluster in the same". (The first of these is a well-known double star; the others are M17 and M11 of Messier's catalogue). Thereza's obvious familiarity with these celestial objects indicates that she was already an experienced observer. She knew her way around the heavens with the greatest ease, and would demonstrate the planets and other interesting sights to family and friends: the Moon; the crescent Venus; Saturn; Jupiter with its four satellites and its belts; the nebula in Andromeda; the Pleiades; various double stars, all identified by their Greek names. A charming entry, for 1858 April 19, reads: "We went to look for Saturn and just got a look at him before he left the range of the telescope, and a beautiful sight it was! We also looked at i Cancri, a beautiful double star, yellow and blue, and at Praesepe [a star cluster] which wonderfully pretty object solicits fresh admiration every time I have the pleasure of gazing into its starry depths." It is evident that her guidebook to the sky was Smyth's *Bedford Catalogue* with its emphasis on double stars and the colours of their components.

Thereza also kept an eye on the Sun and its spots which she would have done by projecting an image on to a screen. The journal, over a period of a year and a half until the end of 1857, contains only 20 astronomical references; but it was not a systematic observing log, and does not cover all the uses to which the telescope was put by Thereza and her father, which included their most impressive achievement, a photograph of the Moon (Fig. 8.2). That historic photograph, preserved among Thereza's archives,[xliv] is labelled *The Moon* with John Llewelyn's initials JWL.[xlv] It ought, however, to be attributed to Thereza, as she was the primary observer who handled the telescope (her father presumably processed the photograph). Her part in the operation, as she records in her *Memoirs*, was to guide the telescope by hand during the exposure. Being equatorially mounted, it required to be moved slowly clockwise to compensate for the rotation of the earth: in Thereza's own words: "as the moon's light is much less than that of direct sunlight, the necessary exposure in the camera must have been a long one, and it was my job to keep the telescope moving steadily as the telescope had no clockwork motion while the photograph was being formed."[xlvi] Unfortunately the exposure time was not recorded. She had a steady hand, as the resulting image of the gibbous Moon (between first quarter and full), shows up the darker areas of the main maria or "seas" remarkably well. It was a considerable feat, and places the Llewelyns among the earliest lunar photographers in Britain.

The first properly successful photographs of the Moon were obtained at Harvard in America in 1850 "with which" to quote Agnes Clerke, "the career of extra-terrestrial photography may be said to have formally opened".[xlvii] Some of these were shown in London at the great Exhibition in 1851, and inspired Warren de la Rue, the telescope-maker and photographer already mentioned, to make his own

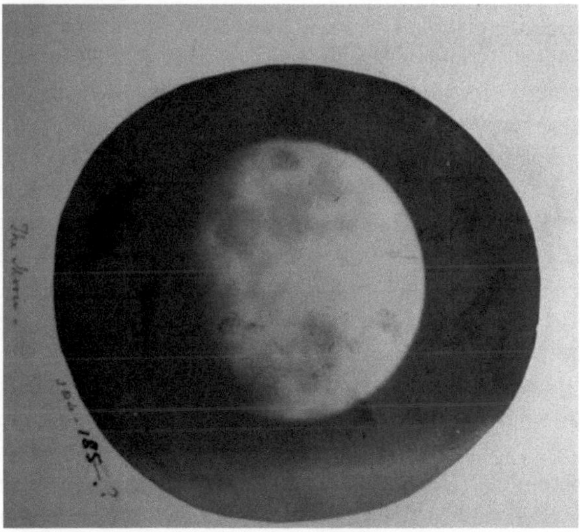

Fig. 8.4 John Llewelyn and Thereza's moon photograph, 1858. (Vanda Morton, in private possession)

attempt at lunar photography with the collodion process. Using the 13-in. speculum mirror telescope which he had constructed entirely himself, in his garden in north London, he secured his first photograph of the Moon in 1852. Realising the disadvantage of having to guide the telescope by hand, he added a driving mechanism and moved his observatory to a clearer site at Cranford, then well outside the city, in 1857. There he dedicated himself with great energy to photography of the Moon and planets (begun in 1857), and later of the Sun.

Thereza's experience with hand-guiding was the same as de la Rue's first attempt five years before. "That photograph", she wrote in her *Memoirs*, "was one of the first ever made of the Moon, though I have since learned … that Mr. de la Rue tried to get photographs of the Moon in 1852, when he found that the labour of keeping the telescope directed was very great and after a few trials, he put aside the work until he got a driving clock attached to the telescope. This was early in 1857".[xlviii]

As to the date of her own photograph, Thereza records in her journal on 12 January 1857: "I had the camera fastened to the end of the telescope"; and on 13 January: "In the evening I went with Papa to try the telescope, but found it did not answer as it was". Vanda Morton, who has made a detailed study of Thereza's writings, considers these entries to represent the Llewelyns' earliest attempt at photographing the Moon. John Llewelyn was away from home on 12 January when Thereza set up the camera; their failure of the following night meant that some further work or adjustment was needed.[xlix] Her *Memoirs* give a further clue. She describes the visit of Nevil Story-Maskelyne (her future husband) to Warren de la Rue at his observatory at Cranford during which he secured a photograph of the Moon for himself. That visit, recounted by Nevil in a gushing letter to Thereza, took place in December 1857.[l] He saw and photographed both Jupiter and Saturn ("a very spectacle of wonder and delight"), and went on: "I secured one of the moon, a capital little negative, about 1 in. by 1 in., taken at 5.55 [p.m.] in 12 seconds."[li] Thereza in her memoirs could not remember "whether it was this letter which made us wish to photograph the moon, or whether we had already made the attempt". Nevil's photograph of the Moon is dated 27 December 1857.[lii] The most likely sequence of events compatible with Thereza's remark is that the Llewelyns, having failed in January 1857 tried again, encouraged by Nevil Story-Maskelyne's experience, and succeeded in January 1858 or soon afterwards. The date was certainly before 29 June 1858, when Thereza married and left home. The limits to the date of the Moon photograph are thus January 1857 and June 1858. It is suggested that the most likely date was early 1858.

Nevil's photograph of the Moon[liii] is also preserved in the family archives, as is a photograph of Saturn, believed to be that taken by him on the same occasion.

It is strange that, while the Llewelyn Moon photograph has earned a place among historians of photography, it was passed over by historians of astronomy. Thereza herself undoubtedly realised this when she commented in her memoirs on the notice taken in London circles of David Gill's much later photograph of 1869.[liv]

Thereza was 24 when she married Nevil Story-Maskelyne (1823–1911),[lv] aged 36. He was already a scientist of considerable standing and had been recently appointed to the newly created prestigious post of Keeper of Minerals at the British Museum, while at the same time retaining his position as Professor of Mineralogy at the University of Oxford. He was the grandson of his namesake, Nevil Maskelyne, Astronomer Royal from 1765 to 1811. His mother, Margaret Maskelyne, was the Astronomer Royal's only child and heir whose inheritance included Basset Down House, a country estate in Wiltshire where she and her husband, Anthony Story, lived from the time of their marriage in 1819. A condition of Margaret's inheritance was that the family should adopt the surname Maskelyne when their eldest son attained his majority. Thus, when that son, Nevil, reached the age of 21 in 1844, the family, hitherto known as Story, became Story-Maskelyne or simply Maskelyne.

Nevil Story-Maskelyne imbibed a taste for science from both his parents. His mother had lived until the age of 24 at the Royal Observatory, Greenwich (a charming portrait of her as a girl of 12, posed against a painting of the Observatory, is reproduced in Vanda Morton's biography of Nevil[lvi] and in Derek Howse's biography of the original Nevil, the Astronomer Royal[lvii]), was well educated, her sphere of interest including, naturally, a knowledge of astronomy which she passed on to her son. One wonders if Margaret attended Mrs. Bryan's school, of which, as we have seen, her father the Astronomer Royal was a supporter (Chapter 2). At her Royal Observatory home in Greenwich Park, Margaret met the famous astronomers of the day, and accompanied her parents on their social rounds. Among her memories was undoubtedly the visit to Greenwich of Caroline Herschel in 1899, who came to give the Astronomer Royal her corrections to Flamsteed's Atlas. Caroline has put that enjoyable visit and the presence of her host's only child on record.[lviii]

Nevil's father, Anthony Story, a lawyer by profession, had specialised in the sciences at Oxford, was a Fellow of the Royal Society and a friend of many of the leading scientists of the day. On his marriage to Margaret, he settled down to his role as landowner, though he never abandoned his scientific interests.

The young Nevil, when a small boy at his boarding school in Somerset, gained as a class prize a copy of Mary Somerville's book, *The Connexion of the Physical Sciences*. Many years later, when he was an established scientist, he recalled in a letter to Mary Somerville, who was then in her eighties, that it was this book that turned him into "a man of science".[lix] His eventual chosen career was in mineralogy which he studied at Oxford, but he was also a successful early photographer. Like John Llewelyn, he was was a founder member of the Photographic Society of London, and in 1845 he introduced himself to Fox Talbot who was kind and generous with advice. In 1846 he purchased a large camera, intending to photograph the Moon, his aim being to record the crater Maskelyne, named after his Astronomer Royal grandfather.[lx] This ambitious idea required a telescope, and he was put in touch with Henry Lawson, an amateur astronomer in Bath[lxi] who owned a large telescope which he allowed Nevil to mount on the roof of his (Lawson's) house.[lxii] The result was a blank, but Nevil Maskelyne's interest in the chemistry of the photographic process continued, and led to the invitation in November 1855 to visit John Llewelyn at his home in Wales.

Nevil visited the Llewelyns on several occasions to discuss and compare photographic techniques and also, no doubt, to compare photographs, as Nevil was a prolific photographer of artistic outdoor scenes and portraits of colleagues.[lxiii] He was sometimes accompanied by one of his three sisters, and the young people in both families soon became friends. A house party at the Llewelyns included the geologist Warrington Smyth, who later married Nevil's sister Antonia, adding an interesting link to the Smyth clan. Smyth was the brother of Charles Piazzi Smyth, another photographic pioneer in Story Maskelyne's circle and a special friend of his. John Llewellyn once took his two eldest daughters to visit Oxford where Nevil showed them his laboratory and demonstrated the phenomenon of light polarisation to the serious-minded Thereza. The young couple's romance blossomed in the darkroom and the dome, and they became engaged in September 1857. Nevil's mother Margaret Maskelyne died shortly before their marriage in June 1858. Marriage separated Thereza from her beloved observatory, her only regret as she left Penllergaer to make her home in London.

The Maskelynes' honeymoon on the Continent was combined with a Grand Tour of scientific laboratories and mineral collections that would be useful for Nevil's work in the British Museum. It brought them to France, Germany, Switzerland and Austria. Looking out of the carriage window one night as they approached Dresden, Thereza, to her great astonishment, spotted a star "where no star ought to be". She looked again through her binoculars, and found that it had a short but bright tail. It was the famous Donati's comet, one of the most spectacular of the century. Thereza had recognised it before its discovery was announced in the newspapers, and continued to observe as it grew in brilliance and size until about the middle of October when it slipped out of sight as it approached the Sun. She sketched its position and appearance on 10 evenings from September 4 to October 11 against the pattern of stars, showing how it moved and how its huge tail developed. Recognising the principal stars by sight, she labelled them by their Greek letters and constellation. Her observations included October 5, when the head of the comet coincided spectacularly with the bright star Arcturus, which is marked on her sketch. Her record of the comet's progress, which survives in the family archives, matches very closely that made by Mary Ward in Ireland (Chapter 7), though her observations, unlike Mary Ward's, were not published.

Thereza, bereft of her telescope and with a growing family, was content thenceforth to admire the sky with the naked eye. She and Nevil spent several hours in an open space watching the great meteor shower of 1866 which materialised as predicted. "Hundreds and thousands sped across the sky towards every point of the compass. We both counted and could scarcely keep count". When the Great Comet of 1882 appeared, a brilliant object with a long tail, Thereza was up before dawn to observe as it rose, sketching its place as it moved through the constellation Hydra.

Her Memoirs also contain a wonderful description of a phenomenon which she had first admired as a child – the Aurora Borealis. This particular one occurred in 1870, a year of exceptional sunspot activity and many auroral displays; "... the sky was absolutely one mass of colour – from Rose colour to the deepest crimson. Arches of white light crossed the sky from which delicate fringes of light played up and form and huge columns of crimson and of white light ran up and down the sky. The country people, both here and elsewhere all over the Continent, thought nothing but the burning of Paris by the German conquerors could explain such a sight!"[lxiv]

Though Thereza did not reach her full potential in astronomy, she was open to other fields of learning. She shared her husband's interests; she helped him catalogue minerals and gems in important collections, and accompanied him on many of his scientific journeys. She moved with him in London's stimulating circle of scientists, that included Charles Lyell, William Flower (the Smyths' brother-in-law), Charles Babbage, Thomas Huxley and John Tyndall, and was a friend of public figures like John Ruskin, Florence Nightingale and Josephine Butler. She was thus very much in the tradition exemplified by her own and Nevil's mother, by Lady Herschel, by Mrs. Smyth, and many others – of the intelligent and sympathetic consort of a successful scientist. In 1897 she wrote a well researched account of the life and family history of the Astronomer Royal Nevil Maskelyne, her husband's grandfather, describing his scientific achievements and published works.[lxv] It is one of the sources of his official biography.

Nevil Story Maskelyne was for many years actively associated with the forward-looking Bedford College for women (founded as a young ladies' school, but established in 1900 as a part of the University of London where, interestingly, astronomy was taught from 1898 and offered as a subject in the BSc. degree in mathematics); his and Thereza's three daughters were educated there.[lxvi] By that time, women were already making their mark in public life, and their daughter Thereza Charlotte (1863–1941), later Lady Rucker, wife of Sir Arthur Rucker FRS, Vice Chancellor of London University, achieved eminence in her own right as a promoter of the serious study of domestic science.[lxvii]

On his father's death in 1885, Nevil Story-Maskelyne took over the responsibilities of the family estate in Wiltshire which thereafter became the couple's home.

He served as a Member of Parliament and took part in local affairs. Astronomy was not entirely forgotten. He had an interest in archeoastronomy, and Thereza accompanied him on visits to the ancient monument of Stonehenge and the standing stones of Carnac in Brittany.

Nevil died in 1911; Thereza lived another 12 years, to the age of almost 90, in the same beautiful home that was once Margaret Maskelyne's, the Astronomer Royal's daughter.

Thereza's observatory at Penllergaer has been partially restored from its dilapidated state by the Lliw Borough Council with the enthusiastic help of Swansea Astronomical Society which hopes eventually to install a telescope and return the observatory to its former glory. It will serve as a fitting memorial to a pioneering woman and an insufficiently recognised early astronomical photographer.

Notes

[i] William H. Smyth. 1986. *The Bedford Catalogue.* (Foreword by George Lovi). (Reprint). Richmond, Virginia: Willman-Bell, Inc.

[ii] Isaac Fletcher. 1865. Obituary of W.H. Smyth. *Monthly Notices of the Royal Astronomical Society*, 121–129.

[iii] Oxford DNB.

[iv] Nevil Story-Maskelyne.1866. Admiral Smyth. *Fraser's Magazine* **73**, 392–98.

[v] Magda Vargha. 2005. *Franz Xaver von Zach (1754–1832), his Life and Times.* Budapest: Konkoly Observatory; Peter Brosche 2001. *Der Astronom der Herzogin*, p 234. Frankfurt am Main: Verlag Harri Deutsch.

[vi] Hartwell House which still stands is now a hotel.

[vii] Allan Chapman. 1998. *The Victorian Amateur Astronomer, Independent Astronomical research in Britain 1820–1920.* Chichester: Praxis Publishing and John Wiley and Sons. Copious records of Hartwell House survive. Chapter 5 in Chapman's book is a delightful and amusing account of the social gatherings there.

[viii] Phebe Mitchell Kendall. 1896. *Life, Letters and Journals of Maria Mitchell.* New York: Houghton Mifflin.

[ix] Henry Albers. 2001. *Maria Mitchell, a life in journals and letters*, p 226. New York: College Avenue Press. The bust was a gift from the British feminist Frances Power Cobbe, after Mary Somerville's death.

[x] *Aedes Hartwellianae*, 1851. London:John Bowler Nichols and Son (printed for private circulation).

[xi] C.F. Bartholomew. 1976. The Discovery of solar granulation. *Quarterly Journal of the Royal Astronomical Society* **17**, 263–289.

[xii] K. Hentschel. 2000. Drawing, engraving, photographing, plotting, printing: Historical studies of visual representations, particularly in astronomy. in P. Klaus Hentschel ans Axel D. Wittmann. *The Role of Visual Representations in Astronomy. Acta Historica Astronomiae vol 9.* Thun and Frankfurt am Main: Harri Deutsch.

[xiii] John Herschel. 1887. *Outlines of Astronomy* (new edition), paragraph 597. London: Longman Green.

[xiv] I am grateful to Mary Creese who provided me with this summary of the careers of the Rümker family.

[xv] Helen Wright. 1959. *Sweeper in the Sky*, p 64. Nantucket: Maria Mitchell Association.

[xvi] Phebe Mitchell Kendall. op. cit.

[xvii] Allan Chapman. op. cit. Part 1 of the book deals with the Grand Amateurs. Part 2 is devoted to obscure and self-taught amateurs, and contains a chapter on the Grand Amateurs' professional assistants. Maria's comments as an eye-witness bear out the picture given by Allan Chapman of the world of astronomy in Victorian Britain.

[xviii] Henry Albers. op. cit. p 95.

[xix] Tim Jeal. 1989. *Baden Powell.* (p 8, quoting Henrietta's diaries). London: Hutchinson.

[xx] Allan Chapman. 1998. The Female Touch. *Astronomy Now*. January 1998. 43–47.

[xxi] Tim Jeal. op. cit., p 7.

[xxii] H.A. and M.T. Brück. 1988. *The Peripatetic Astronomer, the Life of Charles Piazzi Smyth*, p 11. Bristol: Adam Hilger.

[xxiii] Flinders Petrie. 1931. *Seventy Years in Archaeology*, p 4. London: Sampson Low, Marston & Co. Ltd.

[xxiv] Helen Wright. 1959. *Sweeper in the Sky. The Life of Maria Mitchell*, p 113. Nantucket: The Maria Mitchell Association.

[xxv] Agnes M Clerke, *History*, p 323, gives the Harvard observation, but not Maria's prior observation which is recorded in her papers and quoted by Albers op. cit., p 131.

[xxvi] Edna Carter, Professor Emeritus of Physics, Vassar College, told in a Tribute to Mary Whitney. Archives of Vassar College Library, quoted by kind permission.

[xxvii] Vanda Morton. 1987. *Oxford Rebels, the life and friends of Nevil Story Maskelyne* Gloucester: Alan Sutton.

[xxviii] Richard Morris. Oxford DNB.

[xxix] Richard Morris op. cit.

[xxx] John L. Birks. 2005 The Penllergaer Observatory. *The Antiquarian Astronomer*, Issue 2, 3–8.

[xxxi] Llewelyn called it the oxymel process, announced in 1856.

[xxxii] Vanda Morton. 1987. *Oxford Rebels, the life and friends of Nevil Story Maskelyne*, Chapter 9.

[xxxiii] Vanda Morton. op. cit., p 127.

[xxxiv] Thereza Story-Maskelyne. *Ms Memoirs* 1923, op. cit. Thereza was aged 89 when she wrote these memoirs.

[xxxv] Vanda Morton. op. cit., p 126.

[xxxvi] In the Swansea Museum Collection.

[xxxvii] John Birks, op. cit. 2005. A 4.75-inch telescope is believed to have been sold at auction when the contents of Penllergare Mansion were disposed of in 1936. John Llewelyn is reported to have purchased a 4-inch Dollond refractor in 1846, a date which conflicts somewhat with Thereza's remark.

[xxxviii] Lliw Valley Borough Council. *Penllergaer Equatorial Observatory* (pamphlet) n.d. J.

[xxxix] Thereza Story- Maskelyne, *Ms Memoirs*.

[xl] John Birks, op. cit.

[xli] ibid.

[xlii] Mss Journal August 30, 1857. (in private hands, transcribed by Vanda Morton). Perhaps a small reflector. She was showing the moon to her young brother.

[xliii] Ms journal of Thereza Llewelyn 1856–57, in private hands, transcribed by Vanda Morton.

[xliv] In private hands. There is a copy in the possession of the Astronomical Society of Swansea. I thank Dr J.L. Wainright and Dr John Birks for this information and for kindly sending me a reproduction.

[xlv] Exposure time is not given. It is undated; someone later added "185-" which is hardly helpful.

[xlvi] Thereza Story-Maskelyne. *Ms Memoirs*, op. cit.

[xlvii] The words of A.M. Clerke. *History of Astronomy*, p 153. The photograph was a daguerreotype, taken through a 15 inch refractor at Harvard College Observatory.

[xlviii] Thereza Story-Maskelyne. *Ms Memoirs*. From Col Hills, President of the Royal Astronomical Society 1913–14, with whom she was apparently reminiscing in old age. In fact Thereza was aware of de la Rue's history from her husband before she married.

[xlix] The Moon was three days past full on 13 January 1857, a very suitable phase and time of year to view the moon. A winter waxing moon would be quite high up in the evening. I thank Terry Moseley for checking the lunar phase.

[l] Vanda Morton. op. cit., p 137.
[li] Thereza Story-Maskelyne. *Ms Memoirs.*
[lii] The label on the copy in Swansea Museum is dated 27 December 1857 and adds: collodion on glass.
[liii] Reproduced in Vanda Morton. op. cit. It was on display at a Photographic Exhibition in Swansea in the summer of 2005.
[liv] Gill had a drive on his telescope. The photograph was given publicity in Forbes' biography of Gill who says: "This lunar photograph, now historical, came into the possession of the Royal Astronomical Society in December 1913" (George Forbes.1916. *David Gill, Man and Astronomer*, p 40. London: John Murray.
[lv] Vanda Morton, 1987 *Oxford Rebels, the life and friends of Nevil Story Maskelyne* is the principal source for most of what follows.
[lvi] ibid.
[lvii] Derek Howse. 1989. *Nevil Maskelyne, the Seaman's Astronomer*. Cambridge: University Press 1989.
[lviii] Lady Lubbock. *The Herschel Chronicle*, p 258.
[lix] Vanda Morton, op. cit., p 10. The letter written in 1866 is in the Somerville archives, Oxford.
[lx] Vanda Morton. op. cit., p 43–45.
[lxi] Henry Lawson (1774–1855), Oxford DNB.
[lxii] A photograph of this roof-top arrangement is reproduced in *Oxford Rebels.*
[lxiii] Several of Story-Maskelyne's photographs illustrate *Oxford Rebels.*
[lxiv] The siege of Paris by the Prussians 1870.
[lxv] T.S. Maskelyne. Nevil Maskelyne D.D., F.R.S., Astronomer Royal. *Wiltshire Archaeological and Natural History Magazine*, June 1897, 126–139.
[lxvi] Margaret J. Tuke. 1939.*A History of Bedford College for Women 1849–1937*. London: Oxford University Press.
[lxvii] Oxford DNB.

[illegible] 2001.

Chapter 9
Intrepid Travellers

It could be said that Charles Piazzi Smyth,[i] son of William Henry and Annarella, was an astronomer all his life. He was, said his friend and relative Nevil Story-Maskelyne, a son "of a most worthy father, and inherited not a little of the spirit of the gallant admiral".[ii] From the age of 11, when his father set up his renowned observatory in Bedford, he breathed an atmosphere of serious astronomical activity, and at the age of only 16 was considered qualified for the post of chief assistant to his father's friend Thomas Maclear at the Royal Observatory at the Cape, South Africa, where he spent ten happy years. Even more remarkable was his election at the age of 26 to the Regius Chair of Astronomy at the University of Edinburgh, which carried the title Astronomer Royal for Scotland and gave him responsibility for the Royal Observatory on Calton Hill in the centre of Edinburgh. The first decade of his tenure was bogged down with completing a backlog of tedious calculations inherited from his predecessor. Having got rid of that, his mind turned to a field in which he could use his native imagination and energy – a site-testing expedition to the Peak of Teneriffe in 1856, and spectroscopy at home and abroad at later dates. In all these activities he had by his side his devoted wife and amanuensis Jessie (née Duncan) (1815–1896) (Fig. 9.1).

Charles Piazzi Smyth and Jessie, aged respectively 36 and 40, were married on Christmas Eve 1855.[iii] It was the beginning of their 40 years together. Jessie adopted, or perhaps was persuaded to adopt, her husband's affectation by styling herself Mrs. Piazzi Smyth, though of course Piazzi was not really their surname but that given to Charles by his godfather Giuseppi Piazzi.[iv] "Piazzi" was also Charles' preferred forename.

Jessie was the daughter of a lawyer in Aberdeen who died while she was very young. Her mother re-married, and Jessie was brought up from the age of five in Clova, the country estate of her step-father Harry Leith Lumsden of Auchendoir. Leith Lumsden, head of an old Scottish family, was an enlightened landowner who in 1825 created the local village of Lumsden, about 40 miles north-west of the city of Aberdeen.[v] The House of Clova in beautiful and historic countryside is still inhabited. Jessie had a very happy life there: decades later she and her husband were to give the name Clova to the house in England to which they retired and ended their days.[vi]

M. Brück, *Women in Early British and Irish Astronomy*,
DOI: 10.1007/978-90-481-2473-2_9, © Springer Science+Business Media B.V. 2009

Fig. 9.1 Jessie Piazzi Smyth on the site-testing expedition on Tenerife, 1856. (Royal Observatory, Edinburgh)

Jessie, in all likelihood, was educated privately at home. One learns from her later life that she spoke French, possessed a piano, had a keen interest in geology and enjoyed travel. She probably lived in Clova until the death of her step-father in 1844 when she was 29. She and her mother then moved to Edinburgh, to Meggetland House, a mansion which no longer stands.

Jessie's future husband took up his university chair in Edinburgh 2 years after Jessie came to the city. There were various opportunities for the couple to become acquainted. Jessie took a serious interest in geology which she studied in Edinburgh with a well-known tutor Alexander Rose, beginning in 1847. Edinburgh was the cradle of academic geology, the birthplace of James Hutton, and an ideal centre – as it has ever remained – for the practical student, with its dramatic volcanic rock scenery and geologically fascinating surroundings. At that time, Geology in the university came under the aegis of Natural History which also included Zoology and Botany (the chair of Geology was not established until 1871) but tuition was offered by Rose who gave lectures in rooms near the university, and conducted field work in and around Edinburgh. An advantage of Rose's course was that it was open to women.

Rose's classes, laid out in his syllabus,[vii] consisted principally of the classification of strata and rocks, and of the chemical compositions of minerals, all in great detail. The lectures were illustrated by collections of rocks, fossils and minerals, as well as "sections, diagrams, drawings and maps". Particularly impressive was the field work, no different from the programme of modern first year university students – excursions to the famous Edinburgh formations of the Castle Rock, Calton Hill, Salisbury Crags and Arthur's Seat; and to localities further afield, including the Pentland Hills and the Coast of Fife. Jessie attended 43 lectures in Rose's course in 1847–1848 and took part in 10 field trips.[viii] . The trips included one to Burntisland, Mary Somerville's home village, where Mary as a child had marvelled at the fossil ferns picked up in the shale on the shore.

Afterwards, Jessie made extensive geological tours not only in Scotland, England and Ireland, but also in Switzerland and Italy. The first was to Ireland in the summer of 1852 when she attended a meeting of the British Association for the Advancement of Science in Belfast. The British Association, founded in 1831 with the aim of bringing science to a wide public, allowed women to attend. It met in different cities each year, and divided its activities among the various branches of science. Its gatherings were – and remain to the present day – extremely popular, providing an opportunity for social intercourse as well as for reporting scientific progress. Jessie went in the company of friends, Dr. and Mrs. McLagen. Dr. David McLagen, a distinguished Edinburgh surgeon,[ix] was a member of Piazzi Smyth's first Observatory Board of Visitors. Piazzi Smyth was also present at the Belfast meeting, and would have met Jessie there in the McLagens' company. His sketchbook contains a picture of a group of people at that meeting which includes several women – one probably being Jessie, another perhaps Mary Ward (Chapter 6).

The meeting was followed by an excursion to the Giant's Causeway, the famous geological formation on the seacoast in County Antrim. On returning home, Jessie and her mother set off on a three-month tour of England. It was a combination of sightseeing and geologising which took in Bolton Abbey, Fountains Abbey, Harewood House and other attractions. Most significant was a week spent in the vicinity of Aylesbury, 45 miles from London, apparently staying at St John's Lodge, the home of W.H. Smyth, and a focus for astronomical devotees. The week included a visit to nearby Hartwell House, the magnificent home of W.H. Smyth's friend and fellow-enthusiast, Dr. Lee, whose observatory and museum were at the disposal of the Smyth family at all times. The Christmas season was spent on the Isle of Wight.

There was a further visit to Admiral Smyth's at St John's Lodge the following spring, when Jessie's diary records that she went directly there from Edinburgh and stayed a week. It was the start of another extended tour of England. She went to friends in London and in the south of England, and spent some days in Lyme Regis, famous as the home of the late Mary Anning (1799–1847), the great fossil hunter whose discoveries attracted paleontologists in great numbers. Other women fossil-collectors in the district were Elizabeth Philpot (1780–1857) and her sisters who assembled a valuable collection of local specimens, and were known to William Buckland and other important geologists of the day.[x] One wonders if Jessie came to know any these women on her excursions to the Lyme Regis in 1852 and 1853. The area was a geologists' paradise, and Jessie walked the beach from Lyme to Charmouth, no doubt collecting specimens for her collection.

Jessie also made excursions to different parts of Scotland, chiefly for their geological interest, to judge from the addiction to quarries and her occasional notes referring to the work of Murchison and other experts. One trip was through her native Aberdeenshire, including a visit to her old home. In fact, Jessie spent more time away from Edinburgh than at home in the three years between the Belfast meeting and her marriage.

The last and most elaborate of her journeys was a tour of the Continent in August and September 1854. She (probably accompanied by her mother) travelled by boat and train from London to Paris, then to Basel, Baden, Zurich, Milan, Verona, Florence. An unfilled gap in the diary suggests that they continued as far as Naples. There were halts at places on the route, mountains climbed and hot springs visited, with copious descriptions of rocks examined. The entries end with 20 pages about the eruption of Mount Etna which had taken place two years earlier. This ends the extant record of Jessie's geological career. Seven years of her life had been largely devoted to geology; but only her skimpy diaries survive.

The couple, who had no children, were devoted to each other, but Jessie herself remained in the shadow of her flamboyant spouse. Piazzi's own life is well documented. He was a lifetime diarist and an indefatigable recorder of his travels in words and in sketches. Jessie by contrast left practically no personal papers, but as the pair were rarely apart, Jessie's life can be traced from his. Their characters were entirely different: he was impulsive and sometimes distinctly eccentric; she was a woman of gentle personality who did not seek the limelight, a foil to Piazzi's enthusiasms and prejudices. There is no indication that she pursued her own geological investigations in any serious way after her marriage; they we probably crowded out by her husband's demands. She kept her collection of specimens, and perhaps added to it occasionally, but her chief role was as her husband's aide in his diverse projects. The official programme at the observatory, performed by Piazzi and his two long suffering assistants, consisted of positional observations of stars, routine meteorological recordings, and the maintenance of a time service. Original research was done outside the observatory's walls, much of it on foreign trips. Jessie, physically tough and fond of travel, was the ideal companion for a roving scientist.

Within a few months of their marriage, the Smyths had started on the first of their many expeditions. Its destination was the Peak of Tenerife, and its purpose to test the merits for astronomy of making observations from a high altitude site, a long cherished ambition of Piazzi's. Since coming to Edinburgh, with its observatory ludicrously situated in the very heart of a city packed with smoking chimneys,[xi] he pined to observe the stars in a better climate.

Piazzi Smyth's vision of the ideal astronomical site was formed in his South African days, when he spent long spells in the high mountains, often alone, while working on the country's geodetic survey. There, away from all human habitation, he was struck by the brilliance and steadiness of the stars and the clarity of faint objects such as the zodiacal light. He argued forcibly the tremendous advantage of observing from above the clouds and above the densest part of the disturbing atmosphere, views expressed in print to the Royal Astronomical Society as early as 1845. He quoted Isaac Newton who in his "Optiks" stated that telescopes, however perfect, cannot "take away the confusion of rays which arises from the tremors of the atmosphere. The only remedy is a most serene and quiet air, such as may perhaps be found on the tops of high mountains above the grosser clouds." Piazzi Smyth believed that a mountain surrounded by ocean would be the best location. Now, 20 years later, he would put the theory to the test on Tenerife, with its Peak at an elevation of 12,500 ft.

It seems strange that nobody had put Newton's remedy into practice before; but there were reasons. Publicly paid astronomers were expected to provide a time service for their city, or to give university lectures; to some extent, also, observatories were objects of local prestige, like art galleries or cathedrals. Piazzi Smyth's solution was the "peripatetic astronomer". In the Edinburgh case, he envisaged a mountain station where he could take a telescope "into the darkness of the tropical nights" during the summer months when the long twilights made astronomical observations impossible in the north. Observations would be still be done at home during the winter.

Support for Piazzi Smyth's plan came in 1856, in the form of a grant of £500 from the Royal Society in London. Five hundred pounds, even in the mid-nineteenth century, would not have been enough without the additional support of a wealthy well-wisher, the railway engineer Robert Stephenson, son of the pioneer constructor of the first steam locomotive, the Stephenson rocket. Robert Stephenson had worked for many years constructing railways in South America, and had witnessed in the Cordilleras the clear mountain skies described by Piazzi Smyth. Stephenson, now home again and a Member of Parliament, offered the expedition the use (with its crew) of his famous yacht Titania – "that fairy thing of iron, built for speed and for luxury, to skim the waters or to fly like a racer over calm seas".[xii]

This generous gesture brought the Smyth family a new friend in Stephenson himself. Piazzi's sister Henrietta, wife of Baden Powell, named her son, born in February 1857, after him. That favourite son, Robert Stephenson Powell, the future Lord Baden-Powell, was to become considerably more famous than his parents or his godfather, as a war hero and the Boy Scout founder.

On 24 June the Smyths set sail from Cowes with a massive load of scientific instruments. There were two telescopes, with their equatorial mountings – the observatory's 3-in. refractor, called the Sheepshanks after its donor, and a 7-in. refractor lent by an English amateur friend. Auxiliary apparatus, borrowed from various colleagues, included a spectroscope, a polarimeter, an actinometer (to measure the Sun's radiation), a thermomultiplier (for infrared observations of the Moon), a heliometer (to observe double stars), as well as an array of thermometers, barometers and other meteorological equipment. There was no provision in the grant for human assistance of any kind, but Piazzi, fortunately, had the help of his wife who was no stranger to open air activity. The crew of Titania, and willing local helpers on Tenerife, undertook heavy tasks such as carpentry, building, and transporting of provisions. Only Jessie, however, was capable of lending a hand with the multifarious scientific apparatus. She was also needed with Piazzi's own personal project – photography, both scientific and scenic. Piazzi Smyth was a skilled practitioner of that art, which still used the tedious wet process with its messy chemicals and darkroom tents.

After fourteen days at sea the Titania arrived at Santa Cruz on the island of Tenerife. The party entered the town on ponies but changed to mules for the rugged ascent of the mountain. They set up their first camp at a height of 8,900 ft. It had a walled enclosure, built by the Titania crewmen, within which the telescopes and Mrs. Smyth's tent, were housed. Here Jessie worked, according to Piazzi, lending "important assistance, in writing and many other occupations". A Tenerife friend compared her to Caroline Herschel. Fireplaces were built in each wall of the stone room, from which to choose, depending on the wind direction, and in front of the door was placed a "large slab of trachytic lava" to serve as a table. "My wife never regretted that she had herself proposed decreasing the bulk of our camp equipage by declining to adopt either actual tables or chairs".[xiii] After a month's intensive observations, operations were moved to their second station, close to the summit, at a height of 10,000 ft. The transfer, by mules, up the rugged mountain was a marathon task, repeated in reverse a month later. An even larger building, roofed in, to withstand the winds, was erected there. Daily life and weather conditions at this altitude were considerably more severe than at the lower station, but there was no flagging, and all the planned observations including solar spectroscopy were carried out: the scientific data (which Jessie was presumably responsible for keeping track of) occupied 100 printed pages in the official Observatory publications.[xiv] The Smyths also took an interest in the geology of the island, confirming and photographing features recorded by earlier travellers, while Jessie collected specimens. On their very last day at the higher station, their equipment packed and on its way down the mountain, Piazzi and Jessie made a hazardous photographic excursion to the famous ice cavern at 11,000 ft where the snow lingers perpetually. Tourists, they were informed, never entered but contented themselves with standing over the hole and peering in. The helpful yacht carpenter constructed a ladder such that "a lady was handed down with ease", and the enquiring pair with all the "photographic machinery", descended and explored.

The Smyths sailed home after an absence of three months, two of them spent camping on the mountain. Piazzi wrote various accounts of the expedition. The first was his popular *Teneriffe, an Astronomer's experiment*, believed to be the first book ever illustrated by stereophotographs, a delightful travelogue in which the reader is made aware through uneffusive remarks of his wife's collaboration. His formal scientific Report, addressed to the Admiralty (official providers of the £500 grant),[xv] which earned him his Fellowship of the Royal Society, contains no reference whatever to Jessie's presence on the actual expedition. However, she is given credit for her valuable contribution to the illustrations. Piazzi had wished to include some of his striking photographs in the Report, but their reproduction was deemed too expensive by the authorities. So Jessie single-handed (working, it is believed, in her kitchen) produced no fewer than 700 enlarged positive prints which were pasted into place in the volumes. When one looks closely at these, one finds their mounts labelled "C.P.S. phot., J.P.S. pr". They are still in perfect, unfaded, condition. Piazzi held them up as an example of what ought to have been done. "Where is the fault or difficulty?", he wrote in the Report, "my wife has in the course of a very short space of time printed off 350 copies of each [of two negatives]".[xvi]

It was while the Smyths were completing the reports (perhaps they had already finished them) in September 1857, that Maria Mitchell's visit to Scotland took place. Her published diaries and letters record her impressions of the historic city of Edinburgh, and, as was her wont, minute details of the instruments of the Observatory[xvii] and her views on British astronomy.[xviii] She was invited to dinner twice at the Smyths' home where she met some "very cultivated "company – Robert Chambers, the controversial writer whose book *Vestiges of Creation* with its early ideas on evolution had caused a sensation when it was published in 1844, "a noble-looking man", and his good-looking young daughters; and Dr. Charles Maclagen who turned out to be the author of the article *America* in *Encyclopaedia Britannica.*[xix] (In fact, Maclagen was a very conspicuous figure in Edinburgh society. He edited the sixth edition of the *Encyclopaedia* (1823) as well as writing the *America* article which was retained up to and including the ninth edition of 1875. He was one of the founders and the first editor of *The Scotsman* newspaper. He was also a noted amateur geologist.) She also met Robert Stephenson, the sponsor of the Tenerife excursion.[xx] Surprisingly, however, Maria does not appear to have anything to say, at least in her published memoirs, about the Smyths' adventures, or about Stephenson's role as their patron; nor does she mention conversing with her hostess Jessie who, like herself, had experienced an astronomer's life under canvas and had plenty of tales to tell. On the contrary: she took a high-handed view of her. In a letter to her father, written from Edinburgh, Maria had her usual complaint about the ignorance of the British of the great nation of America. "Mr. Airy understands that the Bonds [of Harvard Observatory] are astronomers, but I dare say Mrs. Prof. Smyth never heard of them, tho' of course Prof. Smyth has the transactions" [i.e., the Harvard publications]. One rather suspects that Maria did not care to be displaced from the centre of attention.

To Piazzi Smyth's disappointment (and to the detriment of British astronomy) his recommendations to found a mountain observing station, though praised at the time and earning him his Fellowship of the Royal Society, were not acted upon and were, as so often happens with official reports, filed away and forgotten. The Russian decision to build an observatory in the Caucasus was influenced by the Tenerife results and by direct advice from Piazzi himself when the Smyths (and their photographic apparatus) made a privately-funded trip to Russia by sea across the Baltic three years later. Their experiences as honoured guests at the great Pulkovo Observatory near St Petersburg and of visits to Moscow and Novgorod, were recorded in another delightful book, *Three Cities in Russia*,[xxi] and in a collection of beautiful photographs.

It was most unfortunate that at this juncture Piazzi should suffer a bizarre distraction which delayed his further scientific schemes and dented his academic reputation. This was his ill-advised expedition to Egypt in 1864 to explore the Great Pyramid at Giza for what he firmly believed were urgent scientific reasons. Piazzi Smyth's sudden obsession with the Pyramid arose from many causes that are impossible fully to disentangle.[xxii]

He was not the first to be fascinated by this extraordinary building, the last of the seven wonders of the ancient world, and the most massive on the face of the earth. There was something awesomely mysterious about its size and shape and inner chambers, believed by some to enshrine important secrets of the ancients, if only one could find the key. Its geometrical structure had been investigated by surveyors in the past, including Napoleon's team of experts whose published report inspired new speculations. One such theory which appealed to certain religious minds was that the Great Pyramid was not the tomb of a great Pharaoh, as generally believed, but a monument erected under divine guidance in Old Testament times for some purpose associated with the destiny of humankind. A particular version of this idea arose at a time of public debate on the question of whether Great Britain should replace the traditional system of weights and measures by the French one (the metric system). The old standard rod and the standard weight preserved in the Palace of Westminster had been lost in the fire of 1834, and it was essential in the interest of trade that strict new standards be agreed upon and adopted.[xxiii] Feelings ran high about these possible changes. Scientists on the whole favoured the unified French system. Sir John Herschel, however, took the conservative view, and found arguments demonstrating the advantage of the British inch or multiples thereof over the metre as a unit of length. Piazzi Smyth supported this view.

It was at this juncture that the claim was made that the British inch was related to the sacred cubit of the Bible, and that this same sacred cubit was the unit of measure used by the (reputedly inspired) Pyramid builders. Piazzi, who at first dismissed the idea as nonsense, had second thoughts. It now seemed to him that the Pyramid, with its unique geometry, could not have been designed using the primitive mathematical knowledge of its builders, and hence was part of a higher plan. By extension, so too was the British inch which thus, it would follow, had Biblical authority. A careful metrological survey of the Pyramid would confirm it.

Piazzi Smyth, "unaccompanied by anyone else than my Wife",[xxiv] made elaborate preparations for the expedition. Their baggage of 32 items contained astronomical and surveying instruments and special measuring rods. There was a stereoscopic camera of Piazzi's own design, magnesium for flash photography, and the usual darkroom paraphernalia. There was a complete range of domestic equipment for a planned three months' stay. All this travelled with them by sea to Alexandria at their own expense. Before setting out, Piazzi, a master "spin-doctor", published an exposition of the theory[xxv] that brought converts to the cause even before it was "proved".

The Pyramid expedition was, from Piazzi Smyth's point of view, a great success. A huge volume of data, published and publicised, purported not only to confirm the mystic theory but to provide much additional "evidence" that the Pyramid was, indeed, an inspired structure whose significance was only now being revealed. In a chronicle of Piazzi Smyth's scientific achievements, it is a huge embarrassment which cannot in all honesty be explained away. As for Jessie – the four months spent in Egypt in the winter of 1865 were evidently an enjoyable adventure, as emerges from Piazzi's popular day-to-day account of the expedition,[xxvi] published on their return. Like his books on Tenerife and Russia, it is a highly readable traveller's tale, in which one catches delightful glimpses of Jessie's normally eclipsed presence. She was not the first lady-traveller in Egypt, but it is unlikely that any other woman, before or after, had set up home in a tomb, as she did (Fig. 9.2). They discarded their rents and settled instead into a suite of tomb-rooms – a kitchen with adjoining mummy pits for storing water and a "beautiful room for the lady and gentleman to take their dinner in", chosen by their Arab cook. There was a place to write their daily observations and a bedroom with a bath (they had brought with them a full-sized tub). Piazzi had the services of an experienced guide, and Jessie had her cook and a young male domestic help. Jessie got on splendidly with these and with their friends, who called her affectionately "Mrs. Piazzi".

Jessie was not entirely occupied with domestic affairs, however. She helped Piazzi making astronomical observations and with his photography. They observed the pole star to find orientations of certain structures and trenches. More memorably, she climbed with him the 200 courses of the pyramid's wall to the small flat platform on the top where they stayed an entire night, observing the position of the Moon and of other stars for the purpose of determining longitude.

A genuine achievement of Piazzi Smyth's in Egypt were his miniature stereo photographs. They comprised artistic views of the pyramids and the surrounding landscape, close-up details of the Pyramid's masonry, and magnesium flash photographs of the interior. The last were notable as the first such use of magnesium flash outside a photographer's studio. The Egyptian photographs – like those taken in Tenerife and Russia – belong to the history of photography. Piazzi made lantern slides of many of them to illustrate his lectures on the subject of the sacred Pyramid, and allowed them to be commercially copied and distributed among studious "Pyramidists".[xxvii] The Pyramid cult flourished, with Piazzi Smyth as its interpreter, and to this day has its devotees.

Addressing the Royal Society of Edinburgh, Piazzi described the Pyramid work blatantly as "my own individual labour", with no mention of Jessie or of anyone else. She scarcely appears in *Our Inheritance in the Great Pyramid*, the popular book that first made his name among the public, though she is given brief credit in the Preface: "My wife and self...did...through four months of residence on the Pyramid Hill itself employ a large variety of scientific instruments in obtaining many measures of the mighty monument." With each edition of the book his interpretation of the Pyramid became more fantastic and his followers more enthusiastic. Many reputable citizens were convinced, but most were cranks, some even believing that the Pyramid had prophetic possibilities. One of Piazzi's transatlantic supporters was Henry Mitchell, Maria's brother, a member of the US Coast survey, though there is no indication than the common-sense Maria was attracted.

What Jessie, an intelligent woman with a good understanding of modern geology, thought of these unscientific notions, one can only guess. As a dutiful wife,

Fig. 9.2 Jessie, "the Lady of East Tombs", at the Pyramid of Giza, 1866. (Royal Observatory, Edinburgh)

she probably kept quiet and supported her husband; in the course of time she herself joined in his propaganda, replying to letters from his admirers. There are even hints that she once wrote a pamphlet of her own. Who can tell whether she was truly a convert, or whether her husband's powerful personality had worn her down?

Unfortunately, the Pyramid dabbling cost Piazzi Smyth the respect of colleagues in the London scientific establishment, and caused him rashly to resign from the

Royal Society. Fortunately, at about this same time, he took up a new, fruitful, worthwhile line of research which somewhat redressed the balance: spectroscopy.

Astronomical spectroscopy, known as the "new astronomy", began to take root in the 1860s (Chapter 11). By coincidence, the year 1859, indeed the very month when the Smyths set off on their Russian tour, marked its birth. The German scientists, Gustav Kirchoff and Robert Bunsen, discovered almost by accident the source of the dark lines in the spectrum of the Sun. The laboratory science of spectroscopy could now be applied to luminous celestial bodies, and made to reveal the presence in them of specific chemical elements, something which up to then was considered beyond the power of man. Piazzi Smyth had observed the spectrum of the Sun (its rainbow colours, dispersed by a prism) from Tenerife at a time when its origin was still a mystery. With the application of spectrum analysis, it now became possible to identify the chemical composition of the chief absorbing material, presumed to be gases in the Sun's atmosphere. The elements sodium, calcium and iron were soon revealed; others were gradually discovered.

Piazzi Smyth, with Jessie at his side, was henceforth to put much energy and ingenuity into laboratory and solar spectroscopy. The Smyths had a very large house in an elevated location in Edinburgh (Royal Terrace) from the top floor of which there was an unobstructed view to the north, and an equally fine prospect to the south, ideal for observing the Sun. Here Piazzi assembled his spectroscopes, furnished an experimental laboratory, and made observations, independently of his official duties at the observatory. The sunspot maximum of 1870 was unusually high, and many bright aurorae or northern lights which are correlated with sunspots were visible from Edinburgh over the next few years. The view from the Smyths' north window was particularly favourable, and Piazzi made some first class observations of both the phenomenon itself and of its spectrum. A visitor, a distinguished French spectroscopist, wrote to Piazzi congratulating him on "the most intelligent and agreeable assistant in the person of Madame Piazzi Smyth". In one unusual experiment carried out in the railway yards in Edinburgh, the obliging Railway Superintendent got up a good head of steam in front of a bright gas lamp to allow the astronomers to observe what effect this had on the spectrum. Piazzi, the Observatory's two assistants, and Jessie, each in turn, viewed the spectrum thus produced through a spectroscope. Their result was published in the Edinburgh Astronomical Observations.[xxviii]

In the field of solar spectroscopy, Piazzi Smyth, faithful to his doctrine of "peripatetic astronomy", travelled abroad with Jessie to better climates to procure observations, making four foreign expeditions – to Palermo, to Lisbon twice, and to Madeira – as well as journeys on scientific business to France, Germany and Ireland. (These trips were financed partly from Piazzi's Pyramid books, but also, one suspects, from Jessie's private fortune). The first of these, in 1872, was to Palermo in Sicily, where there was a flourishing observatory, well-known for its leading spectroscopic work. The chief purpose was to observe the spectrum of the zodiacal light – a phenomenon best observed at low geographic latitudes – and compare it with that of the aurora as seen from Edinburgh. The zodiacal light is a faint glow in the sky, caused by a belt of fine interplanetary material spreading out from the

Sun and illuminated by it. Its spectrum is the same as that of the Sun, but at the time in question there were differences of opinion, some spectroscopists claiming that it had the same origin as the aurora. Setting up his spectroscope on one of the Palermo telescopes, Piazzi Smyth could satisfy himself, and his colleagues, that the two were different. Jessie confirmed it. "Jessie declares that the ZL spectrum is as different from the aurora spectrum as night is from day", Piazzi wrote in his journal. His Italian colleagues expressed surprise to find a lady at an observatory. The Smyths' continued their tour of the Mediterranean with visits to the observatories at Trieste and Padua.

This Mediterranean cruise was followed by shopping trips to France to acquire equipment, and included visits to several observatories. Piazzi designed and constructed a series of powerful spectroscopes for use on further expeditions. In Lisbon, in the summers of 1877 and 1878, the Smyths were welcomed by the Director of the Observatory but preferred to set up their apparatus in a rented house outside the city. A room with a large window was the temporary observatory, containing the spectroscope, mounted on a rotatable base, into which sunlight was fed from a heliostat mirror outside the window. Piazzi observed and carefully recorded the positions and strengths of hundred of lines in the spectrum. The heliostat mirror was "managed very steadily during the whole of the operations in both years by Mrs. Piazzi Smyth"[xxix] – so Piazzi wrote in the subsequent publication. As they were interested in observing the spectrum of the Sun at various altitudes above the horizon, the work commenced at dawn each day. The Lisbon solar spectra were better than any achieved by astronomers elsewhere, with wide "bands" resolved into tightly packed lines and certain broad lines revealed for the first time as pairs. Unwilling to trust his own eyes alone, Piazzi asked his wife to look into the telescope and confirm. "Oh, the beautiful double lines!", she exclaimed.[xxx]

The Smyths last foreign expedition was to Madeira in the summer of 1881. They stayed in a hotel where the smoking room was transformed into their solar observatory. As always, Jessie operated the heliostat "with patient enthusiasm and enduring skill",[xxxi] though she was now greatly afflicted with sciatica. It was Jessie's last time in active astronomy.

Piazzi often used a direct-vision spectroscope, a small hand-held instrument which he carried about with him. With this instrument he discovered in 1876 a peculiarity in the spectrum of the daytime sky (i.e., scattered sunlight) which turned out to be due to water vapour in the atmosphere. It was a slight darkening in the yellow part of the spectrum, recognisable by anyone familiar with the normal spectrum of the Sun. He called it the "rainband", and advocated its use for weather forecasting. The "rainband spectroscope" became for a while a popular instrument with amateur weather observers. Jessie made use of one in a meteorological journal, recording twice daily temperatures, humidity, wind speed and direction, cloud cover, hours of sunshine, and – the novel element – strength of the rainband. This journal was maintained without a break for 10 years; when on their travels, Jessie's little meteorological laboratory always accompanied her.

Following their last foreign expedition, to Madeira in 1881, Piazzi wrote up an account of that island, its geography and its flora,[xxxii] in which he compared the meteorological conditions there with those recorded in Lisbon. No definite scientific conclusions were drawn, but it was an opportunity to publish Jessie's meteorological journals in those locations and to draw attention to "Mrs. Piazzi Smyth's long-continued enthusiasm for the idea, as an addition to her daily meteorological journal at home".[xxxiii]

It is a pity – but typical of the prevailing attitude to women – that this contribution to science, slight though it was, was not published by Jessie herself but under Piazzi's name. More regrettably, it was the sole piece of work entirely attributed to her, though she might well have achieved a great deal more in her own right had she continued actively in geology, her first love. But it is clear from Piazzi's copious journals that there were no more geological digs for Jessie after her marriage. She still had her "museum" in the lovely house in Ripon, Yorkshire, where they lived in retirement, but when her health failed, she gave the room over to Piazzi for his spectroscopy. There is no record of what became of her geological collection after her death. She remained ever the docile wife. The American telescope-maker John Brashear, who called on the Smyths in their retirement in 1892, gives his own impression. "I had the honor to break bread not only with a great scientist, but with his wife whom he honored as all men should honor their companions in life's work. There is an adage that if a man becomes truly great, it is usually the help of a devoted wife that is responsible for it."[xxxiv]

Jessie died in 1896, aged 80, 4 years before her husband. She is buried in the country churchyard of Sharow under a low pyramid monument bearing an inscription composed by her husband commemorating his "faithful and sympathetic friend and companion through 40 years of varying scientific experiences by land and sea, abroad as well as at home, at 12,000 ft up in the atmosphere on the wind-swept Peak of Teneriffe as well as underneath and upon the Great Pyramid of Egypt".

Another expedition wife, also from Aberdeenshire, was Isobel (or Bella) Gill (nee Black) (1848–1919) who, as a young woman of 28 accompanied her husband, David Gill, on a historic astronomical expedition to Ascension Island in 1879. It was an operation of great significance to nineteenth century astronomy, as it opened up a new improved method of determining the dimensions of the solar system and the distance of the Earth from the Sun – the same fundamental problem that prompted the Transit of Venus expeditions that preceded it.

The rise of David Gill to become one of the nineteenth century's leading astronomers was inseparably shared by his wife who was by his side from his earliest days in astronomy. Bella was only 16 years of age when she first met her 22-year old future husband on a Sunday morning on his way to Church. "The two loved each other the moment they met", says his biographer,[xxxv] "but they could not impress the fact on their elders"; so they waited 5 years before they married, with approval all round, in 1870. The romance never dimmed: the Gills' undisguisedly happy marriage was famous throughout the astronomical community and ended only with David's death in 1907 (Fig. 9.3).

Bella Black was one of three lovely daughters of a farmer and his wife from near Aberdeen. Lively and intelligent, she was educated at the local school by the parish schoolmaster, one of that breed of scholarly teachers for which Scotland was

Fig. 9.3 Bella Gill before her marriage in 1870. (George Forbes, *David Gill Man and Astronomer*)

renowned. David Gill was a watchmaker in the city of Aberdeen. He had studied physics and mathematics at Aberdeen University, but did not graduate, as he was needed at home to take charge of the family business. He was highly skilled at his profession and perfected it in Germany and Switzerland. His overriding interest, however, was astronomy, into the practical uses of which he was first introduced by Charles Piazzi Smyth when he visited the observatory in Edinburgh seeking advice on how to set up a time service in the city of Aberdeen. He installed a good 12-in. reflecting telescope at his home, mounted and driven by clockwork, which he equipped for photography and took photographs of the Moon that brought him to the attention of the professional astronomers in London. In 1871, the year after his marriage to Bella, came the great opportunity of his life when he joined forces with another young enthusiastic amateur, (Lord) James Ludovic Lindsay, son of the scholarly Earl of Crawford, a nobleman of an ancient Scottish family who had a country estate among picturesque Highland surroundings at Dun Echt in Aberdeenshire. Lord Lindsay, who had studied at Cambridge but, like David

Gill, had not taken a degree, was equally enthusiastic about astronomy and photography. He had a laboratory in his London home, and could afford to employ a professional photographer. He had already successfully observed a total eclipse of the Sun in Spain in 1870 and now had an ambition to pursue astronomy at a really serious level – to set up an observatory which, though "amateur" in name, would be equipped to a professional standard. His father encouraged this idea, and agreed to finance his son's plan to build an observatory in the grounds of Dun Echt.[xxxvi] Lindsay was anxious to get started in time for the famous Transit of Venus of 1874, the first of the nineteenth century pair of that rare phenomenon (mentioned in Chapter 2) which astronomers world-wide were already preparing for. He required a good collaborator, and decided to ask the talented watchmaker from Aberdeen. Gill passed the letter of invitation to his wife, who immediately responded with the words "How glorious!", though it meant a considerable drop in their income and the grave disappointment of his father. It would be not be only time that she would put her husband's passion for astronomy before worldly considerations.

The Gills moved into rooms in the Dun Echt mansion until a pretty house was built for them on the estate. David Gill worked extremely hard, erecting domes and travelling to Russia and Germany to order the best available astronomical instruments, particularly those required for the forthcoming Transit of Venus in December 1874. The purpose of observing the transit of this nearby planet over the Sun's face was to determine its distance from the Earth, and hence, indirectly, the Sun's. The method, simple in principle, required the phenomenon to be observed and timed from widely separated places on the surface of the globe. The official British plans were masterminded at the Royal Observatory at Greenwich, from where several teams, all provided with similar instruments, were sent to locations throughout the Empire. Lord Lindsay, being financially independent, could make his own plans. He and Gill chose the island of Mauritius in the Indian Ocean for their own expedition, and with a small group of helpers set up and operated their station there for several months.[xxxvii] The task of timing the transit turned out, for observers everywhere, to be more difficult in practice than in theory, and the final result on a world-wide scale was disappointing.

However, the Scottish astronomers had not placed all their eggs in one basket. Gill had planned a second, novel, experiment, one that was independent of the Transit but had the same purpose. It employed another planet (in this case the minor planet Juno), which was fortunately so placed in the sky in December 1874 that it could be observed both evening (after sunset) and morning (before dawn) from the same location. Juno's position was measured in each instance with respect to a background of the fixed stars. The angular shift in the interval between the two observations arose from the fact that the observer, carried by the rotation of the Earth, was viewing the planet from two different places in space. The method had never been tried before. Gill observed Juno for several weeks in this way, and his result[xxxviii] from which he succeeded in calculating the distance from the Earth to the Sun (obtained in terms of an angle, the "parallax") was so promising that he was determined to try the method again as soon as possible. The opportunity came three years later, in September 1877, when the planet Mars came particularly close to Earth, its nearest until 2003. (It was during that close approach that Mars' two satellites and its alleged canals were discovered.)

Gill was ready to undertake the mission at his own expense, supported by Bella who once again was willing to take a risk. He resigned his post in Dun Echt at the end of 1875 and moved with her to London to make preparations. He borrowed his favourite instrument (the beautiful heliometer[xxxix] which he had used on Mauritius) from Lord Lindsay and received a modest grant from the Royal Astronomical Society and offers of personal loans from friends, a kindness he never forgot.[xl]

Gill required a site in a good climate, close to the equator. He chose the Island of Ascension, in the South Atlantic (latitude 8 °S), a rocky volcanic island of only 34 sq. mi. in mid-ocean which was entirely uninhabited until early in the nineteenth century. The island was under the control of the British Navy and had no town apart from a small garrison manned by marines. Ships carrying mail called once a month en route from the Cape, South Africa. The Gills sailed in June 1877 and arrived, via the island of St Helena, a month later. In the 10-day stop at St Helena they took the opportunity of visiting, for sentimental reasons, the site where the great Edmund Halley had set up a temporary observatory exactly two centuries earlier (in 1677).

The Ascension expedition – a man-and-wife venture like the Smyths' to Tenerife in 1856 – was a brave undertaking for just two people, but they were young and adventurous: Bella was 28 and David in his thirties. Their twenty tons of baggage included several chronometers and a transit instrument, as well as the precious heliometer, papers, books and household necessities. Their help on this remote island consisted of a few strong men provided by the naval authorities; their only amenity, an empty cottage near the garrison. Bella afterwards described her first task – to fill the larder – which proved to be by no means a simple process. "No butcher! no dairy! no greengrocer! no fishmonger!" The pair took it all in the best of humour. "I hastened home sadly from my foraging expedition with my tale of want and woe;" she recalled, "but so strongly did the comic element prevail in the recital that David and I broke into peal after peal of laughter, and that was almost as good as a meal."[xli] Gill spied a level patch that had once been a croquet ground of some sailors of the past, where he decided to erect the observatory (Fig. 9.4). "Somehow it all came right; and sitting that first evening after sunset in the verandah which looked upon our novel croquet lawn, we could speak of nothing, think of nothing, but the beauty of the heavens. Though Ascension was barren, desolate, formless, flowerless, yet with such a sky she could never be unlovely. The stars shone forth boldly, each like a living fire. Mars was yet behind Cross Hill, but Jupiter literally blazed in the intense blue sky now guiltless of cloud from horizon to zenith; and, thrown across in graceful splendour, the Milky Way seemed like a great streaming veil woven of golden threads and sparkling with gems. The Southern Cross – a poem in the heavens – shone out a bright welcome to us, while our old friend the Great Bear still kept faithful watch in the north over our wanderings. How strengthening and restful after fatigue and petty worry, is such an hour! One forgets to be care-

Fig. 9.4 The Gills' site on Ascension where they spent 6 months observing Mars. (George Forbes, *David Gill Man and Astronomer*)

ful and troubled about many things, and the soul trembles with its load of love and gratitude to Him who made the stars also."

Helped by the marines, Gill soon set up the observatory. It was the same equipment that had been used on Mauritius: the heliometer, mounted on a moving frame within a cage-like canvas dome; the transit instrument fixed in the meridian in a hut with a north-south opening. Astronomical observations had to be made to find the correct orientation and the geographical coordinates, and to keep the chronometers rated. But no sooner were the instruments adjusted and preliminary observations taken than "the face of the heavens darkened". As the days passed and the sky remained cloudy they became worried. Realising that weather conditions varied from place to place, they decided to act. David could not leave his post lest the sky cleared, and Bella offered herself "as a pioneer", in spite of her husband's worries about dangerous gullies and wild cats. But Bella, "having a considerable leaven of Luckier Mucklebackit's[xlii] spirit", had her way. Accompanied by a couple of helpers carrying a lantern, she traversed the island on foot overnight, noting the clouds every half hour until dawn, while her husband did likewise at the original site. She became convinced that the clouds were local, and that conditions would be better further south and near the sea. David accepted her verdict. A new location in a cove at the south-west extremity of the island was decided on, the instruments moved, and work began all over again. They called their camp Mars Bay, the name by which it became officially known and is marked on modern maps.

The Gills now had to live in tents: they declared their "dining-room" at Mars Bay the "nicest, coolest little resting nook in the world." The difficulty of getting from one tent to another over loose clinkers sharp enough to cut through one's shoes was cured by their Kroomen (African) helpers who established paths between the tents and covered them with white sand from the beach; these not only made walking easier but safer at night, showing bright against the black ground.

The actual opposition (nearest point) of Mars occurred in September, and Gill observed it over a period of weeks before and after that date. It was an essential feature of the "evening and morning" method that the observations were all be made by the same observer. Thus Bella did not make any actual astronomical observations, but she had her duties. Her husband took the first watch every evening; she took the early morning watch, calling him when there was the slightest chance of a clear sky so that not single opportunity for observing was lost. She found this duty no hardship. "Happier hours I never spent than those early morning ones under this beautiful heaven", she wrote. When the Mars work was done, Gill made similar observations of the minor planet Melpomene which was in opposition in December, as he had done so successfully in the case of Juno on Mauritius a few years previously.

The second of Bella's duties was to copy the observations daily in duplicate, one set being transmitted to the Royal Astronomical Society in London on each mail boat that called. When news reached the Society that the Mars observations were satisfactorily secured, the President, William Huggins, sent by return mail a letter of congratulations. There was, he said, "a roar of excitement in applause" for Gill himself at the meeting, and an even greater one for Mrs. Gill, "that courageous and enthusiastic lady who at the moments of greatest difficulty and anxiety filled your tent with sunshine and your heart with fresh courage". Undoubtedly, Gill would not have been able to do the work singlehanded. "A considerable part of the success of the expedition was due to [her] unfatigued assistance", said Arthur Auwers, a distinguished German astronomer and later collaborator of Gill.

On their return to London, Mrs. Gill made her own special contribution, a delightful book called *Six months on Ascension*,[xliii] the first published account of the geography and scenery of that little known island. It may well have been inspired by Piazzi Smyth's account of the Tenerife expedition, *An Astronomer's experiment*. Bella described how, as the outbound steamer sailed past Tenerife, their thoughts turned to that charming book, "speculating where Guajara might be, and where the path that the astronomer and his wife had toiled up with their heavy instruments to Alta Vista, the site of their home and Observatory for the time, 11,000 ft above the sea." The book began with a non-technical introduction to the scientific purpose of the expedition and a history of attempts to find the Sun's distance from the Greek Aristarchos onwards. It described the couple's sojourn on the island of St Helena on their way out. Lord Lindsay had arranged for them to be received by the Governor, Hudson Janisch, who, to their surprise and joy, was a descendent of the famous German astronomer J.F. Encke of comet fame. He was himself an enthusiastic amateur astronomer, but the Gills were the first astronomers he had ever actually met. He had been born on the island, and had never been away from it.

Bella went on to describe how, at the spot called Halley's Mount, Gill was charmed to find a bit of wall, overrun with wild pepper, which from its orientation he was sure had been part of that great astronomer's observatory wall of 1677.

The building where Manuel Johnson, later Radcliffe Astronomer at Oxford, had observed a transit of Mars and produced a catalogue of stars in 1832, was easier to find. A sad sight met them there. The observatory was now the artillery mess-room, recorded Bella, "and in the recesses formed for the shutters of the openings through which Johnson's transit used to peep, they stow wineglasses and decanters, and under the dome they play billiards!". Philip Gosse, well-known naturalist and traveller, who visited St Helena in 1937, followed the Gills' path to Johnson's observatory. Half the roof had fallen in, and the transit instrument was discovered supporting an ant-eaten bookshelf in the Supreme Court.[xliv]

On Ascension, as well as the scientific record and their domestic adventures, Bella recounted vivid descriptions of the geological appearance of the island, its flora and exotic wild life, the turtles and birds. The book, full of interest and humour, was greatly admired and is still read and enjoyed.[xlv] Bella dedicated it to their friend Samuel Smiles, famous author of *Lives of the Engineers*[xlvi] and other tales of nineteenth century science. The first copy to come off the press was presented to Lord Lindsay.[xlvii]

Ascension Island today has military uses as a telecommunications centre in the modern world. Thomas Cave, an American scientist whose work takes him to the telemetry tracking station there, revisited the Gills' observing camps in 1992 and followed closely Mrs. Gill's footsteps in her search for a new site, as described in her book. He hiked across the lava plain, with its hard lava formation and cinders, and was amazed that she had been able to make her way through such inhospitable ground at night. He was also surprised that anyone could survive for several months in such conditions, as the Gills had done, enduring sun, wind and blowing cinders. He compares the terrain on Ascension with the planet they had come to study. Remarkably, he found the layout of the camp at Mars Bay exceptionally well preserved. The paths covered with white sand which connected the tents and buildings, as described by Bella, are still there, as are the walls of small rocks outlining them.[xlviii]

David Gill, having completed his lengthy calculations, derived a value of the parallax of the Sun (and hence its distance) which was correct to a percent and accepted as the most accurate to date. He published the result in 1881 and was awarded the Gold Medal of the Royal Astronomical Society in 1882. But he did not have to wait until then for a tangible reward. He was appointed in 1879 Director of the Royal Observatory at the Cape, South Africa, a post that had become vacant following the previous director's move to the Chair at Oxford. He took up his new role with the warmest goodwill of astronomers at home and abroad. He did not disappoint them. He became one of the most admired and best-liked astronomers in the world and a leading expert on the parallaxes of minor planets and other fields of research.

The new surroundings which the Gills found at the Cape were very different from the "dismal swamp" experienced by the Fallows (Chapter 4). Their home was within the observatory building, and visiting astronomers and collaborators from abroad were their personal guests, often for long periods. Gill, enthusiastic and energetic, was determined to make his observatory the leading one of the southern hemisphere.

He initiated new, modern, programmes of research, and battled for funds with the holders of the purse-strings in London. His ambitious and forward-looking project to photograph the entire southern sky met with considerable opposition in that quarter. "After thinking the matter well over", he and his wife made up their minds to finance it out of their own resources. It was a repetition of situations they had confronted before, and, as before, Bella was prepared to make the sacrifice, which in this case was considerable – equivalent to half the husband's salary for some years. It proved scientifically well worth while, resulting in a valuable catalogue of stars published in 1900 and providing copious material for new researches.[xlix]

The Gills were very happy at the Cape. Following the example he had seen at the Pulkovo Observatory in St Petersburg, headed over by the great William Struve, Gill ran the observatory as "a happy, enthusiastic, patriarchal colony". Unlike Piazzi Smyth's wife, Mrs. Gill forsook active astronomy after her one experience on Ascension but, like other astronomical wives of her generation,[l] this "highly gifted wife" (as she was described by another of Gill's distinguished associates, the Dutch Jacobus Kapteyn[li]) was well-informed about her husband's work. One must regret, however, that she did not continue (or that her husband did not persuade her) to use her talent as a travel writer and as an expositor of astronomy. *Six Months on Ascension*, like Piazzi Smyth's account of the Tenerife expedition, is a truly delightful book; and Bella had the whole of Cape Province in front of her from which to obtain further inspiration.

The couple had no children of their own, but they adopted three young orphan nephews of David's to whom they were devoted. Their home at the observatory was renowned for its hospitality, a port of call for visiting dignitaries to the Cape. Bella, with her "dignity and fun, captivated the hearts of the Colony", and was active in charitable and Church affairs[lii] There was only one shadow on their lives in the later years – Bella's health, which began to give cause for concern in 1895, after 16 years in Africa. The nature of her illness is somewhat vague; her husband described it as "nervous", an obituary notice said she was "a martyr to persistent headaches".[liii] Gill's letters to friends such as Agnes Clerke (Chapter 12) contain constant references to Bella's improvements and disimprovements, her good and bad spells, her visits to health resorts. He attended to her every wish, and she basked in his attention. "He loved his wife and he loved astronomy".

David Gill was knighted for his services to science in 1900, and Bella became Lady Gill. He retired, full of honours, in 1906. The couple settled in London, where he continued an active scientific life, she a quiet one. Gill died unexpectedly in January 1914 at the age of seventy of pneumonia, developed from a cold caught at the funeral of an old astronomer friend Sir Robert Ball. Bella, notwithstanding her reputedly delicate constitution, survived him by more than five years. She spent her last years in her native Scotland where she died in September 1919. She is buried next to her husband in the plot that had chosen in his lifetime at St Machar's Cathedral, Aberdeen.

Bella was, like Jessie, a satellite to her star. Yet there is no doubt that the high-altitude astronomy tests on Tenerife in 1856 and the observations of Mars on Ascension in 1877 – two pioneering enterprises on these Atlantic islands – would never have been undertaken let alone accomplished without the stoutheartedness and encouragement of these intelligent skilful collaborator-wives.

Notes

i H.A. and M.T. Brück. 1988. *The Peripatetic Astronomer, the Life of Charles Piazzi Smyth*. Bristol: Adam HIlger-IOP.

ii NSM [Nevil Story-Maskelyne]. July 1858. Tenerife. *Fraser's Magazine* **58**, 35–43.

iii They were married on Christmas Eve 1855 at her mother's home, Meggetland House, Edinburgh, by her relative, Reverend Harry Leith of Rothiemay, a Church of Scotland minister. (*Scotsman*, 26 December 1855).

iv Her sister-in-law, Mrs Powell, after her husband's death, followed their example by elevating her own and her children's surname from plain Powell to Baden Powell. The wife of the French astronomer Camille Flammarion similarly styled herself Gabrielle Camille Flammarion.

v Moray McLaren. 1965. *The Shell Guide to Scotland*. London: Ebury Press.

vi Now the Clova Nursing Home, in Ripon, Yorkshire.

vii Alexander Rose. *Lectures and Conversations in Geology and Mineralogy* by Mr Alexander Rose, Fellow of the Royal Scottish Society of Arts, Hon. member of the Jena Geological Society, etc" Printed Syllabus. Piazzi Smyth Collection, Royal Observatory Edinburgh.

viii J. Duncan's geological notes. Piazzi Smyth Collection, Royal Observatory Edinburgh.

ix David McLagen had a distinguished career in the Royal Navy, and served in the Peninsular War. He was then President of the Royal College of Surgeons. *Dictionary of National Biography*: entry on his son W.D. McLagen, Archbishop of York.

x Mary R.S. Creese and Thomas M. Creese. 1994. British Women who contributed to the geological sciences in the nineteenth century. *JBHS* **27**, 23–52.

xi Edinburgh, known as Auld Reekie (i.e., smoky), was worse than most cities in this respect, as it was densely built in tenements (multistory dwellinghouses), with some streets on two levels.

xii Story-Maskelyne. op. cit.

xiii C. Piazzi Smyth. 1858. *Teneriffe, An Astronomer's Experiment*. London: Lovell Reeve. All quotations come from this book.

xiv C. Piazzi Smyth. 1863. The Teneriffe Experiment of 1856. *Edinburgh Observations* **12**, 401–514.

xv C. Piazzi Smyth. 1859. *Report on the Teneriffe Astronomical Experiment of 1856*. London and Edinburgh: Neill and Co.

xvi ibid. p 574.

xvii Phebe Mitchell Kendall. 1896. *Life, Letters and Journals of Maria Mitchell*, 130–32. New York: Houghton Mifflin.

xviii Henry Albers. 2001. *Maria Mitchell, A Life in Journals and Letters*. New York: College Avenue Press, p 107.

xix The name is mis-spelt in Maria's letter which is quoted in Henry Albers op. cit., p 97.

xx Helen Wright. 1959. *Sweeper in the Sky, the Life of Maria Mitchell, First Woman Astronomer in America*. p 112. Nantucket: The Nantucket Maria Mitchell Association.

xxi C. Piazzi Smyth. 1862. *Three Cities in Russia* (2 volumes). London: Lovell Reeve and Co.

xxii Piazzi Smyth's biographers devote two chapters to this problem.

xxiii Simon Schaffer. 1997. Metrology, Metrification and Victorian values. in B. Lightman (ed.). *Victorian Science in Context*. Chicago: University of Chicago Press.

[xxiv] C.P. Smyth. 1871. *Edinburgh Astronomical Observations*, volume 13, Appendix 1, p 3.

[xxv] xxv C.P. Smyth. 1874. (4th edition). *Our Inheritance in the Great Pyramid*. London: Isbister.

[xxvi] C. Piazzi Smyth F.R.SS.L. and E. 1867. *Life and Work at the Great Pyramid*, volume 1. Edinburgh: Edmonton and Douglas.

[xxvii] Piazzi Smyth's original photographs and lantern slides are preserved in the Piazzi Smyth Collection at the Royal Observatory Edinburgh. One of his miniature stereo cameras is also preserved there.

[xxviii] *Edinburgh Astronomical Observations* **13**, Appendix 2, 1871.

[xxix] Piazzi Smyth. The Solar Spectrum in 1877–78. *Transactions of the Royal Society of Edinburgh* **29**, 1878. 385–342.

[xxx] P. Smyth. 1878. Measures of the great B Line. *Monthly Notices of the Royal Astronomical Society* **31**, 38–43.

[xxxi] Quoted in H.A. and M.T. Brück, op. cit., p 231, from Piazzi Smyth's journal. The observations were published in C. Piazzi Smyth 1882. *Madeira Spectroscopic*. Edinburgh: W. and A.K. Johnston.

[xxxii] C. Piazzi Smyth. 1882. *Madeira Meteorologic*. Edinburgh: David Douglas.

[xxxiii] ibid. p 25 and Appendix 1.

[xxxiv] John Brashear. 1988. *A Man who loved the stars*. (reprint) Pittsburgh: University of Pittsburgh Press, p 125–6.

[xxxv] George Forbes FRS. 1916. *David Gill, Man and Astronomer*, p 42. London: John Murray.

[xxxvi] H.A. Brück. 1992. Lord Crawford's Observatory at Dun Echt 1872–1892. *Vistas in Astronomy* **35**, 81–138.

[xxxvii] M.T. Brück. 2005. Lord Lindsay's expedition to Mauritius in 1874. in D.W. Kurtz and G.E. Bromage (eds.) *Transit of Venus: new views on the solar system. Proceedings IAU Colloquium No. 196, 2004*. 119–126. Cambridge University Press.

[xxxviii] Published in 1877 in the *Publications of Dun Echt Observatory*.

[xxxix] A heliometer is a refracting telescope in which the lens is divided in two across its diameter. When the two halves are displaced relative to each other, two images are observed. By bringing an image of a particular object in coincidence with the image of another nearby object, the angular distance apart of the two objects may be calculated in terms of that displacement. It was called a heliometer because it was first designed to measure the diameter of the sun, but it became famous as the instrument with which Bessel measured the parallax of a star in 1834.

[xl] Gill left £250 in his will to the Royal Astronomical Society "in grateful remembrance of a like sum paid out of the funds of the society in aid of my expedition to Ascension".

[xli] Mrs Isobel Gill. 1878. *Six Months on Ascension*: an unscientific account of a scientific expedition. London: John Murray.

[xlii] A character from Sir WalterScott.

[xliii] I. Gill. op. cit. The book is quoted in Encyclopaedia Britannica, 11th edition, 1910.

[xliv] Philip Gosse. 1989 (first published 1938). *St Helena 1502–1938*. p 331–35. (Reprint) Oswestry: Anthony Nelson Ltd.

[xlv] The complete text of *Six Months on Ascension* may be read on the web, thanks to Dr Barry Weaver, University of Oklahoma.

[xlvi] Samuel Smiles. 1861. *Lives of the Engineers, 2 volumes*. London: John Murray. He also wrote the popular *Self Help*, 1859. London: John Murray.

[xlvii] This copy is in the Crawford Library, Royal Observatory Edinburgh.

[xlviii] Thomas Cave. Ascension Island Astronomy, *Sky and Telescope*. February 1994. Also on Cave's website.

[xlix] It was called the CPD (Cape Photographic Durchmusterung), the combined work of Gill and J. Kapteyn.

[l] Allan Chapman. 1998. Women in Astronomy 1780–1940: Summary of an RAS Special Discussion Meeting. *Observatory* **118**, p 270.

[li] J.C. Kapteyn. 1914. Obituary, Sir David Gill. *Astrophysical Journal* **50**, 161–72.

[lii] *Cape Times*. 10 December 1919.

[liii] *Cape Argus*. 8 October 1919. Bella's nervous symptoms had visited her at an early stage in her married life in Dun Echt. In December 1873, writing to Lord Lindsay, David reported that "Mrs Gill is getting all right, not the slightest return of the symptoms" and a few days later "My wife keeps all right – no trace of return of hysteria". (Gill letter book, Dun Echt archives A26.184, Royal Observatory Edinburgh).

Chapter 10
Adventurous Amateurs

Astronomy in the last decades of the nineteenth century spread in popularity beyond the sphere of affluent gentlemen and, as Allan Chapman describes, could be taken up by "people of more modest fortune and ambitions."[i] These lovers of astronomy were not content to listen passively to the expositions and demonstrations of an earlier generation, but preferred to be active in societies where they could compare their knowledge and experiences. The first truly successful popular astronomical society in Britain was founded in Liverpool in 1881. Liverpool, one of England's great industrial cities and one of the world's major seaports, had many astronomical connections. It had a busy municipal observatory where chronometers were regulated for seagoing vessels. Its suburbs had housed "Starfield", the home and observatory of William Lassell, successful telescope maker, celestial discoverer, and honoured member of Britain's scientific elite, whose talented daughters have already been mentioned (Chapter 7). The city had a thriving cultural life: its beautiful St George's Hall, a library, art gallery and museum, its concert hall and music festival. In 1881 Liverpool University College received its charter – the same year in which a small group of local amateur astronomers got together to form an association which would coordinate their efforts and, more ambitiously, publish their results.[ii]

The Liverpool Astronomical Society thus bridged the gap between ordinary enthusiasts and the professionals and well-to-do Fellows of the Royal Astronomical Society. An important element of its constitution was that women were welcome – unlike the Royal Astronomical Society which was open only to men. The new society solved one of its problems – that of suitable instruments – by acquiring a supply of lenses and brass tubing, from which members constructed their own telescopes. Keen observers, who aimed to do useful work and not merely have fun, organised themselves into sections according to their favourite fields of interest, and either reported their observations at meetings, or sent them in to be read out. Starting from a handful of local people, the Liverpool society flourished, and attracted members from throughout the country and even corresponding members from abroad. Observations made by members, coordinated by heads of the various sections, were published in the Society's *Proceedings* (re-named the *Journal of the Liverpool Astronomical Society* from 1884).

An early member of this excellent society was Elizabeth Brown (1830–1899), the first woman observational astronomer in Britain to enjoy a career of her own

DOI: 10.1007/978-90-481-2473-2_10, © Springer Science+Business Media B.V. 2009

Fig. 10.1 Elizabeth Brown at her private observatory c 1890. (Royal Astronomical Society)

making, not as an appendage to or a dependent on a male family member (Fig. 10.1).[iii] Her home was in Cirencester in Gloucestershire, many hours by train from Liverpool, which did not deter her from attending meetings regularly. She was the elder of two daughters of a prosperous wine merchant with an established family business, a scholarly man with a keen interest in science. He was a Fellow of the Geological Society and an amateur meteorologist who maintained a weather station and collected data for the Meteorological Society and the British Association for the Advancement of Science. His wife died when the girls were quite young, and they were educated in the manner usual in their milieu – at home by a governess. Elizabeth in particular attained a high standard of knowledge in science, in literature, and in art; like others before her, she took an interest from childhood in her father's scientific activities.

Her personal fate in life, as her biographer Mary Creese remarks,[iv] was that of many spinsters of the Victorian age – the care of an aging parent. From the age of about forty she devoted herself to that filial duty and also took over, and supplied to the Royal Meteorological Society, the rainfall, temperature and thunderstorm records which her father had kept for the previous 26 years. She herself was eventually elected a Fellow of that Society, one of the first women members, in 1893. With other amateur meteorologists, she kept watch for and reported on unusual phenomena such as aurorae and meteors. After the huge volcanic eruption on Krakatoa in 1883, spectacular sunsets and dramatic colour effects in the sky, attributed to ejected fine dust spread through the atmosphere, were seen world wide, and observers were encouraged to report on them. Observations of such phenomena including those of Elizabeth Brown were published in *Nature*[v] and elsewhere and afterwards assembled in a special volume.

In the field of astronomy, Elizabeth began with her father's hand-held 3-in. refractor, with which she scrutinised the Moon, the Sun and the planets. This was replaced by a slightly larger, 3.5 in. one, equatorially mounted with a clock drive in a proper dome. A separate building housed the meteorological instruments. Her early situation was similar to that of Thereza Story-Maskelyne (Chapter 8), a woman of the same age group, who also assisted her father in meteorological observations and possessed her very own astronomical observatory but whose circumstances directed her life on a different course. When her father died in 1882, Elizabeth Brown, now 52 years of age, could at last emerge in the wider world. She was free to travel, and found scientific comradeship in the Liverpool Astronomical Society.

Elizabeth Brown was proposed and elected to the Liverpool society in December 1883, apparently the first woman to join. She presented a paper at the same meeting on the movements of a recent large sunspots, referring also to observations accumulated in preceding years.[vi] Her chosen speciality was the Sun, and from then onwards she was the director of the Society's Solar Section, collecting and collating observations of the members of her group.[vii] Her own method of observing the Sun was the straightforward one, dating back to Galileo, of projecting its enlarged image from her telescope on a white card and making drawings of any sunspots present. This she could do efficiently using her clock-driven larger telescope, which held the image of the Sun steady while the spots were being traced. Many of her meticulous drawings, from her daily watch on the Sun, were published by the Society.

Elizabeth also observed variable stars, a field particularly suitable for amateur observers which still flourishes among them. She was active in another specialist field cultivated within the Liverpool society – that of so-called sidereal chromatics. Even from a casual glance at the sky with the naked eye it is evident that stars have different hues. Some gleam very white, others, such as the bright Betelgeuse in the constellation of Orion, are distinctly red. Admiral W. H. Smyth, the noted amateur observer of an earlier generation (Chapter 8), placed great store on this fact, and recommended that stars be classified according to their colours in a range of subtle shades which he defined himself. However, with the growing use of the new technique of spectroscopy, the study of stellar spectra came to offer a better way of understanding the nature of stars as hot luminous bodies, and the usefulness of visually estimated chromatics was ousted.[viii]

Her major contribution to astronomy, however, was as a solar observer. She concentrated on two aspects of the subject – the detailed structures of individual sunspots or groups of spots (one of her beautiful drawings of a complex group of spots may be seen in Agnes Clerke's *Problems in Astrophysics*[ix]); and their positions and movements on the face of the Sun. Spots grow and develop with time, sometimes singly but more often in pairs, and persist for anything from days to months before fading and disappearing. They cross the face of the Sun once in 27 days as the Sun rotates on its axis; but, in addition to being carried along in this way, they tend to drift slowly relative to the Sun itself – a phenomenon studied very particularly by her contemporaries at the Royal Observatory, Greenwich. It was already well established that the numbers of sunspots vary in a cycle of 11 years. An unusually high maximum fell around 1890, which fortunately allowed Elizabeth Brown to capture a complete cycle of rise and fall from the time she took up her serious systematic observations.

Elizabeth's sharp eye for sunspots was put to another good use in a search for a possible planet orbiting close to the Sun. There was reason to suspect the existence of such an object. The nearest known planet to the Sun, Mercury, was known to perform an elliptical orbit which ought to have been – but was not – compatible with Newton's Law of Gravitation. The discrepancy showed up as precession: the perihelion (the point in orbit nearest the Sun) seemed to swing in space, with the planet returning to a different spot on each circuit. The cause was put down to the gravitational influence of an invisible object, postulated to be an unseen planet. The problem seemed to be the same as that which had been met and solved in Mary Somerville's time, when the perturbations in the motion of the planet Uranus proved to be caused by an unknown planet, Neptune, which was duly discovered as predicted in 1846 (Chapter 6). The astronomer who on that occasion had correctly calculated the position of Neptune was the brilliant Urban Leverrier of Paris. Now, 40 years later, Leverrier applied himself to the problem of Mercury's precession, and surmised the existence of another unknown planet between Mercury and the Sun which, he predicted, would at certain intervals be on the near side of the Sun, and would pass directly in front it. The hypothetical planet was given the name Vulcan, and astronomers were asked by Leverrier to look out for it quite particularly in 1886. It was expected to show up as a dark spot on the Sun – as Venus and Mercury do when in transit – distinguishable from a sunspot by the rate of its motion across the Sun's face. Stephen Perry, President of the Liverpool Astronomical Society, recruited Elizabeth Brown, with her experience and diligence, to join in the search for this tiny object. As she reported in the Liverpool Journal, nothing was found.[x] The true explanation for Mercury's erratic behaviour had to wait until Einstein explained it in terms of general relativity in 1915.

Perry, a Jesuit priest and Fellow of the Royal Society, who had joined the Liverpool Society soon after Elizabeth Brown, was a talented astronomer at Stonyhurst College in Lancashire which for many years maintained an observatory equipped with excellent telescopes and modern equipment for solar work and solar spectroscopy. He belonged to a circle of active solar workers in Britain, and was an official British observer of solar eclipses. Elizabeth Brown also took part in eclipses expeditions, her journeys being partly motivated by her great love of travel. Her first foreign trip was in 1884 to Montreal, when the British Association for the

Advancement of Science took the bold step of holding a summer meeting overseas. Many leading astronomers and physicists were among the 700 participants who gathered at the McGill University where proceedings were chaired by the great Sir William Thompson (later Lord Kelvin). They included the famous popular writer and lecturer Sir Robert Ball, Director of Dublin's Dunsink Observatory, who kept an amusing diary of his adventures[xi] which, however, does not mention lady participants. It is a pity that Elizabeth did not record her impressions of the illustrious company at that meeting, or of her subsequent tour of North America. She did, however, publish popular accounts of her travels on two total solar eclipse journeys, in 1887 and 1889.[xii]

Expeditions to foreign lands became an important feature of observational astronomy in the course of the nineteenth century. Elaborate efforts and considerable expense went into observing phenomena which could not be observed at home – such as the transits of Venus of 1874 and 1882, and several total eclipses of the Sun. The Transit teams, needless to say, did not include women. Some eclipses, however, were watched by ladies who were present as spectators. A large entourage of the British expedition at the total eclipse of July 1860 in Spain included the wife and daughter of the Astronomer Royal and the daughter of the Russian astronomer Otto Struve.[xiii] However, Elizabeth Brown was the first woman to take part in such a venture entirely on her own account when she travelled to Russia to observe the total eclipse of 1887.

The eclipse of 19 August 1887 had a long shadow path extending from Germany through Russia to Japan, which attracted many expeditions. Feodor Bredechin,[xiv] Director of the Moscow Observatory, owned a private estate at Kineshma on the river Volga, a place which was fortunately on the eclipse path. He offered hospitality there to the official observers from the Royal Astronomical Society. These were Stephen Perry and Ralph Copeland, director of Lord Crawford's observatory at Dun Echt in Scotland who was accompanied by his engineer. These two astronomers equipped themselves elaborately with spectroscopes and with coronographs (telescopes for photographing the corona),[xv] and after a long voyage were installed two weeks before the eclipse in their Russian host's country house, with every facility and human assistance. Bredichin's invitation extended to Father Perry's amateur colleagues and friends. He reported: "Most fortunately we were joined a few days before the eclipse by Ms. Brown of the Liverpool Astronomical Society and her cousin, Ms. Jeffreys, and the latter most obligingly and most efficiently took charge of the cameras of my spectroscopes during totality, thus leaving me free to attend to the two coronographs". Elizabeth had planned her own observations with her own telescope. Unfortunately, though the weather was generally good beforehand, the astronomers were frustrated by clouds during the actual eclipse – a piece of bad luck experienced by many an observer before and since. Nevertheless, Elizabeth Brown and her friend deserve credit for being the first British women to be part (though unofficially) in a scientific eclipse team.

The eclipse of 22 December 1889 was visible in the Caribbean and South America. Elizabeth Brown and her companion chose Trinidad where they set up a telescope and had an impressive view (when the sky cleared) of the last moments of the ethereally beautiful corona, without attempting serious astronomical observations. They travelled on the outward voyage in the company of Father Perry, who was again an official observer of the Royal Astronomical Society, bound for his chosen site on the Island of Salut off French Guiana. Perry gave a lecture about the eclipse on the outboard journey. It was to be the last time that Elizabeth would see that colleague and friend. He contracted malaria on the island, and, having struggled to obtain his photographs of the corona, "crawled to the hospital as soon as the eclipse was over".[xvi] He died tragically at sea on the way home and was buried in Barbados.[xvii]

The Liverpool sunspot work brought Elizabeth Brown into collaboration with a fellow solar worker, (Edward) Walter Maunder of the Royal Observatory, Greenwich. Maunder occupied a special position at Greenwich, having been recruited in 1873 to a newly constituted post of Photographic and Spectroscopic Assistant which he was to occupy for 40 years. The "new astronomy" or astrophysics (next chapter), involving spectroscopy and photography, did not fall within the Royal Observatory's traditional domain; but the Astronomer Royal of the day, Sir George Airy, felt that the Observatory could not entirely ignore this new field, and he therefore set up a special department for these new techniques. Maunder, chosen to take charge of this venture, was not a university graduate; in the scientific hierarchy he was not to be ranked with the high flying mathematicians who dominated the British astronomical scene. His duty was a routine one, to photograph the Sun daily, and to note the sunspots, supplementing the observations by others made with similar photographic telescopes in far flung parts of the Empire. It was thought that sunspots influenced the weather, and that such records might provide clues to the causes of the devastating famines in India. In the course of time the highly competent Maunder's unique experience and his naturally enquiring mind made him the most knowledgeable solar worker in the country, and under Airy's less autocratic successor, Sir William Christie, he became a respected official observer of solar eclipses.

Maunder, who was also a keen popular speaker and writer on astronomy, was the leading spirit in the foundation in 1890 of the British Astronomical Association, a society modelled on the Liverpool Astronomical Society, to be open to all lovers of astronomy including women. Elizabeth Brown's contacts with Maunder, and her practical experience with the Liverpool society, meant that she became closely involved in the setting up of the new London-based organisation. She was a founder member of its Council and was made Director of its Solar Section, filling these places with great efficiency until her death. She published a steady flow of her own detailed observations of spots, as well as papers on aurorae and other phenomena, for a whole 15 years. The British Astronomical Association went from strength to strength, and women joined in considerable numbers. Nevertheless, it was still felt that women were unfairly excluded from the ranks of the quasi-academic organisation, the Royal Astronomical Society: for one thing, work done by amateur societies tended to be overlooked by the scientific establishment.

The Royal Astronomical Society, founded in 1820 with Sir William Herschel as President, at no time required prospective Fellows to leap hurdles of academic qualifications; nor has it ever become an exclusive gathering like the Royal Societies of London or Edinburgh to which distinguished achievers are elected as a mark of recognition. Equally, it is not, and has never been, a union of professionals such the Royal Colleges of Surgeons or Physicians to which practitioners are expected to belong. The Royal Astronomical Society is open to anyone with a genuine enthusiasm for astronomy, provided he/she is duly proposed and approved by the Fellows, and pays the required subscription. In the nineteenth century, when few salaried posts existed, the Fellows were overwhelmingly amateur – and were all male.

An attempt to remove this barrier had been made once before in 1886 when Ms. Elizabeth Isis Pogson of Madras Observatory in India was nominated by a group of Fellows.[xviii] Isis Pogson was working in India as assistant to her father, Norman Pogson, Director of the Madras Observatory. Pogson will ever be remembered as the originator of the familiar scale of magnitude which defines a difference of 5 magnitudes as representing a factor of 100 in brightness of stars which he first proposed in 1856.[xix] He was also a successful discoverer of asteroids, for which he was awarded the French Lalande Prize: one of them he named Isis (1856), after his daughter.

Pogson established his career in his native England in the private observatories of wealthy patrons – the last being at Hartwell House with John Lee – before gaining what appeared to be a promising colonial appointment, to which he was recommended by the Smyths, in 1861.[xx] The post did not fulfill his expectations. He was short of staff, poorly paid, and could carry on his work (in classical meridian astronomy) only with the help of his family.[xxi] His daughter, though in delicate health, was given a post as a computer in 1873 with the salary of "a cook or a coachman". She worked continuously at the observatory for 25 years, continuing even after her father's death in 1891. She eventually married and returned to England.[xxii] The unhappy situation in the Pogson family is reminiscent of that of the Fallows at the Cape – but with the difference, that Mary Fallows worked willingly for love, whereas Isis Pogson appears to have worked mainly from financial necessity. Many years later, in 1920, H.H. Turner, the professor at Oxford who played a quiet role in the careers of several women astronomers, rectified the Royal Astronomical Society's rejection of 1886 by proposing her (she was now Mrs. Kent) for election and having her duly made a Fellow, though she took no further part in astronomy.

The next attempt, in 1892, to have women elected to the Society, was largely if not entirely at the instigation of Walter Maunder, ever a champion of women and of amateurs. Among seven candidates for Fellowship in that year, three were women – Elizabeth Brown, proposed by Captain William Noble, President of the British Astronomical Association, and two young members of staff of the Royal Observatory. The two young women, whose interesting careers are reported later (Chapter 13), had studied mathematics together at Cambridge and had taken the Tripos (degree) examination with distinction: one of them, Annie Russell, who would later become his wife, was proposed by Maunder himself. They were well qualified academically to be Fellows of the society; Ms. Brown was equally so in consideration of her recognised contributions to solar physics. All three had reason to expect that their election would follow without difficulty; yet when the secret ballot of Fellows took place at the next meeting of the Society (May 1892), they failed to secure the required three quarters majority. In the debate that preceded the election, the President, speaking on behalf of the Council, decided to leave the matter in the hands of the Fellows. A Fellow responded that it was "practically a proposal to introduce into these dull meetings a social element, and all we shall require is a piano and a fiddle" and "to lay down a parquet flooring, and I am sure many of my young friends will be glad to dance through most of the papers".[xxiii] The facetious speaker was John Brett, a professional artist but only a fringe astronomer. The rejection is known to have rankled for decades in the mind of Mrs. Maunder. Though Elizabeth Brown did not openly complain, the rebuff was especially ungracious in her case, a woman of 62 with a significant record of practical contributions to astronomy to her name. There is no doubt that her skill as an observer of sunspots was insufficiently appreciated in her lifetime due to the exclusion policy of the Royal Astronomical Society.

A few months later that Society offered (in what seems to a modern reader somewhat flippantly expressed terms) "a mild plaister to the possibly wounded feelings of the ladies".[xxiv] The Council resolved that the President be authorised to issue admission cards "to such persons as it may be thought desirable to admit". Even this concession was hedged. The card was valid for one season at a time; the President would submit a list to the Council, and one third of its members were "sufficient to veto a name". The system remained in operation until 1915 when the Fellowship was at last opened to women. Several women, including the rejected candidates, and Agnes Clerke and her sister (Chapter 12), availed themselves of the cards of admission and attended meetings of the Society regularly thereafter, though they were not eligible to speak, to vote, or to be Council members.

Elizabeth Brown continued to devote herself assiduously to the work of the British Astronomical Association, mainly in the Solar Section which she directed, collating observations and publishing annual reports in its Journal.[xxv] She recommended solar work particularly to her women members. "The Sun is always at hand", she wrote in the first volume of the journal, "No exposure to the night air is involved, nor is there any need for a costly array of instruments".[xxvi] Nor, she might have added, did it involve venturing out of doors in darkness unescorted. She repeated her plea on behalf of solar work, having noticed temporary bright patches or "faculae" near sunspots on two occasions in 1892 (a year of exceptionally high sunspot activity), and recommended that "the sun should be much more constantly observed"[xxvii] – an idea ahead of its time. Such continuous monitoring was organised internationally during the International Geophysical Year of 1958–1959.

In the summer of 1896 the British Astronomical Association took part in its first total eclipse expedition, to a site at Vadsö in Finnmark, northern Norway. The eclipse path extended from northern Europe to Japan, and several expeditions were attracted to Norwegian locations. The ambitious venture was organised by Walter Maunder, now married to his young colleague, who organised a party of 58 members and their friends, including several women, to travel by sea from Tilbury, carrying their various instruments, some planning to team together for joint observations. They travelled in the company of the official British Government party, the whole making a lively company on the week-long journey. The day of the eclipse was unfortunately cloudy, but the month-long adventure was a resounding social success and the first of further expeditions.

Elizabeth Brown was one of the British Astronomical Association's participants on this eclipse. Sadly, it was to be her last expedition. Three years later, on 5 March 1899 as she was planning her fourth one, to the Mediterranean in 1900, she died suddenly

Notes

[i] Allan Chapman. 1998. *The Victorian Amateur Astronomer*, p 221. Chichester: Wiley-Praxis.

[ii] Gerard Gilligan et al., 1996. *The History of the Liverpool Astronomical Society*. Liverpool Astronomical Society, Liverpool.

[iii] Mary Creese is the principal authority on the life and work of Elizabeth Brown. The present account is based on her publications: Mary Creese. 1998. Elizabeth Brown, solar astronomer. *Journal of the British Astronomical Association* **108**, 4, 193–7; Mary R.S. Creese. 1998. *Ladies in the Laboratory? American and British Women in Science 1800–1900*, 236–7. Lanham Md. and London: The Scarecrow Press.

[iv] Mary Creese, op. cit. *Journal of the British Astronomical Association.*

[v] *Nature*, 6 December 1883, reported an observation made in Cirencester on November 26 which was surely by Elizabeth Brown.

[vi] *Proceedings of the Liverpool Astronomical Society*, Volume 2, 1883–4, p 18–19.

[vii] E. Brown. 1884. *A Short Review of the Sunspots of 1882 and 1883*, presented March 1884. She published similar reviews each year.

[viii] Later, a more efficient version of sidereal chromatics became a standard method of analysing starlight, known as photometry. The a star's radiation in various bands of colour or wavelength was measured by comparing its brightness on a photograph (which responded to blue and violet light) with that visually observed by eye (which is more sensitive to yellow and green). Later again, the same purpose was achieved by recording its brightness through various colour filters.

[ix] Agnes M. Clerke.1903. *Problems in Astrophysics*. p 79. London: Adam and Charles Black.

[x] Elizabeth Brown. 1886. The Search for Vulcan. *Liverpool Astronomical Society Journal* **5**, p 146.

[xi] P.A. Wayman. 1986. A Visit to Canada in 1884 by Sir Robert Ball. *Irish Astronomical Journal* **17**, 185–96.

[xii] They are described by Mary Creese, op. cit. The books are *In Pursuit of a Shadow*, London 1887, and *Caught in the Tropics*, London 1889.

[xiii] Peter D. Hingley. 2001. The first photographic eclipse? *Astronomy and Geophysics* **42**, 18–22.

[xiv] Alan H. Batten 1988. *Resolute and Undertaking Characters: the lives of Wilhelm and Otto Struve*. p 243. Bredichin succeeded Otto Struve as Director of Pulkovo Observatory.

[xv] R. Copeland.1887. *Monthly Notices of the Royal Astronomical Society* **48**, 49–51; S.J. Perry, *Monthly Noticesof the Royal Astronomical Society* **48**, 51–54.

[xvi] Agnes Clerke. 1902. *History ofAstronomy during the Nineteenth Century*, 187. (reprint edition Sattre Press 2003).

[xvii] Agnes Mary Clerke. *DNB*. Entry on Stephen Perry.

[xviii] J.L.E. Dreyer and H.H. Turner (eds.) 1923. *History of the Royal Astronomical Society, Volume 1 1820–1920*, p 233. London: Royal Astronomical Society (reprinted for the Society, Oxford: Blackwell 1987).

[xix] John B. Hearnshaw. Origins of the Stellar Magnitude System. *Sky and Telescope*, November 1992, 494–499.

[xx] Obituary. 1892. *Monthly Notices of the Royal Astronomical Society* **52**, 235–238; Vishnu Reddy, Keith Snedegar and Balasubramanian Ram Kumar. 2007. Scaling the magnitude: the fall and rise of N.R. Pogson. *Journal of the BAA* **117**, 237–245.

[xxi] R.K. Kochhar. 1991. The Growth of Modern Astronomy in India 1651–1960. *Vistas in Astronomy* **34**, 69–105. Allan Chapman 1998. *The Victorian Amateur Astronomer* op. cit. p 151.

[xxii] Roger Hutchins. Entry on Isis Pogson. *Oxford DNB*.

[xxiii] Report of the Meeting of the Royal Astronomical Society, 1892. *Observatory* **15**, 217. The remark is quoted by J.L.E. Dreyer in the *History of the Royal Astronomical Society*, volume 1, Chapter 7, p 234. (J.L.E. Dreyer and H.H. Turner (eds.) op. cit.

[xxiv] ibid. p 234.

[xxv] A full list may be found in Mary Creese's *Ladies in the Laboratory*. op. cit.

[xxvi] Elizabeth Brown. 1890–91. *Journal of the BAA* **1**, 58–60.

[xxvii] Elizabeth Brown. 1892. *Astronomy and Astro Physics* **12**, 74.

Chapter 11
The New Astronomy

One of the great dramatic events in the history of astronomy occurred in 1859 (a momentous year that also marked the publication of Charles Darwin's *Origin of Species*), when Robert Bunsen and Gustav Kirchhoff in a laboratory in Germany resolved the long-standing mystery of the dark gaps in the spectrum of the Sun. The first to observe such gaps in the Sun's rainbow colours was the experimental chemist and polymath William Hyde Wollaston, already mentioned as a dear friend of Mary Somerville to whom he demonstrated the phenomenon in his home soon after she arrived in London.

In the course of time many hundreds of gaps or lines were identified in the solar spectrum. Their true cause, however, remained a puzzle: all that could be said was that they were caused by some absorbing material either on the Sun itself or in the Earth's atmosphere, or both. (The Piazzi Smyths had made observations pertinent to this problem on Teneriffe in 1856 (Chapter 8)).

Meantime, the art of spectroscopy flourished in the laboratory. It was found that the light of each chemical element, when set to glow, produced its own particular set of bright coloured lines when seen through a spectroscope. Sodium, to give the most conspicuous example, was found to display a pair of yellow lines which are unique to that element; conversely, to observe these lines in a spectrum pointed unmistakably to the presence of sodium. Other elements, in the same way, had their own signature spectra. The immensely useful technique of spectrum analysis evolved whereby the composition of an unknown substance could be discovered by examining its spectrum and comparing it with the spectra of known elements.

In Kirchhoff and Bunsen's experiment, white light from a bright hot lamp, when sent through a cooler gas, showed a spectrum in which dark lines replaced the bright ones which that gas would normally produce on its own. It was like a photographic negative. For astronomers in particular the discovery was a revelation. It explained at a stroke the origin of dark lines in the Sun's spectrum: that they denoted the presence of identifiable gases somewhere between the Sun's luminous surface and the eye of the observer, chiefly in the Sun's own atmosphere. It opened up a "new astronomy", later known as astrophysics.

There was a surge of activity among astronomers and spectroscopists as they tried to identify the chemical elements present in the Sun by matching the dark lines

M. Brück, *Women in Early British and Irish Astronomy*,
DOI: 10.1007/978-90-481-2473-2_11, © Springer Science+Business Media B.V. 2009

in its spectrum with those produced by known chemical elements in the laboratory. To do the same for the stars was a far greater challenge: it involved mounting a spectroscope at the end of a moving telescope, directing the starlight onto its slit, and making the delicate adjustments needed to view its faint spread-out rainbow colours. A few attempts had been made earlier than 1859, but it was only after that date that such efforts gained a definite purpose. The first in the field was G. B. Donati in Florence (the same astronomer who had discovered the famous comet of 1868 observed by Mary Ward and Thereza Maskelyne (Chapters 7 and 8)), who examined the spectra of some fifteen stars in 1860. Within a few years others had also succeeded, chief among them Angelo Secchi in Rome and the Englishman William Huggins whose wife Margaret was later to join him in the new astronomy (Fig. 11.1).

William Huggins was one of the outstanding figures of nineteenth century science in Britain.[i] He had no formal academic training, never went to university or held a salaried post, yet he became recognised as one of the founders of astronomical spectroscopy. He was born in 1824, the only child of parents who owned a drapery business in London. He attended school for only a few years, and from the age of fifteen was educated at home by private tutors, developing wide intellectual tastes and a special inclination towards science. On receiving a modest inheritance he decided to abandon the world of commerce and to devote himself entirely to astronomy. He bought a house at Tulse Hill in Clapham, at that time on the outskirts of London, and built in the garden a small observatory connected to the house by a covered passage. In this building he soon afterwards installed a beautiful telescope

Fig. 11.1 Portrait of Lady Huggins in later life. (Royal Astronomical Society)

Fig. 11.2 The Huggins home and observatory at Tulse Hill, London. The building no longer stands. (Royal Observatory, Edinburgh)

of 8 in. diameter equipped with a clock drive. The house and observatory were to be his home for the rest of his life (Fig. 11.2).

Huggins was now 30 years of age. The first observations he made in his observatory were of planets, favourite targets of amateur astronomers. He longed to do something more original, to make some discovery – but how? The answer came with Bunsen and Kirchhoff's discovery. Huggins described his reaction to this announcement as like "coming upon a spring of water in a dry and thirsty land". "A feeling as of inspiration seized me", he wrote. "I felt as if I had it in my power to lift a veil which had never before been lifted; as if a key had been put into my hands which would unlock a door which had been regarded as ever closed to man – the veil and door behind which lay the unknown mystery of the true nature of the heavenly bodies".[ii]

These high-flown and oft-quoted sentiments, written when he was in his seventies, are probably a romanticised recollection of his actual experience,[iii] but the fact is that Huggins did indeed turn to astronomical spectroscopy at an early stage. He had no experience of spectroscopy; "a star spectroscope was an instrument unknown to the optician", he recalled. However, chancing to make the acquaintance of an experienced spectroscopist, William Allen Miller, Professor of Chemistry at King's College, London University, he found in him a colleague whose skill complemented his own. Together they constructed a small spectroscope which could be attached to the end of Huggins' telescope and which produced a tiny spectrum or rainbow, less than a centimetre long from red to violet. Their real triumph was to see dark lines or gaps in the spectra of a number of bright stars including Sirius, Aldebaran and Betelgeuse. They compared these with the spectra of various gases in the laboratory and succeeded in identifying in these distant stars earthly elements such as hydrogen, sodium, magnesium and iron. Huggins and Miller published their first results in 1863; Angelo Secchi published his at about the same time.[iv] When Miller returned to his own laboratory work, the field was virtually left to Huggins and Secchi, the latter specialising in comparisons of large numbers of stellar spectra and dividing them into recognisable types, while Huggins concentrated on detailed spectra of individual bright stars. The veil had indeed been lifted and the door unlocked. Secchi died in 1868. Huggins continued for over 30 years, becoming the acclaimed sage of stellar spectroscopy (Fig. 11.3).

Huggins now turned his attention to the nebulae, those objects which William Herschel (not forgetting Caroline's help) had studied so carefully and about which he had pondered so deeply. The nebulae were generally fuzzy objects, but there were others – the globular star clusters – which when examined closely with good telescopes could be resolved, or nearly resolved, into individual stars. Others, however, remained stubbornly hazy; and the unanswered question was whether these, too, were assemblies of stars or whether they were what Herschel described as "a shining fluid of a nature unknown". The problem was further complicated by the existence of peculiar objects which Herschel named "planetary nebulae" on account of their appearance, with small extended disks reminiscent of planets, but obviously not real planets because they were stationary with respect to their stellar neighbours in the sky. John Herschel called them "enigmatic", but we now know that a planetary nebula is a star surrounded by a balloon or shell of gas ejected from it at a certain stage in its evolution.

Huggins' reputation was greatly enhanced in 1864 when he made the startling observation that the spectra of certain nebulae such as the Orion Nebula exhibit individual emission lines, of the same kind as those given off by fluorescent gas-filled tubes in the laboratory. The phenomenon was correctly interpreted by him and Millar as showing these objects to be composed of "enormous masses of luminous gas or vapour". The light of the well-known planetary nebula (in the constellation Draco) revealed its gaseous component in the same way. Other nebulae like the Great Nebula in Andromeda, to be later recognised as external galaxies composed

Fig. 11.3 Sir William Huggins and his telescope at Tulse Hill. (Royal Observatory, Edinburgh)

of millions of stars, showed a continuous spectrum. Still collaborating with Miller, Huggins found that the spectrum of the nova of 1866 in the constellation of Corona Borealis, discovered by the Irish amateur astronomer, John Birmingham, who had the good sense to communicate directly with him, was also of a gaseous nature. It was the first nova (a suddenly brightening star) to be observed spectroscopically. These spectacular discoveries earned for Huggins the Fellowship of the Royal Society and its Royal medal in the same year, as well as, jointly with Miller, the Gold medal of the Royal Astronomical Society in 1867.

In 1869 Huggins – always an imaginative thinker – was the first to introduce the use of the Doppler shift (slight change in wavelength) in a star's spectrum as a means of determining its movement in the line of sight. Though his first result (for Sirius) was crude, the method was right, and opened up an entirely new and huge field of astrophysics.

Astronomical research in nineteenth century Britain was dominated by independent gifted scientists whom Allan Chapman labels the "Grand Amateurs".[v] Sir John Herschel was the prime example of such astronomers, who believed that discoveries were best made in freedom by non-salaried individual practitioners rather than in centrally directed institutions. Huggins, an independent astronomer whose work was of professional standing and who moved in the highest scientific circles, fell into the Grand Amateur category, but his financial means were modest and he could not realistically aspire to acquiring a larger telescope. The Royal Society therefore took the unusual step of providing a substantial grant to allow him to equip his private observatory at Tulse Hill with new first class instruments, to be supplied to him on loan. These were an interchangeable 18-in. reflector and a 15-in. refractor, with suitable spectroscopic and photographic attachments, constructed by the firm of Howard Grubb of Dublin, which were set up in 1871 in a new dome in his garden.

The beautiful new instruments brought with them considerable responsibility and effort for a lone observer. Huggins – unlike wealthy amateurs such as Lord Rosse, Lord Lindsay or William Lassell – could not afford to hire an assistant. "It would cripple me", he told a friend. His dilemma may have influenced his decision to "abandon over half a century of bachelorhood"in 1875.[vi]

John Birmingham, discoverer of the nova that led to one of Huggins' discoveries, imparted the news from Ireland in March 1876 in a letter to the other great pioneer, Angelo Secchi.[vii] "Mr. Huggins has just got married to an Irish lady. I do not know what effect this may have on his astronomy". He cannot have been the only one to wonder: Huggins, aged 51, was to all appearances a confirmed bachelor wedded exclusively to astronomy and about to embark on a new ambitious phase in his scientific career. Birmingham need not have worried: Huggins' Irish wife, far from being a distraction, was to be his greatest asset for the rest of his life.

The unexpected bride was Margaret Lindsay Murray (1848–1915), aged 27, a highly talented and artistic young woman, already a skilled amateur astronomer and a keen admirer of her future husband's work. She was born in Upper Gardiner Street, Dublin, on 14 August 1848, the year of Caroline Herschel's death.[viii] She was of Scottish stock. Her father, John Murray, a solicitor with a lucrative private practice, was born in Scotland but lived in Ireland from the age of three. Her mother, Helen Lindsay, was Scottish by birth and upbringing; she died when Margaret and her only brother were eight and five respectively. The father re-married – and had three more children – and moved the family to a house on the newly built stately Longford Terrace, overlooking Dublin Bay at Dun Laoghaire (then called Kingstown), homes of affluent citizens who were moving out of the city's overflowing Georgian streets and squares. The house, on four floors, with its lofty rooms and its breathtaking view, still stands unspoiled, though now divided into apartments. It is marked with a commemorative plaque erected by the Irish scientific community in 2002. Here Margaret lived until her marriage to William Huggins in 1875.

Margaret came to love astronomy at an early age. She recounted in later life that she owed this interest to her Highland grandfather who used to take her out in the evenings to show her the constellations. The grandfather, however, was no humble shepherd like James Ferguson, but the Chief officer of the Provincial Bank in Ireland, recruited from Scotland when the Bank was first established there, whose signature appeared on its Irish banknotes. Scottish culture was nurtured in the family. Margaret's father had been sent back to Scotland for his education at the Edinburgh Academy, and Margaret's only brother Robert was in his turn sent to the same famous school.[ix]

Like most girls of her social class, Margaret was educated privately at home and later attended a finishing school in Brighton, on the south coast of England, a popular location for boarding schools on account of the reputed health-enhancing qualities of the sea air. She learned languages becoming fluent in French, acquired a good knowledge of the classics, music and art. Though science and mathematics were not part of the normal curriculum for young ladies, Margaret may well have received encouragement from her father, who would have received a thorough grounding in mathematics to an advanced level at Edinburgh Academy from the legendary Mr. Gloag who taught a generation of brilliant scholars including James Clerk Maxwell and Peter Guthrie Tait.[x]

When she was thirteen Margaret was given Mary Ward's charming book "Telescope Teachings" (Chapter 7) by an aunt.[xi] She went on to read the popular books of Thomas Dick and John Herschel's *Outlines of Astronomy*. She found further material in *Good Words*, a monthly evangelical family magazine emanating from Edinburgh, supplied, perhaps, by her grandfather. *Good Words* was edited by Norman MacLeod, a well-known Presbyterian clergyman and scholar, and contained articles on secular as well as religious topics, including - unusually for its time – contributions by leading scientists such as John Herschel and David Brewster. Here Margaret could learn about chemistry, physics, and geology as well as her favourite astronomy. Articles on spectroscopy in this magazine which would have come her way include an account of Birmingham's nova of 1866 and its spectrum. The author of this article, published in *Good Words* in 1867 when Margaret was 19, was Charles Pritchard, President of the Royal Astronomical Society, who was in the chair when Huggins read his paper on the subject, describing the nova dramatically as "a world on fire".[xii] In 1872 the President of the British Association for the Advancement of Science chose for his Presidential address the topical subject of spectroscopy in which he traced the history of that technique, with the astronomical discoveries of William Huggins forming an inspiring climax. The address was published in instalments in *Good Words* where Margaret would have read it. She was an ardent admirer of Dr. Huggins (as he was then, with his honorary degrees) before she met him in person.

Margaret also tried to do simple scientific experiments. With her small terrestrial telescope she looked at the stars and was able to project an image of the Sun on which she studied sunspots of which there were unusually high numbers in the years around the sunspot maximum of 1870. She did some photography, then quite a fashionable hobby with artistic ladies,[xiii] which was later to become one of her principal areas of scientific activity. Remarkably, she also succeeded in producing a spectrum of the Sun showing some of the dark lines, and when she married, a friend – unnamed - gave her a spectroscope as a wedding present which is preserved in Wellesley College, USA.[xiv] Like many serious-minded and conscientious young women of her background she also taught in Sunday schools.[xv]

Margaret Murray's interest in astronomy was not pursued entirely in a vacuum. Ireland, though a small country on the map, was nevertheless well-known in the astronomical world. Lord Rosse's 6-ft telescope at Birr erected in 1845 was one of the scientific wonders of the age. In the mid nineteenth century Dublin also saw the establishment of the Grubb firm of telescope makers which soon became internationally renowned, providing instruments for some of the world's major observatories.[xvi] Its founder, Thomas Grubb, was succeeded in 1872 by his son Howard, a contemporary and personal friend of Margaret Murray's. Indeed, it was through Howard Grubb that Margaret Murray became friendly with the astronomer William Huggins, perhaps as early as 1870 when Grubb was given the commission to build the new instruments for Huggins' observatory at Tulse Hill. In after years, Huggins used to joke with Grubb that however correctly they squared their business accounts, he still remained indebted to him for introducing him to his wife.

Margaret Murray and William Huggins were married on 8 September 1875 in the Parish Church at Monkstown, near her home. It appears to have been an unostentatious wedding: there was no bridesmaid, the witnesses being Margaret's brother Robert and an uncle. Whatever her father's misgivings may have been at her choice of husband, he made a marriage settlement on his daughter in the form of a trust to ensure her future.[xvii]

Despite the age disparity, the marriage was not only "one of the most successful husband and wife partnerships in astronomy"[xviii] but apparently extremely happy.[xix] Huggins' former home life had been a lonely one. His mother, to whom he was extraordinarily devoted, had died seven years previously. So shattered was he by her death that he gave up all work for some months. At this period also, Huggins appears to have dabbled for a while in spiritualism. However, his interest in the spirits vanished when he married Margaret. Indeed his life was totally transformed.

The Hugginses had no children, and Margaret dedicated herself ardently to him, and to astronomy. She enthusiastically compared their marriage to the Brownings', the poets whose romantic elopement in 1846 had captured the popular imagination, and spoke of William as her "beloved" and her "Dearest". The romantic theme was extended to the home and work, where the engraved wooden mantelpiece eulogised their chosen vocation in verse:

Stars above that shine and glow
Have their image here below
Flames that from the Earth arise
Still aspiring, seek the skies.
Star-dust glimmers in the sky:
Hearthstone sparks are flying high.
Love and light glow evermore
Upward with the flame we soar.

And framed on William's desk was his favourite "life-verse" "which he delighted to give, especially to younger friends"[xx] :

> Then away with Longing, and Ho for Labour!
> And ho for Love - each one for his neighbour,
> For a life of Labour and Study and Love
> Is a life that fits for the joy above.

Margaret was herself an extraordinarily interesting and amusing character, a woman of strong individuality, vivacious, determined. She wore her hair bobbed for convenience when observing, and dressed in flowing pre-Raphaelite garments of her own design. She filled the house with her sketches and woodcuts. She bought and restored antique furniture and objects d'art while William collected old astronomical instruments. She tended the garden which was adorned by a sundial inscribed *Nil nisi Caelesti Radio*. She installed a piano and an organ and persuaded her husband to take up his violin again which he used to play in his younger days.[xxi] They played together and held musical evenings in their house with friends and colleagues. John Brashear, a well-known American telescope maker, described visiting Tulse Hill for tea and observatory viewing, to find another astronomical guest who had come, explained Huggins, "not to talk astronomy, but to talk violin".[xxii] Piazzi Smyth, who with his wife visited the couple when they were not long married, has left an account of their home-cum-observatory.[xxiii] "A small house, small garden in front, large behind; small rooms and low and narrow staircases but all filled with the most exaggerated ideas of medieval furniture; the painted glass at the door reproducing the group from the Bayeaux tapestry admiring the comet; a Sun with bright prominences and nebulae with their respective spectra. Fernery and palm-house though small, grandly successful." The stained glass panels were installed after their marriage, no doubt at Margaret's instigation. One, illustrating William's fields of work (the Sun, a comet, a spectrum), was designed by William himself. The second was copied from the Bayeau tapestry and represents people including King Harald observing the comet of 1066 (a former apparition of Halley's Comet). A third, a stained glass window, depicting "The Shining Ones saluting Christian" from Pilgrim's Progress, was added in 1900 to commemorate the silver jubilee of their marriage.[xxiv]

Access to the observatory dome was through the house. It had a carpeted floor, with a space for a laboratory, induction coils, spectroscopes, and other apparatus. There was no dividing line between home and place of work. Even the dogs were part of the astronomical paradise, one a big yellow mastiff called Kepler, claimed by William to be able to do sums, and the other a little black terrier called Tycho.[xxv] Kepler dated back to William's bachelor days: the physicist James Clerk Maxwell, also a dog lover, sent his friend a photograph of his Toby and himself "with regards to you and Kepler", and when Kepler died, Clerk Maxwell wrote: "We can well enter into the feeling of the melancholy home it makes when a dear doggie dies. Of course you have buried him in the garden".[xxvi] No doubt they did: Margaret in fact wrote his biography, and Huggins himself, having discovered that Kepler had inherited certain aggressive characteristics from his forbears, communicated the information to Charles Darwin.

On the material level the couple's tastes were simple, almost spartan. William himself ate mainly vegetables and drank nothing stronger than weak tea and milk coffee. It was his chosen life-style: Margaret liked to claim that their life was simple "in a great measure, for we are poor in our position - very poor", surely an exaggeration.[xxvii] Holidays in the early years of their marriage were trips to the continent of Europe, with visits to instrument makers in Paris, but when William got older, he preferred to go fishing nearer home, while Margaret did her sketching.

However, the Hugginses' main purpose in life was astronomy. In his recollections written when he was quite old, William referred to his "great happiness in having secured an able and enthusiastic assistant" in his marriage. Margaret echoed the same sentiment after his death when she declared that she looked upon herself "first and last as William Huggins' assistant". The remark is reminiscent of Caroline Herschel's even more humble description of herself as no more than a trained puppy dog. But the Tulse Hill Observatory diaries reveal that Margaret Huggins was far more than a mere assistant; as Barbara Becker's researches prove, she was her husband's close collaborator, without whom it is unlikely that he would have maintained his former level of research. Within six months of marriage, she was designing and adding apparatus.[xxviii]

The success of the partnership was the result not only of their shared love of astronomy but of the gift they both had of patience and meticulousness. William Huggins has been described as having "an enthusiastic heart and a cool head". Margaret also had an enthusiastic heart, and an artist's eye for perfection. Their educational backgrounds were not dissimilar. William was a self-taught astronomer, and his intelligent wife was an apt student who would have quickly shared his store of knowledge - as well as adding her own. Their work was observational and experimental, Margaret's skilful hands and excellent eyesight being a special advantage. They worked hard, and experienced the joys as well as the frustrations of night observing, vividly described by Margaret herself: "One of the compensations was the glorious beauty of the midnight sky, of the skies in the early morning, and no imagination can fail to be struck with the wonders of the heavens sweeping round in perfect peace within 5 mi. of the greatest city in the world". It reminds one of Wordsworth's thoughts on Westminster Bridge.

It was a happy coincidence that the Hugginses, early in their partnership, could pioneer the use of the dry plate in photographic astronomy. The invention of dry gelatin plates in the 1870s brought about a revolution in photography. Previously, photographers made use of the wet collodion process, in which glass plates were coated with light-sensitive material just before use. It was a messy and a tricky job; but its chief disadvantage as far as astronomy was concerned was that exposure times had to be relatively short so as to be completed before the collodion dried out. Astronomical photography had therefore been more or less confined to the Sun and Moon, while stellar spectra continued to be observed visually. Huggins had attempted in his early days to photograph the spectrum of the bright star Vega but succeeded only in recording a faint smudge. Henry Draper, the versatile American astronomer and pioneer photographer, obtained the first photograph on wet collodion of the same spectrum showing some dark lines, in 1872.

With the new process, the plates, pre-coated with dry emulsion and chiefly sensitive to blue and violet wavelengths, were supplied ready for use essentially as with modern films. Exposures could be made as long as one wished and could even extend over more than one night. With the spectrograph in place, the light of the star under investigation had to be directed on to the narrow slit of the spectrograph, and the observer had to ensure that the image of the star remained in that position for the duration of the exposure. This was not an easy task. The telescope was driven by a clock which, in ideal circumstances, would keep the star correctly centred indefinitely. In practice, it was necessary to watch continuously, and to make small adjustments as required. Huggins placed a small viewing telescope in such a position that the observer could see the image of the star on the shiny metal edges or faces of the slit. To reach this guider usually meant perching on a ladder. Margaret became an expert observer, and it was she, being the more agile, who climbed the ladder in the dark and guided the telescope, often for hours on end. "As I observe", she said in a magazine interview, "I direct William as to what I need and he moves me bodily on my ladder, so that I am not disturbed more than is necessary. Such work as this, to do well, requires very considerable skill".[xxix] She also did a great deal of the actual photography and took charge of operations from which one deduces that she already had experience in these techniques before her marriage. During the day the photographs were developed and studied. Enlarged paper prints of the spectra in published papers were made by her and labelled in her neat handwriting.

The Hugginses soon succeeded in photographing the spectra of a number of bright stars (the results were published by Huggins in 1880), using the 18-in. diameter reflector and a spectroscope which had quartz lenses and a set of prisms made of calcite. These materials, unlike ordinary glass, transmit ultra-violet light which, though invisible to the human eye, can be recorded on photographic emulsion which is sensitive to these wavelengths. The photographed spectrum of Vega in the constellation of the Lyre, the third brightest star in the sky, showed a beautiful series of dark lines of which two were familiar from the visual spectrum and known to be produced by hydrogen gas. The lines in the invisible ultra-violet were a new discovery, but their regular pattern left no doubt that they all had a common origin. This famous group of lines, later named the "Balmer series of hydrogen" after the physicist who reproduced them in the laboratory, are found in many stars and are especially conspicuous in hot white stars such as Vega. The writer Agnes Clerke (next chapter) always maintained that the series ought to be called the "Huggins series" after the discoverers.

In 1881 a bright comet appeared, the brightest in 20 years and the first since the introduction of the dry photographic plate. It was Tebbutt's Comet, discovered in Australia by John Tebbutt and later visible in the northern hemisphere where it remained for two months, a spectacular object with a 20 degree long tail. The Hugginses grasped the opportunity of photographing its spectrum, the first ever such observation of a comet. Many years earlier Huggins had visually observed the spectrum of a comet, and had identified bright bands (regular close groups of lines) due to molecular gases such as carbon monoxide, rather than to individual atoms. The photographed spectrum of Tebbutt's comet showed several extra molecular bands in the violet and ultra-violet part of the spectrum; these, too, were due to carbon compounds. The spectrum of the comet also showed the presence of sunlight, reflected from dust in the comet's composition.

The second of the Huggins telescopes, the refractor, was afterwards mounted next to the reflector so that the two telescopes could be used simultaneously rather than interchangeably, one for photography and the other visually. The two astronomers now had a powerful instrument that could cover the whole range of spectrum.

Huggins continued to publish papers in his own name with no acknowledgement to his helper. The first researches to be published by the couple jointly - studies of the spectrum of the Nebula in Orion and of the planets – appeared only in 1889, 15 years after their marriage. Agnes Clerke, reporting the Orion work, in which Margaret had been actively involved since 1882, wrote of the "invaluable cooperation that all lovers of astronomy must rejoice to see publicly recognised".[xxx] It is somewhat strange that William, should suddenly have waited so long before giving her credit.[xxxi] The Orion Nebula showed a spectrum line which had no counterpart in the laboratory, and was attributed by the Hugginses to an unknown element, "nebulium" (a name launched by Margaret herself), which existed only in the nebulae. Norman Lockyer, another leading spectroscopist and a bitter rival of Huggins, claimed that the disputed line was due to magnesium. In fact, the Huggins wavelength proved correct, but nebulium remained a mystery until 1927 when it was identified as ionised oxygen, found in this form under extremely rarified hot cosmic conditions.

The Orion Nebula remained a favourite object with William and Margaret Huggins, though it was very difficult to observe under the poor skies of a London suburb. They also did valuable work on the spectrum of Nova Aurigae, a star that flared up in 1892. Margaret began visual work on the evening after its discovery and continued for four successive evenings, recording every detail of the spectrum from red to blue. She and William together then photographed the spectrum from blue to ultra-violet over a period of several nights. By combining the two sets of observations they were able to produce a map of the entire spectrum, where they recorded two sets of hydrogen lines and were able to say for sure that these were displaced relative to each other, which indicated the outer envelope of the star blowing outwards. Another field of study were the Wolf-Rayet stars (stars which have conspicuous bright lines in their spectra due to their extended hot atmospheres).

The progress of astronomical spectroscopy revealed that stars come in many types or "classes", depending mainly on their temperature. The Hugginses assembled samples of typical spectra as a guide for future observers in an elegantly produced *Atlas of Representative Stellar Spectra*, with wavelength scales and identifications of the principal spectral features. The Atlas was published in 1897 and earned for the authors the coveted Actonian Prize of the Royal Institution. It was the first ever photographic atlas of spectra, the precursor of many, and was widely used.

These various researches, made and published over a period of 20 years, kept the Hugginses in the forefront of astronomical spectroscopy. There were naturally others who emulated them, as other schools of astronomical spectroscopy grew up in Europe and the United States of America. In the 1890s the younger generation of American astronomers, especially in places like California, began to challenge the dominant position of Huggins and other Europeans "with their small telescopes and primitive spectroscopes in a poor climate". The Hugginses now confined their work to the laboratory. They examined the spectra of elements conspicuous in the spectrum of the Sun and sun-like stars. The relative strengths of the individual lines in these spectra are not constant but vary with the physical conditions under which they are produced according to theories which still lay far in the future. The Hugginses, in their laboratory experiments, varied temperature and pressure, and noted the changes. The interpretation of such variations had to await the Boltzmann-Saha law, formulated only in 1920.

Apart from their joint scientific papers, published in the *Proceedings of the Royal Society*, Margaret Huggins contributed independently to the widely-read *Observatory* magazine which carried reports of meetings, notices of books, and astronomical news generally. Writing as an artist she published an article in 1883 on the subject of astronomical drawing, invoking the artist Albrecht Dürer as a model. At a time when it was still usual for astronomers to make drawings of what they saw through their telescopes, she stressed the need for practice in the art of drawing and for training the eye to see and the hand to set down what it saw. "Feeling for the beautiful should not be absent from the scientific artist, for few forms are destitute of beauty; but any tendency to idealise must be anxiously kept in check. The scientific artist must beware of jumping to the conclusion that certain appearances necessarily indicate certain states of things." This valuable paper received less than its due recognition when it was first written, but it is of considerable interest today in the discussion of visual representation in the history of science.[xxxii] The Hugginses were both interested in pre-telescopic astronomical instruments which they collected. Margaret made a study of their history and became something of an authority on astrolabes and armillaries. She wrote a paper on the subject in the American journal *Astronomy and Astro-Physics* in 1894, and contributed articles on these two subjects in the eleventh edition of *Encyclopaedia Britannica*.

Margaret also found time for scholarly research in the field of music. As already mentioned, she introduced music-making as a feature of home life when she married. William extolled the benefits of his prized Stradivarius violin in one of his public lectures and deemed £500 for such an instrument good value in return for the joy and relaxation it provided. He studied his violin from the scientific point of view and, with his wife's assistance, did some experiments on the part played by the soundboard. These results were elevated to publication in the *Proceedings of the Royal Society*. Margaret's own contribution to music was more substantial. She studied the history of the early Italian instrument makers and wrote a monograph on the Brescian violin maker Gio Paulo Maggini (1581–1632), a predecessor of the more famous Stradivarius and Guarneri, based on material collected by the London violin makers W.E. Hill and his sons of New Bond Street.[xxxiii] She included background historical and technical information, and an illustrated list of extant examples of Maggini's instruments.[xxxiv]

Among Margaret Huggins's other activities was educational work on one of the London School Boards which supervised the running of public elementary education and operated from 1870 to 1902.

The Diamond Jubilee of Queen Victoria (she was 60 years on the throne in 1896) was celebrated with great pomp throughout the land in 1897. Illustrious citizens were honoured, including William Huggins who became a Knight Commander of the Order of the Bath. His citation was "for the great contributions which, with the collaboration of his gifted wife, he had made to the new science of astrophysics". As was pointed out by Margaret's friend, the American astronomer Sarah Whiting, this reference to the wife of the new knight made her the only woman even remotely mentioned in the honours list.

Sir William's honours did not rest there. In 1901 Queen Victoria died and was succeeded by her son King Edward VII. On the occasion of his coronation the King founded the Order of Merit on the lines of the Prussian "Pour le Merite" to be awarded to men of distinction in military and naval services in the empire and to men who had made great names in the fields of literature and science. The number of holders was limited to 24, and the first list contained twelve names that included just four scientists – Lord Lister, Lord Rayleigh, Lord Kelvin and Sir William Huggins. The Order of Merit was the highest accolade that William Huggins could have received from his country. He had already been awarded the greatest honour that his peers could bestow – the Presidency of the Royal Society – which he held from 1900–1905. It was the peak of a long and successful career.[xxxv]

Margaret, now Lady Huggins, enjoyed the reflection of her husband's glory as she performed her duties as hostess at Royal Society social functions. It was during this period that she was at last to receive recognition in her own right. In 1903, at the age of 55, she was made an honorary Member of the Royal Astronomical Society in company with her dear "companion in astronomy" Agnes Clerke.[xxxvi] Only three women had previously been so honoured by that all-male society – Caroline Herschel and Mary Somerville, elected in 1835 (Chapters 3 and 6) and Anne Sheepshanks, elected in 1862 in gratitude for her financial beneficence to astronomy. The President of the Society, in welcoming the two women, made the inevitable comparison of Lady Huggins' role with that of Caroline Herschel as assistant to a great man.

A historic scientific event during William Huggins' term as President of the Royal Society was the award of the Society's Davy medal in 1903 to the celebrated husband and wife, Pierre and Marie Curie. The Curies had earlier visited London at the invitation of the Royal Institution when Pierre gave a lecture on "Radium" attended by a bevy of great British physicists and met the Hugginses for the first time. On the occasion of the Davy medal, Marie was ill and Pierre came alone, staying with the Hugginses at Tulse Hill. Marie wrote a letter of thanks to Margaret, saying that her husband "was very moved to see you two, Mr. Huggins and you, in the place where you have spent your life work. It is truly a beautiful existence which serves as a worthy example, and it is wonderful to devote one's entire life to the accomplishment of a beautiful piece of work."[xxxvii]

The Curies' discoveries, for which they (with Henri Becquerel) were awarded the Nobel Prize that same year, were an inspiration to the Hugginses who now turned their talents to devising ways of observing radioactive radiation spectroscopically. "In radium", they wrote, "we have a body which appears to be spontaneously and without ceasing giving off energy in several forms". They recorded how a box of unopened photographic plates stored inside a drawer in which they kept a sample of radium bromide had become fogged. Such casual handling of radioactive material shows how little the danger of these penetrating rays were appreciated in the early days. The experiments consisted in photographing the spectra of the glow emanating from radioactive samples. The gist of their conclusions was that the radioactive radiation was as energetic as electrical high voltage in its power to activate ultra-violet spectrum lines. Their work in 1903–1905 on the spectra of radioactive substances, described in four papers published in the *Proceedings of the Royal Society*, was their very last piece of research.

Huggins' term of office as President of the Royal Society closed on St Andrew's Day (November 30) 1905. His wife was present to hear his closing address and attend the banquet that followed. She recorded: "Sir William has closed his 5 years of office brilliantly .... And so a new chapter of life which he began on December 1 must be a full and interesting one." She added: "I have been Sir William's sole Privy Counsellor, all through. I am aware that the Royal Society has been good enough to do me honour. It recognises me as Sir William's faithful and sole and, I trust, pretty capable assistant." It was a notable feature of the Huggins regime that they worked entirely alone. Visitors came aplenty, but none was invited to share in their night observing sessions.

Margaret lost no time in providing a permanent memorial of her husband's Presidency. A book, *The Royal Society, or Science in the State and in the Schools* edited by Lady Huggins which appeared in 1906 is a collection of William Huggins' Presidential addresses including some which were then of topical interest such as a discussion of the future of education in Britain. Huggins was an excellent speaker; his lectures, adorned with quotations from English and classical literature, are highly readable. The volume, suitably illustrated by Margaret's woodcuts, was a labour of love.

To round off their astronomical life's work the Hugginses collected and edited William's and their joint papers in another volume entitled *The Scientific Papers of Sir William Huggins and Lady Huggins*, published in 1909. It was handsomely produced and beautifully illustrated, and assembled on a most attractive plan. When Huggins was knighted in 1897 he had been prevailed upon to write his personal recollections for a London magazine in which he described his quest for knowledge from the day he decided to devote himself to astronomy.[xxxviii] Instead of reprinting the papers chronologically, they were arranged by topic introduced by a relevant passage from these memoirs, providing an illuminating insight into the way a creative scientist develops his ideas. It is a demonstration of Margaret's devotion her husband's genius, that although all of William's publications - including the one on the soundboard of the violin - find a place, she herself is represented only by their joint papers.

The observatory and home at Tulse Hill continued to be a place of scientific pilgrimage. One of the many visitors recalled "the hearty greeting received from the venerable sage and his devoted wife, and the unostentatious simplicity of their entertainment".[xxxix] "For the visiting astronomer", wrote Simon Newcomb, doyen of American classical astronomy, "scarcely a place in London has more attractions than the modest little observatory and dwelling house on Upper Tulse Hill, in which William Huggins has done so much to develop the spectroscopy of the fixed stars.... The charm of sentiment is added to the cold atmosphere of science by the collaboration of Lady Huggins".[xl] "[Her] striking and attractive personality expressed itself in her appearance and manner", wrote an obituarist, more fully described by her American astronomer friend Sarah Whiting on her first visit in 1896: "A lovely lady, who was not at all ordinary, but like an embodied poem or piece of music met me at the door. She had a spirituelle face set with grey hair, she wore a trailing satin skirt and red velvet loose jacket of quaint cut."[xli] Photographs taken in both youth and old age, and the portrait which hangs in the rooms of the Royal Astronomical Society, show Margaret with her chosen bobbed hair, a most individualistic style for a nineteenth century lady.

In 1908, William and Margaret Huggins felt that they had contributed as much as they could to astronomy and decided to retire. He was now 86; he had worked in his little observatory for well over 40 years, she for 30. They could look back on an amazingly successful and fulfilled life. They had the use of beautiful instruments, but had no formal duties. Their income from William's private resources were modest: it was only in 1890, when he was already 66 years of age that the Government granted him a civil list pension of £150 a year, half a normal professorial salary. They worked for love of astronomy, with no motive of material gain.

The Royal Society arranged that the instruments which they had on loan should be given to the University of Cambridge, where H.F. Newall, the recently created Professor of Astrophysics, was ready to make use of them. When the time came for the telescopes to be dismantled Howard Grubb, who had made and installed them, came to supervise the packing for the move. The two Hugginses watched with sadness. Howard Grubb himself left a moving account of the scene. As the lid of the case was about to close on the object glass, the "faithful collaboratrice" took her husband by the hand and led him across the room to take a last look "at their very old friend". The telescopes were re-erected at the Cambridge Observatory in the "Huggins Dome" which was given a plaque inscribed:

> "1870–1908 These telescopes were used by Sir William and Lady Huggins in their observatory at Tulse Hill which formed the foundation of the Science of Astrophysics".

There had probably never been a scientific institution which produced so much original research at so little cost to the taxpayer as the observatory at Tulse Hill. The only disadvantage from the point of view of the scientific community was that no students were trained there to benefit from the experience of the two masters and carry on their work. As far as the Hugginses themselves were concerned, the arrangement was surely idyllic: they had no teaching or administrative duties, they worked when and as it suited them; they were in the truest sense amateur astronomers.

William Huggins died two years later (1910) at the age of 88. He fortunately lived to see the publication of the beautiful volume of their collected scientific papers.

Margaret went through the remaining instruments accumulated in a long life and sent many of them to Cambridge to join the telescopes. She moved to an apartment in Chelsea near the river Thames. It stood on the reputed site of the house of Sir Thomas More, Lord Chancellor of England who was executed by King Henry VIII, a historical connection that appealed to her sense of romance. As she was now bereft of her husband's Civil List pension, the Royal Society arranged that she should receive a pension of £100 a year. She accepted, only on the understanding that it was in recognition of her contributions to science, not as a form of charity.[xlii] One final task she now set herself – to write a worthy record of her venerated husband's life and work. "I hope to work every forenoon on my Dearest's life", she wrote to an American friend. Sadly, her own health began to fail, and to her disappointment she was unable to complete the promised biography. She survived her husband by less than five years. She died after a long illness on 25 March 1915, aged 66. Her ashes were laid next to her husband's in Golders Green Cemetery, London.

In her last years she made careful provision for the safety of her papers and the valuable relics of the Tulse Hill Observatory. The preliminary outline which she had written of her husband's life was left to a lifelong friend[xliii] on whose death it came into the hands of the friend's solicitor. He and a collaborator undertook to edit the manuscript to the best of their ability, and had it privately printed[xliv] in 1936 under the title *A Sketch of the Life of Sir William Huggins KCB OM*.[xlv] Thus Margaret's version of the story of her famous husband did not appear until a quarter of a century after his death. Inevitably, with Margaret's hand so closely involved, the picture is somewhat hagiographic, but a primary source of information survives in the day-to-day observing diaries kept at Tulse Hill, first by Huggins himself, and after their marriage, mainly by Margaret. These were presented to the observatory of Wellesley College in the United States by Margaret before her death.[xlvi] She also donated to it the remainder of her and her husband's scientific and artistic treasures, including their antique astronomical instruments, art books and pictures, as well as the lovely stained glass panels that had adorned the Tulse Hill residence. These are lovingly preserved together with the lively correspondence between Margaret and Sarah Whiting, Professor of Physics at Wellesley College and Director of the College's Whitin Observatory.[xlvii]

Wellesley College in Massachusetts, founded in 1875, was a women's college intended from the start to include the sciences alongside the classics in its curriculum. A chair of Physics was instituted and filled by Sarah Whiting, a teacher of classics and mathematics who had studied at colleges in New York. Special arrangements were made to allow her to learn practical physics – then the province of men only – before taking up her appointment in 1879. These took the form of attendance at the classes of Edward C. Pickering at Harvard, the famous astronomical spectroscopist and founder of the great school of astronomy at that university. She was thus introduced to the new science of photographic astronomical spectroscopy at about the same time that the newly married Margaret Huggins was starting her own career in London. Sarah Whiting later travelled to Europe to gain further experience in Physics, and spent some time in Glasgow with Lord Kelvin. In 1896 she studied for a while in Edinburgh with Professor Peter Guthrie Tait[xlviii] and paid frequent visits to the Hugginses, forming a warm friendship with Margaret who was of about the same age. Wellesley College, though well equipped with Physics apparatus, was still without astronomical instruments, but eventually a benefactor, Mrs. Whitin, came forward who financed the building and equipment of a fine observatory, completed in 1910. This was the very year of William Huggins' death, after which Margaret moved out of her old home and was considering how to dispose of her possessions. The new Whitin observatory was the perfect choice for her gift.

The education of women was a much debated subject in Margaret Huggins' day. She herself was born too soon to benefit from the first wave of university education for women, but she was a friend of two of the old campaigners for women's rights, Anna Swanwick and Hannah Pipe, and contributed after their deaths to their published Life and Letters.[xlix] Anna Swanwick, a lady of independent means, studied in Germany and was one of the earliest to introduce in translation the works of the great dramatists Goethe and Schiller to the English-speaking world. In her mature years she promoted female education and was one of the founders of Queen's College and Bedford College, advanced secondary schools for women in London. Hannah Pipe, headmistress of Laleham School for girls in Clapham, was – like the more celebrated Ms. Buss of North London Collegiate College - a supporter of Emily Davies' scheme to get women students into Cambridge University; indeed the very first student to enrol at Hitcham, the forerunner of Girton College, was one of her pupils.

At the same time, Margaret was not belligerent in the cause of women's rights, but preferred rather that women should be judged by their achievements than by their demands. When asked about this she is quoted as replying (in 1905): "I find that men welcome women scientists provided they have the proper knowledge. It is absurd to suppose that anyone can have useful knowledge without a great deal of study. When women have really taken the pains thoroughly to assist or to do original work, scientific men are willing to treat them as equals. It is a matter of sufficient knowledge. That there is any wish to throw hindrances in the way of women who wish to pursue science I do not for a moment believe."[l] She did, however, also say that in her youth "intellectual justice was denied to women".[li]

There is something ambiguous about this view of the place of women in the scientific world. William Huggins himself did not "throw hindrances" in Margaret's way, but neither did he strew many tributes, at least in public. His oft-quoted remark (in 1897) that on his marriage he gained "an able and enthusiastic assistant" is his only recorded reference to her status, apart from his entry in Who's Who (in his last years) where he mentions her as his collaborator. They were 15 years married before her name was included as co-author on their published papers. In a list of the world's astronomers, compiled in 1903, to which all astronomers including modest amateurs and popular writers were invited to enrol, there is only one name registered under the heading "Private Observatory at Tulse Hill", viz. "Owner and Observer, Sir William Huggins".[lii] Barbara Becker, chief authority on the life and work of the Hugginses, concludes that the "myth of the wife as the able assistant" was largely of the couple's own creation. Conforming with the idealised notions of the Victorian era, they liked to project themselves in the roles of the man as "the doer, the creator, the discoverer" and the woman as sweet and submissive[liii] – a situation reminiscent of Martha Somerville's attempt to soften the image of her mother, except that in the Huggins case they were themselves the perpetuators of the myth.[liv]

In his last Presidential Address to the Royal Society on the subject "Science and Education", included in the collected *Scientific Papers*, Huggins made a strong case for the teaching of science in schools. He was against early cramming of Latin and Greek and in favour of allowing young minds to use their faculties of "wonder and imagination". He listed the advantages of scientific training - but quite clearly only for boys. His vision was to develop the faculty of observation "in a boy's early years" and to "educate young men in public affairs". It is strange that, after 30 years of marriage to a scientifically talented woman who most certainly had a gift for "wonder and imagination", and having been responsible for receiving the Nobel Prizewinning Madame Curie, he could yet outline a scheme for the education of the young that ignored girls.

It is also known that Huggins – in common with the majority of Fellows – was opposed to the election of women to the Royal Society. He was President of the society at the time when Hertha Ayrton, a remarkable woman and a gifted physicist, was proposed for the award of the Society's Hughes medal for 1906. However, on the day of the Council meeting at which the decision was taken, he was deemed by Margaret to be too ill to attend and was thus prevented from voting against it, as had been his intention.[lv] Mrs. Ayrton received the award, and Margaret wrote congratulating her. The gesture was sincerely appreciated; Hertha wrote to a friend: "It is particularly generous of her, because she has done some splendid work in astronomy herself, with her husband, and has not had a bit of recognition for it".[lvi]

The story of William Huggins' reaction to the Ayrton award is told by the historian Joan Mason in her account of Hertha Ayrton's career and of the efforts made by certain enlightened Fellows of the Royal Society to have her elected to their number.[lvii] . After her nomination for the medal, Huggins, in a letter to the Secretary of the Royal Society (which one hopes Margaret was not shown), wrote sarcastically: "There will be great joy and rejoicing in H.M.'s gaol, among the women in prison [suffragists who were serving sentences for their part in a recent demonstration, whom Hertha Ayrton had publicly supported in the newspapers]. I suppose Girton and Newnham [the women's colleges in Cambridge] will get up a night of orgies…in honour of the event!…. The papers will teem with publications from all the advanced women! I suppose the P[resident] will invite her to the dinner, and ask her to make a speech. As the only lady – I should say woman – present, the P. will have to take her in, and seat her on his right hand!…. Can we now refuse the Fellowship to a Medalist?"

Huggins was at the end of his five-year presidency, and the President in question was the incoming one, Lord Rayleigh, the famous physicist and Cavendish Professor at Cambridge, who, unlike Huggins, was a sincere supporter of women in science. Huggins had one consolation: Mrs. Ayrton was not elected a Fellow of the Royal Society, nor were any women admitted until 1945.

One concludes that William Huggins was a conformist with the times in which he lived. As a representative of the scientific establishment (and what higher place could one hold in that Establishment than President of the Royal Society?), he held rigid traditional views on the place of women in society, while privately admiring individual women such as his wife and her clever friends (who knew their place).

Margaret, we assume, went along with this. As far as she was concerned, her mission in life, first and last, was to promote William, and after his death to keep the flame alive. She left in her Will a substantial sum of money to the City of London School for Boys to endow a "Sir William Huggins scholarship" for the study of astronomy at Cambridge. She left a similar sum to Bedford College for Women, which was affiliated to the University of London, to endow a postgraduate scholarship for research in Sociology.[lviii] Sociology at Bedford College encompassed social work, and it is remarkable that the scholarship should have been directed to an essentially charitable purpose rather than to science. The Lady Huggins scholarship was awarded every few years until 1952, but in the 1960s when the fund may have been running out it appears to have been amalgamated with other scholarships.[lix]

Margaret's final tribute to her "Dearest" was a bequest to have a memorial erected to his memory in St. Paul's Cathedral (Fig. 11.4). She wished the task to be given to the Royal and the Royal Astronomical Societies. Her friends ensured that she, too, would be remembered. The lovely memorial in the crypt of the Cathedral takes the

Fig. 11.4 Huggins Memorial in St Paul's Cathedral. (Photographed with permission)

form of a pair of medallions with the profiles of husband and wife,[lx] one inscribed "Sir William Huggins, astronomer 1824–1910" and the other "Margaret Lindsay Huggins 1848–1915, his wife and fellow worker".

Hers was indeed "a life of Labour and Study and Love".

Notes

[i] Barbara Becker. *William Huggins*. Oxford DNB, The present account is greatly indebted to the researches of Dr Becker, the foremost authority on the Huggins partnership.

[ii] Ian Glass. *2006. Revolutionaries of the Cosmos: the Astrophysicists*, Chapter 5. Oxford University Press; William Huggins. 1897. The New Astronomy: a Personal Retrospect. *Nineteenth Century* **41**, 907–29.

[iii] Barbara Becker. 1993. *Eclecticism, Opportunism, and the Evolution of a New Research Agenda: William and Margaret Huggins and the Origins of Astrophysics*. Doctorate Dissertation, Volume 1, p 62.

[iv] Another pioneer was an American amateur, Lewis Rutherfurd.

[v] Allan Chapman.1998. *The Victorian Amateur Astronomer. Independent Astronomical Research in Britain 1820–1920*. Chichester: John Wiley and Sons in association with Praxis Publishng Ltd.

[vi] Barbara Becker. 1993. op. cit. vol. 1, p 222.

[vii] Birmingham to Secchi. Letter reproduced in: Paul Mohr. 1995. *Irish Astronomical Journal* **22**(2), 204.

[viii] I. Elliott. 1999. The Huggins Sesquicentenary. *Irish Astronomical Journal* **26**(1), 65–68; M.T. Brück and I. Elliott. 1992. The family background of Lady Huggins (Margaret Lindsay Murray). *Irish Astronomical Journal* **20**(3), 210–211.

[ix] Robert Murray (1852–1934) became a barrister, and was a senior legal figure becoming Father (i.e., the most senior member) of the Irish bar.

[x] John Murray was a few years ahead of Clerk Maxwell and Tait, but was taught by the same masters.

[xi] Ian Glass. 2006. op. cit.

[xii] Barbara Becker. 1993. op. cit., volume 2, p 251, suggests that it was this article that chiefly inspired the young Margaret Murray.

[xiii] There was an active Photographic Society in Dublin, but its records are incomplete for the relevant years. Howard Grubb, a friend of hers, was a member. The basement of the spacious home in Longford Terrace would have provided ample space for photography.

[xiv] Sarah Whiting. 1915. An International Gift. *Popular Astronomy* **23**, 698; Ian Glass. 2006. op. cit.

[xv] Lady Margaret Huggins' entry in *Whos Who* (1897–1916) contains this information, supplied by herself.

[xvi] I.S. Glass. 1997. *Victorian Telescope Makers, the lives and letters of Thomas and Howard Grubb*. Bristol and Philadelphia: Institute of Physics Publishing. The book has a full account of the work done for Huggins with a picture of the Huggins Twin Equatorial.

[xvii] John Murray died in 1893, but the list of mourners at his funeral in Dublin (*Irish Times*, 8 December 1893) did not include either of the Hugginses. I thank Dr Ian Elliott for this reference.

[xviii] J.B. Hearnshaw. 1986. *The Analysis of Starlight*. Cambridge University Press. p 74.

[xix] C.E. Mills and C.F. Brooke. 1936. *A Sketch of the Life of Sir William Huggins KCB, OM*, London (privately printed).

[xx] Whiting, S.F. 1914. Priceless accessions to Whitin Observatory Wellesley College. *Popular Astronomy* **22**, No. 218, 487–85. The wooden mantelpiece and the framed verse are in the

Huggins Collection at Wellesley College. The words have been copied from Whiting's article (except for the American spelling).

[xxi] Huggins' violin, known as the Huggins Stradivarius, is now owned by the Nippon Music Foundation in Japan. I thank Ian Glass for this interesting information. The violin is mentioned in the Nippon Foundation website.

[xxii] John Brashear. 1988 (new editiona). *A Man Who Loved the Stars, Autobiography of John A. Brashear*. University of Pittsburgh Press, p 120.

[xxiii] Mary T. Brück. 2003. An astronomer calls: extracts from the diaries of C. Harles Piazzi Smyth. *Journal of Astronomical History and Heritage* **6**, 1, 37–45. Also quoted in H.A. Brück and M.T. Brück, 1988. *The Peripatetic Astronomer, the life of Charles Piazzi Smyth*, p 191. Bristol: Hilger. The house and dome no longer stand.

[xxiv] It was designed by John Dawson Watson (1832–88), a painter and watercolourist, and taken from his illustrations for Pilgrim's Progress published in 1860.

[xxv] Tycho Brahe was the great sixteenth century Danish astronomer; Kepler, his pupil, discovered the famous laws of planetary motion.

[xxvi] Letters from J. Clerk Maxwell to Huggins, 1872 and 1877. Quoted by Robert H. Kargon in the New Preface to Lewis Campbell and William Garnett. (reprint) 1969. *The Life of James Clerk Maxwell*. New York and London: Johnson Reprint Corporation.

[xxvii] Quoted from a letter of hers in Mills and Brooke's *Sketch of the Life of Sir William Huggins*, op. cit., p 37. Her estate at her death was valued at more than £12,000.

[xxviii] Tulse Hill Observatory notebooks in Wellesley College, are quoted by Barbara Becker, distinguished historian of science and expert on the work of the Hugginses, in several publications (opera cit.).

[xxix] Quoted in Mills and Brooke. op. cit.

[xxx] Mary Brück. 2002. *Agnes Mary Clerke and the Rise of Astrophysics*. p 77. Cambridge University Press.

[xxxi] Barbara Becker suspects that his motive may have been less than chivalrous. The observation involved a crucial determination of the wavelength of a certain unidentified spectrum line and William Huggins (ever cautious, and with a glittering reputation to uphold) was anxious that, should their measurement prove inaccurate, he would have an alibi.

[xxxii] Klaus Hentschel. 2002. *Mapping the Spectrum. Techniques of Visual Representation in Research and Teaching*. Oxford: University Press.

[xxxiii] Grove's *Dictionary of Music and Musicians*, 3rd edition (H.C. Colles ed.) 1928. Vol. 2. London: Macmillan.

[xxxiv] M.L. Huggins. 1892. *Gio Paulo Maggini, His Life and Work*. London: W.E. Hill. The book was reprinted in the mid-twentieth century.

[xxxv] He had been awarded medals by the Royal and the Royal Astronomical Societies, the French Janssen Prize and the American Draper Prize, numerous honorary university doctorates and membership of several scientific societies at home and abroad.

[xxxvi] Lady Huggins. 1907. *Agnes Mary Clerke and Ellen Mary Clerke, an Appreciation*. London, privately printed.

[xxxvii] Susan Quinn.1996. *Marie Curie, a Life*, p 185. London: Mandarin Paperback. The Curies and Henri Becquerel received the Nobel Prize later that same year.

[xxxviii] Sir William Huggins. 1897. The New Astronomy: a Personal Retrospect. *The Nineteenth Century* **41**, 907–29.

[xxxix] T.J.J. See. 1910. Tribute to the memory of Sir William Huggins. *Popular Astronomy* **18**, 387–99.

[xl] Simon Newcomb. 1903. *The Reminiscences of an Astronomer*. p 279. London and New York: Harper.

[xli] Sarah F. Whiting. 1914. Priceless accessions to Whitin Observatory Wellesley College. *Popular Astronomy* **22**, 487–502.

[xlii] Letter from Margaret Huggins to the secretary of the Royal Society, 1910, quoted by B.J. Becker. 1996. Dispelling the Myth of the Able Assistant, Margaret Huggins and the Work

of the Tulse Hill Observatory, in H.M. Pycor et al. (eds.), *Creative Couples in the Sciences.* New Brunswick: Rutgers University Press.

[xliii] The friend was Miss Julia Montefiore, whose brother John was to be the writer but who died before he could accomplish it.

[xliv] Lady Huggins had left £300 in her Will towards the cost of publication.

[xlv] C.E. Mills and C.F. Brooke. 1936. *A Sketch of the Life of Sir William Huggins KCB OM.* London.

[xlvi] They have been studied extensively by the historian Dr Barbara Becker.

[xlvii] Annie J. Cannon. 1927. Sarah Frances Whiting (obituary notice). *Popular Astronomy* **35**, 1–7.

[xlviii] No record of Miss Whiting's stay in Edinburgh exists in the archives of the Physics Department at the University. She was probably a non-graduating student. I owe this information to the late Professor William Cochran.

[xlix] Susan M.P. McKenna-Lawlor. 1998. *Whatever shines should be Observed.* Dublin: Samton Ltd. Chapter 5 gives extracts from Margaret Huggins' characteristically gushing correspondence with Miss Pipe.

[l] H.F. Newall. 1916. Obituary. *Monthly Notices of the Royal Astronomical Society* **76**, 278–282.

[li] Sarah F. Whiting. 1915. Lady Huggins. *Astrophysical Journal* **42**, 1–3.

[lii] P. Sroobant et al. 1907. *Les Observatoires et les Astronomes*, p 129. Brussels: Observatoire Royal de Belgique.

[liii] John Ruskin's description, quoted by B. Becker. A similar image, the Angel in the House, was created by Ruskin's friend, Coventry Patmore, in a popular poem eulogising marriage.

[liv] It is interesting, in this connection, that when Margaret was given an entry of her own in Whos Who after William's death, she should have provided a long flowery account of her early achievements.

[lv] Barbara Becker. 1993. op. cit. One may well wonder, as Dr Becker does, whether Margaret's intervention was "out of innocent concern for her husband's well-being or personal interest in the outcome of the Council's vote".

[lvi] Joan Mason. 1991. Hertha Ayrton (1854–1923) and the admission of Women to the Royal Society. *Notes and Records of the Royal Society of London* **45**(2), 201–220.

[lvii] Joan Mason. ibid.

[lviii] The endowments were worth £1000 each.

[lix] I thank Dr Laura Bentley, archivist at Royal Holloway and Bedford College for this information (letter 18/02/1992).

[lx] The artist was Henry Pegram.

Chapter 12
The Scholarly Sisters

Among her wide social circle, Margaret Huggins had one very special close friend. She was Agnes Clerke (1842–1907), an expert in the "new astronomy" who played an influential role as an expositor of astrophysics in the last quarter of the nineteenth century not only in Britain but internationally (Fig. 12.1). Like Margaret Huggins, she was born and raised in Ireland, though it was in London that the two first came to know each other and to become, as Margaret put it, "companions in astronomy".[i]

Agnes Clerke (1842–1907) was born in Skibbereen, Co Cork, a small market town on the picturesque south coast about 50 miles from the city of Cork. Her father, John William Clerke, manager of the town's Provincial Bank, came from a long established and highly respected family in the district, originally of English stock. Agnes' mother, Catherine Mary Deasy, was a daughter of Rickard Deasy, a wealthy brewer and shipbuilder in the nearby town of Clonakilty and an influential member of the local community. The Deasys, an old Irish family with a colourful history, hold an honoured place in the annals of their native county: the museum at Clonakilty has a section devoted to the history of the family over several generations. Agnes was the second of the Clerkes' three children. Her sister Ellen Mary (1840–1906) made her own successful career as a writer and journalist; their only brother, Aubrey St John (1843–1923) became a barrister.

John Clerke was educated at Trinity College Dublin where he was a scholar in classics, attended classes in mathematics and took a keen interest in the sciences. He was a shy, studious person, who devoted his leisure to "a close study of the arts and sciences". Catherine, a talented and cultivated woman with a gift for music, was a woman of strong character and high ideals. These influences fostered in Agnes Clerke her capacity for sustained study, her devotion to her family and her unostentatious lifestyle.

On account of Agnes' delicate health, the sisters never attended school but were educated at home by their scholarly parents. In a record of their early life Aubrey paid special tribute to the father's "painstaking teaching."[ii] The study of the classical languages and mathematics, as is evident from Agnes' later work, was serious and undiluted; it is significant that those popular feminine occupations, sketching and needlework, did not figure in the Clerke parents' curriculum. The girls' studies began early: by the age of eleven Agnes had mastered Herschel's *Outlines of Astronomy*,[iii] and Ellen in her 14th year had composed a ballad which she published in

DOI: 10.1007/978-90-481-2473-2_12, © Springer Science+Business Media B.V. 2009

Fig. 12.1 Agnes Clerke. (Royal Observatory, Edinburgh)

later life. Agnes' favourite subjects were astronomy and music. A collection of science books which she used as she grew up is still preserved in the family. It includes well-known expositions such as J.P. Nichol's *Architecture of the Heavens* and Humboldt's *Cosmos*, and some advanced works on spherical astronomy. In the realm of practical science, the father had a chemistry laboratory and possessed a 4-in. telescope through which the children were shown Saturn's rings and Jupiter's satellites. The telescope was probably a portable transit instrument, equipped with a chronograph to provide a time service for the town which at that time was not connected to the wider world by either railway or telegraph.

In 1861 when Agnes was nineteen the family moved to Dublin where John Clerke took up a new profession as Registrar at the court of his brother-in-law Rickard Morgan Deasy, a lawyer and former Member of Parliament at Westminster, who had been appointed a High Court Judge. Aubrey was a student at Trinity College Dublin where he took mathematics and natural sciences. He had a brilliant academic record. He was elected a scholar in 1862 and graduated in 1865, carrying all the prizes before him, including gold medals in mathematics and natural sciences. He was then awarded a post-graduate studentship in science and won the much valued McCullough Prize in 1867 for an essay on theoretical mechanics. His undergraduate courses of study included astronomy; the set text was Brinkley's *Astronomy* (a standard work from the early decades of the century), and, at a more advanced level,

Laplace's *Méchanique Céleste*, that great work that Mary Somerville had mastered alone. One can imagine that Agnes would have done her best to keep up with her brother's studies in private during those Dublin years: such activity would have been in the spirit of Aubrey's lifelong devotion to his sisters, and it is on record that his quiet support was forthcoming behind the scenes on many occasions in later life when a mathematical problem arose in Agnes' own studies. Here was a case of role-reversal: Agnes' career at every stage was encouraged by her father and brother who remained in the shadows.

Academic posts in the sciences were few, and Aubrey, in spite of his strong scientific bent, chose a career in law and eventually took up practice as a barrister in London.

Meantime Agnes and Ellen, aged respectively 25 and 27, went to complete their education in Italy in 1867, where they stayed for 10 years. Their base was Florence, where they frequented the city libraries which were open freely to all readers. Their interests ranged widely. Agnes' chief interest was the philosophy and science of the Renaissance, while Ellen's was Italian history and literature. Both became fluent linguists. Agnes was entirely at home with written and spoken Italian and German. She taught herself Portuguese on one occasion, while Ellen was also a specialist in Spanish. They were both good classical scholars, able to read Latin and Greek. Agnes, who first considered making music her career, studied music seriously at the Conservatoire in Florence and became an accomplished pianist who once played for Franz Liszt in Rome. In late 1877, the father being now retired, the sisters returned from Italy and the closely bound family was reunited in London. None of the three Clerke children ever married. They lived together in perfect harmony at Redcliffe Square, Chelsea, until death separated them one by one: the brother, the last of the family, died in 1923.

Agnes Clerke's first serious published work appeared in 1877 in the intellectual quarterly *Edinburgh Review*.[iv] Founded in Edinburgh by the circle of liberal literary figures among whom Mary Somerville moved, the journal was at this time published in London. Agnes Clerke's introduction to the *Edinburgh Review* was through Janet Ross, a leading hostess in the British literary circle in Florence who was a cousin of Henry Reeve, the journal's famous editor. Writing was one of the few careers open to women, and at this stage Agnes aimed simply at making a career as a writer. She offered the journal essays on two very different topics – Italian politics and history of science – hoping that one or other would succeed. In the event, the editor accepted both. Reeve wrote to her to Florence: "My dear Ms. Clerke, It gives me very sincere pleasure to have introduced you to your first literary success. I hope it may be the prelude to many more."[v] Agnes could not have had a happier contact in the literary world than Reeve. For the rest of her life she provided him with two contributions yearly, and became a personal friend. When he died in 1895 it was she who wrote his entry in the *Dictionary of National Biography*.

The first article in the *Edinburgh Review* entitled "Brigandage in Sicily" dealt with a much debated political topic, the rise of the Mafia. The second, "Copernicus in Italy", was based on volumes published in connection with the 350th anniversary (1873) of Copernicus' *De revolutionibus*, the famous work postulating the sun-centred solar system. Most of Agnes Clerke's articles in the *Edinburgh Review* during the first few years dealt with the science of the Renaissance – Copernicus, Campanella, Bacon, Harvey, Newton, Galileo, Descartes. As time went on she dealt with other interesting and varied topics not met with in her books. Of a total of 53 articles written over a period of 30 years, 29 were on scientific subjects. The books reviewed might be in Latin, Greek, German, French, Italian or Spanish, with all of which languages she was conversant.

The publishers Adam and Charles Black of Edinburgh were at the time bringing out the restyled scholarly ninth edition of the *Encyclopaedia Britannica*. Agnes was invited to provide articles on the history of science for this edition. The early volumes had already been published, so it was not until volume 10 (1879) at the letter G that the first of her contributions, a long and erudite dissertation on Galileo, appeared. This brilliant article remained unchanged through several subsequent editions of the Encyclopaedia: in fact, the only alterations made by later contributors were in reducing the length, as the encyclopaedia perforce increased the number of its entries. Other major biographies which Agnes Clerke wrote and which survived many editions were of the giant figures of science – Alexander von Humboldt, the optician Huygens, the astronomers Kepler and Leverrier, the mathematicians Lagrange and Laplace, the chemist Lavoisier. It was this work for the encyclopaedia that first brought Agnes Clerke to public notice when one weekly magazine commented: "It is worthy of remark that the lives of Lagrange and Laplace have been entrusted to a lady, A. M. Clerke, who seems desirous to emulate the acquirements of Mrs. Somerville."[vi] The essay on Laplace – which included a reference to Mary Somerville's *Mechanism of the Heavens* – probably had its origin in Aubrey's prize essay as a graduate student in Dublin.

The first indication of Agnes Clerke's interest in modern developments in astronomy was an article on "The Chemistry of the Stars" in the Autumn 1880 number of the *Edinburgh Review* which related the story of astronomical spectroscopy from the beginning of Huggins' work in 1864 to the time of writing. This was an entirely new subject for her: astronomical spectroscopy came too late to figure in her brother's university courses. This excellent essay was the seed of her first book, *A Popular History of Astronomy during the Nineteenth Century*, which she began to write immediately that article was published.[vii] As encouragement, her sister Ellen composed a poem extolling the joys of astronomy.

Earlier histories of astronomy tended to begin with the Greeks and end with Newton, whose work had once been deemed the culmination of knowledge. The so-called "new astronomy" was at that time almost entirely non-mathematical, and Agnes Clerke decided that the best way to deal with the subject was as a chronological account of its development. To put it into proper context, she took as starting point William Herschel's ideas on the construction of the heavens at the end of the eighteenth century and went on to cover the birth and development of astronomical spectroscopy and studies of the Sun. The book, of almost 500 pages of print, was a model of accuracy and meticulousness. It lived up to the author's claim of completeness and clarity, with a "full and authentic system of references to the sources of information" that took "as little as possible at second-hand." She called it "popular", by which she meant without the use of mathematical formulae. She had the gift of explaining quite advanced ideas in words, without sacrificing the meaning. Her writing was no mere entertainment but, as Margaret Huggins wrote, "ministered to those who long to know, not to people who are too lazy to do more than go to lectures to have their ears tickled". Agnes spent four years on the book, doing the research entirely alone in the British Museum. It appeared in December 1885. Agnes Clerke was then almost 44 years of age and virtually unknown in the astronomical world.

The book took the scientific community by surprise. The fact that its author was a woman added to its fascination in the eyes of the readers. The astute Margaret Huggins immediately identified the mysterious "dark horse" as the "Unknown" who had contributed the *Edinburgh Review* essay on astronomical spectroscopy six years earlier. She soon made her acquaintance. The two women became intimate friends, sharing a deep interest in music as well as astronomy.

Professor Robert Stawell Ball, the book's first reviewer and himself a well-known writer on astronomy wrote[viii] : "We have read this book with very great interest and no little pleasure. The authoress (for this learned volume is indeed the product of a lady's pen) has modestly described her *History of Astronomy* as a *popular* work. We certainly hope that the book will be as popular as it deserves, and that it will be widely and extensively read. We think, however, that few men of science who use this book will think that it ought to be classed as a popular work in the ordinary acceptation. It might be more correctly described as a masterly exposition of the results of modern astronomy in those departments now usually characterised as physical'. The *History* was no less well received in non-scientific circles. The *Edinburgh Review* described it as "written – not by a Fellow of the Royal Society – but by a gifted member of a scarcely less interesting association – a sisterhood, be it remarked – for the narrative is traced by the pen of a lady on whom the mantle of Mary Somerville seems to have descended."[ix] A second edition appeared within eighteen months of the first.

In the last stages of writing, an enthusiastic American astronomer was put in touch with Agnes by Richard Garnett, the learned librarian at the British Museum where Agnes did her research. He was Edward S. Holden,[x] Director of Washburn Observatory, Wisconsin, who generously took it upon himself to advise her, sending her copies of his own scientific papers and popular writings. He became a good friend of the Clerke family by correspondence, though they were never to meet in person. In July 1885 Holden was appointed President of the University of California and director-designate of the planned Lick Observatory, the first great mountain-top observatory in the world. He publicised the book in America with effusive warm reviews and recommendations. The *History* soon reached other leading American astronomers – Charles A Young of the University of Yale and Edward C Pickering, Director of Harvard College Observatory who became lifelong correspondents. "Students in the Old World are deeply indebted to American investigators both for their brilliant results and for their generosity in communicating them", Agnes wrote in thanks.[xi] It was part of her charm – and her success – that she never failed to express her appreciation in an unfailingly courteous fashion.

The first astronomer personally known to Agnes Clerke in London was Norman (later Sir Norman) Lockyer who was introduced to her by Garnett at the British Museum as the reader who always asked for books about astronomy. Lockyer, like Agnes Clerke herself, did not have a formal scientific education and had begun his career as a popular writer. He came to fame in 1868 when he devised a method of viewing prominences on the Sun outside a total eclipse by the use of a spectroscope. He also discovered in the spectrum of the Sun during the total eclipse of 1868 a chemical element unknown on Earth, to which he gave the name "helium", the sun element. Helium was not isolated in the laboratory until 1895. Lockyer was also the founder and editor of the journal *Nature* which went on to become the most widely read scientific publication in the world. He generously invited her to compose her own publicity of the second edition of her book for *Nature*, a gesture which, she wrote to him, was "of a piece with all your friendliness to my book and myself."[xii]

Lockyer was at this time attached to the Science Museum in South Kensington where he had a spectroscopic laboratory and an observatory for solar work. In addition, he built in 1884 a private observatory at his home at Westgate-on-Sea where he used to organise parties for astronomical evenings. "Among the many visitors to the observatory at this time", wrote Lady Lockyer in the biography of her husband, "none saw the wonders of the heavens with a more lively and intelligent interest than Ms. Agnes M Clerke, now well known for her excellent books on the history and problems of astronomy, but at that time a beginner".[xiii] There Agnes made some more good friends. They included well-known independent astronomers and astronomical photographers, Andrew Ainslie Common and Isaac Roberts, who provided her with beautiful illustrations for her later books.

Thanks to Lockyer, Agnes Clerke became a regular contributor to *Nature*, with articles on new developments in astronomy and book reviews, especially of foreign language publications. She sometimes contributed original articles; she wrote a paper on *Homeric Astronomy* in 1887, about astronomical allusions in the Iliad and the Odyssey which was republished as a chapter in a charming book, *Familiar Studies in Homer*, in 1892.[xiv]

Lockyer formed a high opinion of Agnes Clerke's scientific as well as literary talents. He also favoured the idea of women in science, and at the turn of the century was well-known as a leading advocate of women's rights in this field. He employed a young woman helper at his observatory who, according to norms of respectability of the day, was attended by a chaperone. The chaperone's only duty presumably was to be present, but Lockyer, who was far from wealthy, no doubt paid her wages as well as those of the assistant.

In 1886 Lockyer proposed to Agnes Clerke that she should write another book on Spectrum Analysis. She was encouraged by this, and having studied further, she produced her second book, *The System of the Stars*, in 1890.

That book would owe a great deal to another influential and sympathetic astronomer who entered into Agnes Clerke's life in 1887 and became a dear friend as well as a vital influence – David Gill, whom we have already met with his wife Isabel (Chapter 9). Gill who was now well established as Her Majesty's Astronomer at the Cape of Good Hope was in Europe in the spring of that year taking part in the Astrographic Congress which launched the huge international project of the Carte du Ciel or the Chart of the Heavens. This ambitious undertaking was destined to play an important part in the entry of women into professional astronomy, albeit at a lowly level (Chapter 13).

After the congress Gill visited London and among other engagements delivered a lecture at the Royal Institution on the subject of the astronomical photography. Agnes Clerke, being a regular member of the audience at the Institution, heard the lecture and was deeply impressed by it. Mrs. Gill was introduced to Agnes Clerke afterwards (probably by the Hugginses), and was "charmed by her artistic temperament".[xv] She read the *History* and persuaded her husband to read it too. Agnes Clerke's account of this episode was that David Gill, seeing that the book was written by a lady, gave it to his wife with the remark: "This, my dear, will probably suit you". She, however, returned it to him with an emphatic: "I think it will suit *you*". Having read it, "he was convinced of the intellectual power and originality of the authoress". He promptly invited her to visit the Cape. "In your *History of Astronomy*", he wrote to her, "the one weak point was your want of critical knowledge of practical work, and that can only be gained by experience". Agnes hesitated. It was not easy to break away from her writing engagements for three or four months and, as she wrote to Holden, she had great doubts as to her capacity for learning anything practical. "I am not at home with instruments, and I am very short-sighted. So that I have every type of disqualification for observational astronomy". However, Gill kept urging her and in August 1888, she sailed to South Africa. It was the best decision she ever made.

Agnes Clerke spent two months – September and October 1888 – at the Cape (Fig. 12.2). It was a time of particular excitement for Gill. His two great astronomical projects were well in hand. One was his Cape Photographic Durchmusterung (the German name for catalogue), which was to include all stars to 9th magnitude in the southern sky. The other was the observation of the parallax of minor planets, using a heliometer, a continuation of the work begun on Ascension Island in 1877 (Chapter 9). When Agnes Clerke arrived at the Cape she had the opportunity of watching the work in action. She lived with the Gills in their beautiful residence which was a wing of the main observatory building, and in this way became a close personal friend of Bella, whose problems, including health problems, real or imaginary, she always responded to sympathetically.

Gill planned a practical project specially for Agnes. It was the spectroscopy of stars in the southern Heavens which, he said, was "absolutely virgin soil". While in the northern hemisphere, observers like the Hugginses and Angelo Secchi were busily engaged in examining spectra of numerous stars visible to them, little work of the kind had been done in the southern. All that was required, said Gill, was a telescope with a prism attached to the eyepiece. He gave her the use of a very good 7-in. equatorial telescope, which was housed in a dome close to the main observatory building. A small prism attached to the eye piece spread the star's light out into a tiny spectrum or rainbow. She was shown how to handle the telescope by the light of a dim lamp, and in spite of poor weather, and after some mishaps,

Fig. 12.2 Agnes Clerke and the Gills at the Cape Observatory 1888. (George Forbes, *David Gill Man and Astronomer*)

she succeeded in observing a number of unusual-looking red and variable stars, noting their spectra and colours. She also made observations of the mysterious variable nebula, Eta Carinae, one of the glories of the southern sky. The results of this piece of research were published in two papers in the British magazine *Observatory*.[xvi]

The visit confirmed Gill in his high opinion of his disciple's abilities. He declared her to be "one of the ablest women and most original of thinkers that I have ever met". He wrote to the secretary of the Royal Astronomical Society in London, proposing that her *History* deserved recognition by the Society – perhaps even the Gold Medal, but at the very least the honour bestowed upon Caroline Herschel – honorary membership of the society. The practical value to her of such a step, he pointed out, would be access of right to the society's library instead of as a favour. Nothing came of this suggestion until many years later.

On the personal side, Agnes Clerke's visit was a very happy time for her. Gill, who was about her own age, was warmhearted and good-humoured, a great conversationalist and a music lover. He was also a great letter writer, and the lively correspondence between him and Agnes which continued to the end of her life is a wonderful source of illumination on both their personalities. One might almost imagine that they were in love with each other, were it not for Gill's fervent devotion to his wife Bella, which he expressed over and over again in every letter to his friend.

Agnes began research on her new book at the Cape, where Gill provided her with a quiet place to work, and gave her the benefit of his authoritative guidance. When she returned home, she sent drafts of chapters to him according as they were written for his approval. He returned them with his "scribblings in the margin, written in the delighted and at the same time critical attitude of first perusal".[xvii] Her other devoted adviser, Edward Holden, kept her informed of the latest results from the great new Lick Observatory, saying that he found "pleasure in speaking of astronomical subjects to you who appreciate so thoroughly". She illustrated the book with some recent sensational photographs, freely given to her: some famous wide-angle photographs of Milky Way fields by the young Lick astronomer Edward E Barnard, Isaac Roberts' beautiful photographs of nebulae, and the photographic map of the Pleiades star cluster by the French astronomers Paul and Prosper Henry. On its title page Agnes chose as motto a lovely line from Dante's *Divine Comedy*: Io vidi de cose belle que porta'l ciel (I saw the fair things that heaven holds).

The final stages of writing were overshadowed for Agnes by the death of her adored father in February 1890, aged 76. "It is with a heavy heart that I am now working on at the book he would so keenly have enjoyed the publication of", she wrote to Edward Holden, "but no doubt it will be made all right for us in a better way than the one we would choose." She dedicated the book to his memory.

The System of the Stars[xviii] came out in November 1890, and had the same success as the *History*. It began with accounts of the various varieties of stars and then went on to discussing star clusters, nebulae and, finally, "the Construction of the Heavens". It presented what was the definitive picture of the universe, as inferred from the observational facts of the time. In that model the entire universe – stars, nebulae and everything else – was enclosed in one vast agglomeration, a single "System". Her summing up was: 'A practical certainty has been attained that the entire contents, stellar and nebular, of the sphere belong to one mighty aggregation, and stand in ordered mutual relations within the limits of one all-embracing scheme – all-embracing, that is to say, as far as our capacities of knowledge extend. With the infinite possibilities beyond, science has no concern.' This one-system universe was not toppled for 30 years; it eventually (long after Agnes Clerke's time) gave way to the expanding universe and the modern Big Bang.

In 1889, while she was writing the *System of the Stars*, the possibility was put to her of an appointment at the Royal Observatory, Greenwich. It was a remarkable suggestion, as the observatory had never before employed women in any capacity; indeed the employment of women in scientific work was practically unheard-of. The attraction of the offer, as she wrote to Gill, was that she was to have the use of a powerful telescope for observing "according to any plan I fancied". The instrument had once belonged to the celebrated amateur astronomer William Lassell whose daughters donated it to the Royal Observatory (Chapter 7) where it was installed and maintained, but not used because of shortage of staff. She was naturally attracted to the idea; but when the formal proposal came she had doubts. The post offered was a "supernumerary computership" at £8 a month, the lowest rank in the staff. She feared that the job would turn out to be wasteful routine and also reflected on the "almost insurmountable difficulties from the fact that Greenwich Park is said to be unsafe for ladies at night, so that a good deal of the glamour disappeared". The Hugginses encouraged her to accept, but she decided not to abandon her existing situation for one so uncertain. "Nevertheless I feel somewhat sore and sorry at having refused, and so shut out finally a prospect that was not without its attractions for me", she wrote to Gill. It was, in her case, a wise decision. It is hard to imagine how she would have coped with a large telescope without help. Had she accepted the post she would have been the first woman on the staff of the Royal Observatory. Some young women were actually appointed later, but the scheme had mixed results, and after some years was dropped (Chapter 13).

Meanwhile Holden tried to secure a professional appointment for her in the United States. It was to the Chair of Astronomy at Vassar College to succeed Maria Mitchell (Chapter 8). He put Agnes Clerke's name forward, with a warm recommendation: "No woman has ever rendered the service to Astronomy that Ms. Clerke has already given; and there are very few living men who have her philosophical grasp of the whole science and very few of equal erudition. It would be an honor to America should you invite her to your Chair of Astronomy". Agnes Clerke appreciated Holden's action, but even if the offer were made, she would not have wished to be parted from her family.[xix]

When the British Astronomical Association was founded in 1890 to cater for the interests of amateur astronomers, including women (Chapter 10), Agnes, Ellen and Aubrey Clerke all joined. Agnes reluctantly allowed herself to be elected to the council. "I am totally useless", she told Gill, "but must attend a few of the meetings for form's sake". She was already since 1885 a member of the Liverpool Astronomical Society, and a corresponding member of the Astronomical Society of the Pacific – the first non-American to be elected – and of the Astronomical and Physical Society of Canada in Toronto.

In 1892 the question of allowing women to belong to the Royal Astronomical Society came up but was rejected by the Fellows (Chapter 10). Agnes Clerke, never an activist, had not sought election, but when, as a compromise, cards of admission to meetings were issued to women by the President, she and her sister took advantage of them, and attended meetings of the society regularly thereafter. She was happy to listen in silence; when, on one occasion she was invited by the President to say a few words she was tongue-tied. However, she was frequently seen at meetings "surrounded by leading astronomers, genuinely keen to hear her opinion on some knotty point."[xx] and enjoyed the company of friends and colleagues privately in her own home.

In the summer of 1892 Agnes Clerke and her sister attended a meeting of the British Association for the Advancement of Science in Cardiff to hear the President, William Huggins, give his statutory Address. There she had the good fortune of meeting the young genius George Ellery Hale of Chicago. Only 22 years old, he was already the inventor of the spectroheliograph for observing prominences and other changing features in the Sun's outer atmosphere. In the course of his subsequent brilliant career he founded the famous Mount Wilson Observatory in California, and installed the 100-in. telescope, the world's largest. Agnes and he developed a close friendship which continued until her death. She was also on terms of warm and mutually beneficial friendship with Holden's young successors at Lick Observatory which, by the end of the century, led the world in astronomical photography and spectroscopy.[xxi]

Agnes Clerke's third and last major book was *Problems in Astrophysics*, a sort of sequel to her *History* in which she brought the advances in astronomy up to date and discussed unresolved problems and how they might be solved. The idea of such a book had been "haunting" her for years, and she began writing after the death of her mother in 1897. She had definite opinions as to what astronomers ought to be doing with their powerful instruments. "The globe is studded with observatories, variously and admirably equipped. Yet innumerable objects in the sidereal heavens remain neglected, mainly through inadvertence to the extraordinary interest of the questions pending with respect to them". The book discussed the Sun, stars and nebulae; planetary astronomy was to be left for another volume, "should my powers last long enough to reach it", which they sadly did not. Her "magnum opus", as she called it, was published in 1903, and dedicated Sir David Gill.

The new book brought its reward. According to the astronomical correspondent of the *Times*, "*Problems in Astrophysics* was of such great scientific value that the [Royal] Astronomical Society could no longer ignore her claims to public recognition".[xxii] At the May 1903 meeting of the society Agnes Clerke and Lady Huggins were made honorary Members – a long overdue recognition: Agnes Clerke was over 60 years of age. The Society's President declared: "It is a pleasure to think that there is a considerable resemblance between the claims of these ladies and those of our original honorary members. Lady Huggins has been associated with the work of her husband as Ms. Caroline Herschel was associated with the work of her brother. The work of Ms. Agnes Clerke is similar to that of Mrs. Somerville, lying in the domain of scientific writing, and, I may say, with reference to her last work, it is not merely an astronomical history, but a work of actual constructive thinking in our science".[xxiii]

This was the universal opinion among her contemporaries and her readers. She appears to have had only one enemy in her entire career – the assistant editor and book reviewer for *Nature*, Richard (later Sir Richard) Gregory, who criticised *Problems in Astrophysics* in strong terms, claiming that she was not qualified to have an opinion on matters in which she was not an active investigator, and also accusing her of being a partisan of Huggins in a feud between him and his great rival Sir Norman Lockyer concerning theories of the evolution of stars. He coupled his criticism with an implication that being a woman meant poor judgment: "A cynic has said that it is a characteristic of women to make rash assertions, and in the absence of contradiction to accept them as true. Ms. Clerke is apparently not free from this weakness of her sex,"[xxiv] an outrageous sentiment repeated a few years later in his review of the second edition of *System of the Stars*. "The intuitive instinct of a woman is a safer guide to follow than her reasoning faculties; and though in these days it is considered ungracious to make these suggestions, evidence of its truth is not difficult to discover in most literary products of the feminine mind. It is no disparagement to Ms. Clerke that even she shares the characteristics of her sex." It was the only instance of gender prejudice that she encountered in her entire career. She ignored it.[xxv]

When *Problems in Astrophysics* was published Agnes updated *The System of the Stars* (1903). With the latest edition of the *History* (1902), the books thus made up a trilogy providing a complete picture of the state of astronomy at the opening of the new century. In 1905 she published a shorter but much praised book, *Modern Cosmogonies*, a popular account of theories of the origin of the universe from the ancient Greeks to her own day in which she made plain her antipathy to current agnostic interpretations of science.[xxvi] Margaret Huggins claimed that it was in some ways her greatest achievement on account of its spiritual dimension, the "clues helping to sustainment of soul in the midst of the majestic mysteries surrounding us."[xxvii]

Agnes Clerke's prolific output went beyond her books, and her scientific interest went beyond astronomy. She was one of 100 core contributors to the *Dictionary of National Biography*, and the only woman in that group, with responsibility for the biographies of 150 astronomers and scientists in allied fields. These meticulously documented essays were rightly described by Margaret Huggins as "models of painstaking inquiry and of clear, concise statement". To them may be added some 30 biographies in *Encyclopaedia Britannica*, among them those of William, Caroline and John Herschel, who were also the subject of a book (*The Herschels and Modern Astronomy*, 1895). The *Encyclopaedia* also contains a long essay on the history of Astronomy, and another on ancient cosmologies, under the title Zodiac. Further essays that she was working on were taken over by others after her death. It was a tragedy that she did not live to see the monumental eleventh edition of the *Encyclopaedia* in print.

Agnes was a familiar attendant at the Royal Institution's lectures, and met many illustrious scientists there. She kept track of advances in the new atomic physics to provide some excellent essays for the *Edinburgh Review*. The Institution's septennial Actonian Prize for an essay celebrating 'the Beneficence of the Almighty' as illustrated by discoveries in science, was awarded to her in 1893, and in 1901 she was commissioned to write the Hodgkins Trust Essay on the subject of Low Temperature Research at the Royal Institution. It was a scientific report on the experiments to freeze gases that had been going on for 10 years in the Institution's laboratory under James Dewar, a distinguished physicist and laboratory spectroscopist who was a friend of hers. A year later, in 1902, she was honoured by being elected a member of the Institution.

While still working on her *Encyclopaedia* articles, Agnes suffered the loss her beloved sister and constant companion Ellen. She survivor her by just 1 year. She died on 20 January 1907, of pneumonia, following a chill. Her friend Margaret Huggins gave the sad news to George Ellery Hale, and no doubt to other friends: "She was conscious almost to the last, and went to the God she had loved and served, with a sweet and perfect trust and faith". A review of a paper by Hale on which she was working lay on her desk. Her very last article in the *Edinburgh Review*, on radioactivity, was published after her death.

In her lifetime, Agnes Clerke was the leading writer and commentator on astronomy in the English language. Her friend Lady Huggins saw her as a new type of scholar, with a mission 'to collect, collate, correlate, and digest the mass of observations and papers – to chronicle, in short, on one hand; and on the other, to discuss and suggest, and to expound... and at the same time to inform and interest the general public.' A century later, her books, particularly her *History of Astronomy during the Nineteenth Century* (reprinted in 2003[xxviii]), remain indispensable sources of accurate and readable information. "Of all the works on the history of astronomy, those by Agnes Clerke must surely be the most enduring". says the modern expert Michael Hoskin. She was an indefatigable worker and a meticulous researcher. She lived a simple life, rarely travelled, and though a most hospitable hostess in her own home, did not seek a hectic social life outside it. After her labours in writing the *System of the Stars*, her friend Henry Reeve arranged or her to take a recuperating cruise on the Baltic with a friend who owned a yacht; but when they arrived in St Petersburg, Agnes was too shy to make herself known at the famous Pulkovo Observatory, much though she longed to visit it. She and Ellen were invited by Edward Pickering and his wife to visit Harvard as their personal guests at their home, but though greatly appreciating the honour, they made their excuses. She was happy as long as she had access to knowledge, which she gladly shared with other lovers of astronomy. She was uninterested in women's emancipation or in campaigning for entry into the male establishment. In her Will, she left all her worldly wealth to charities – to the Church and to the poor of London, among whom she had worked unostentatiously. 'No purer, loftier, and yet sweetly unselfish and human soul has lived' was Lady Huggins' summary of her character.[xxix]

Agnes Clerke's sister, Ellen Mary, was her devoted companion from childhood until death. So inseparable were they that to those who did not know them well, Ellen may have seemed to be her sister's satellite. But in fact, Ellen had a very successful independent career as a professional journalist. From the time she arrived in London she was a regular contributor to the *Dublin Review*, a Roman Catholic quarterly periodical similar in style to the *Edinburgh Review* for which Agnes wrote, specialising in geography and anthropology. She joined the Manchester Geographical Society which admitted women, and through this and other connections could be relied upon to provide up-to-date information on discoveries and explorations in distant lands as well as political commentary. Mary Creese, in her major study of nineteenth century women scientists,[xxx] places Ellen Clerke in the category of substantial "geographers, explorers and travellers" though in fact she never visited the places she described. In addition to this, Ellen was for 20 years on the editorial staff of the weekly *Tablet*, reporting on European politics. Agnes probably envied Ellen's status as a salaried woman, and one can sympathise with her regret at turning down the post that she was offered at Greenwich.

As far as her own personal taste was concerned, Ellen's favourite field of study was Italian literature. She came to love all things Italian, and on settling in London wrote some charming essays on Italian life and customs, as well making translations of early Italian poetry.[xxxi] She also published a collection of her own poems and ballads, some of which appeared in anthologies.[xxxii] As a writer and poet, she is described by a modern literary scholar Susan Elkin as possessing a "formidable breadth and depth of learning".[xxxiii]

With such a busy life, one wonders how Ellen had time left for astronomy. But she and Agnes had shared each other's thoughts and interests from childhood. With Agnes, she was an early member of the British Astronomical Association and occasionally contributed short pieces to its journal and to other periodicals. She was among the women made welcome at meetings of the all-male Royal Astronomical Society from 1892 onwards, and used to accompany her sister there. She published two small books for the general reader on Jupiter and Venus[xxxiv], respectively, described by a reviewer as "popular yet accurate". Among her other writings were the entries for Mary Somerville and her husband in the original *Dictionary of National Biography*. At the age of 62 she wrote her sole novel, *Flowers of Fire*,[xxxv] a romance set mainly in Italy which includes a dramatic account of an eruption of the Vesuvius volcano in 1872, something that the Clerkes had witnessed and which the aged Mary Somerville also saw.

Ellen died of pneumonia, after only a few days' illness, in March 1906. After Agnes' death the following year, Margaret Huggins who had known them for well over two decades, wrote at their brother's request a moving personal account of the lives and labours of Agnes and Ellen. She ended: "These sisters were lovely and pleasant in their lives, and in death they were but little divided."[xxxvi] It is appropriate that they are jointly commemorated by portraits on a plaque on the house in Skibbereen where they were born, erected in 1999 by the Irish scientific and the local communities (Fig. 12.3). Agnes is also commemorated by a crater on the Moon, near the Sea of Tranquillity.

Their brother lived on, unmarried, until 1923. All three are buried with their parents under a plain monument in the peaceful Brompton Cemetery near their home in London.

At the time when the Hugginses were at the height of their powers, and when Agnes Clerke was already established as the chief writer on astronomy in the English language, another name entered the ranks of women astronomers. This was Williamina Fleming (1857–1911), initiator and mistress for 30 years of one of the greatest projects ever undertaken in an observatory – the massive spectroscopic survey of stars carried out at the Harvard College Observatory in the USA which culminated in the great Henry Draper Catalogue of stellar spectra. Britain's small claim to this magnificent opus rests with the fact that Mrs. Fleming was born and brought up in Scotland, a country which is proud to have given her birth,[xxxvii] though she herself had no idea when she left home that her career would be in the stars.

She was born Williamina Paton Stevens in 1857, in Dundee, a thriving industrial city and sea port on Scotland's the east coast. Her father, Robert Stevens, a carver and gilder with his own business, was also a photographer, the first to bring that new exciting trade to his city. Williamina was a school teacher, having trained in the customary way as a pupil-teacher from the age of 14. At the age of 22 she married a widower, James Orr Fleming, aged 35, an accountant,[xxxviii] and a year later, in 1878, emigrated with him to the United States. The marriage did not last, and Williamina found herself obliged to fend for herself and their little son. Her association with the Harvard College Observatory began in 1881 when she was appointed by the

Fig. 12.3 Commemorative Plaque at the Clerke sisters' birthplace in Skibbereen, Co. Cork. (Mary Bruck, *Agnes Mary Clerke and the Rise of Astrophysics*)

Director, Edward C Pickering, to a humble post, copying and routine computing. By some accounts she had begun as an employee in his private home; but the implication that she was no more than an untutored domestic servant[xxxix] does not do justice to her Scottish background and her educational qualification. It is also likely that her father's professional activities meant that she was not entirely ignorant of the new techniques of photographic observations being introduced at Harvard.

That work expanded significantly with the founding in 1886 of the Henry Draper Memorial, a fund set up by Mrs. Anna Palmer Draper, widow and collaborator of the early astronomical spectroscopist Henry Draper, for a project to record on photographs the spectra of stars in large numbers. This was achieved by placing a large prism, known as an objective prism, in front of the telescope, so causing the image of each star to be drawn out into a small spectrum or rainbow. Each photograph would therefore record not one but hundreds of spectra. It was a complementary approach to that used by spectroscopists like the Hugginses, who photographed one spectrum at a time in considerable detail. Stars in the northern hemisphere were observed from Harvard itself; the work was extended to the southern hemisphere when Harvard set up its own station in Arequipa, Peru. The result, over time, was a huge supply of photographs, containing spectra by the thousand, which required to be analysed, organised and classified.

Pickering employed women specifically for this work, and Mrs. Fleming was assigned the task of devising an empirical system to classify the stars according to their spectra. Her classification, with stars labelled according to letters of the alphabet, was the basis of all future systems. In the first four years she catalogued over ten thousand stars and discovered hundreds of unusual objects. She published some important papers in her own name, but, like Agnes Clerke, she did not speak in public. When Pickering read a paper on her behalf at a conference in Harvard in 1898, he pointed out at the end that the author had omitted to mention that all the stars described there had been discovered by herself, "whereupon Mrs. Fleming was compelled by a spontaneous burst of applause to come forward and supplement the paper by responding to questions elicited by it".[xl] The bulk of the famous catalogue work was published in the official publications of the Harvard Observatory for which she did the proof reading. The library of photographs, each one labelled, sorted and stored, reached two hundred thousand: a veritable gold mine for future researchers. A team of women under Mrs. Fleming, a stern and efficient supervisor, expanded (21 were recruited up to 1900) to become one of the most famous and successful ventures in the history of astronomy. It also provided one of the earliest opportunities for women in science, and launched the careers of other talented women astronomers that continued well beyond its founder's own 30 years of service and management.[xli]

In 1899 Williamina Fleming was elevated to a post of Curator of Astronomical Photographs, the first woman to hold a formal appointment at Harvard. She devoted much time after this to cataloguing all known stars with unusual spectra. listing the various types and giving each object its discoverer, year of discovery and all references in the literature. The catalogue is very valuable source in the history of stellar spectroscopy.[xlii] She was elected an Honorary Member of the Royal Astronomical Society in 1906, a rare accolade in that then all-male organisation, joining Margaret Huggins and Agnes Clerke there. Williamina Fleming's talent as an astronomer and team leader was, according to her colleague and eventual successor Annie Cannon, coupled with "a large-hearted and sympathetic nature", proving that "a life spent in the routine of science need not destroy the attractive human element of a woman's nature". She was still in office at her death at the age of only 54.[xliii]

The team of women whom Mrs. Fleming gathered about her included products of the American women's colleges, by now well established: Antonia Maury, a graduate of Vassar College in 1887, Maria Mitchell's last year; Annie Cannon, a graduate of Wellesley College where she was a student of Margaret Huggins' friend Sarah Whiting and herself a correspondent of Lady Huggins; and Henrietta Leavitt, a more recent (1893) graduate of Radcliffe College, a women's College affiliated to Harvard. The work of all these members of the Harvard team was followed closely by Agnes Clerke, and discussed in her books. Opportunities for women in professional astronomy in America had far outraced those available to their sisters on the east side of the Atlantic.

Notes

[i] Mary Brück. 2002. *Agnes Mary Clerke and the Rise of Astrophysics*. Cambridge University Press; Mary T. Brück. 1991. Companions in Astronomy. *Irish Astronomical Journal* **20**, 70–77.
[ii] A. St J. Clerke. 1907. Foreword to Lady Huggins. *Agnes Mary Clerke and Ellen Mary Clerke, an Appreciation*. London, privately printed.
[iii] Hector MacPherson. 1905. *Astronomers of Today*. London and Edinburgh: Gall and Inglis.
[iv] A.M. Clerke. Entry on Henry Reeve, DNB.
[v] J.K. Laughton. 1998. *Memoirs of the Life and Correspondence of Henry Reeve, CB, DCL*, vol. 2 p 250. London: Longman.
[vi] *Athenaeum* **60**, 330, 1882.
[vii] A.M. Clerke. 1885. *A Popular History of Astronomy during the Nineteenth Century*. Edinburgh: Adam and Charles Black.
[viii] R.S. Ball. 1886. *Nature* **33**, 313.
[ix] *Edinburgh Review* **163**, 372, 1886. The *Dublin Review* **15**, 219, 1886, also compared Agnes Clerke to Mary Somerville.
[x] D.E. Osterbrock. 1984. The Rise and Fall of Edward S. Holden, Parts 1 and 2, *Journal for the History of Astronomy* **15**, 81 and 151.
[xi] A.M. Clerke to E.C. Pickering, 26 October 1886. Harvard College Observatory archives.
[xii] A.M. Clerke to Lockyer, 6 May 1887. Exeter University archives.
[xiii] T. Mary Lockyer and Winifred L. Lockyer. 1928. *Life and Work of Sir Norman of Lockyer*, p 117. London; A Fowler, appendix to the same biography.
[xiv] A.M. Clerke. 1892. *Familiar Studies in Homer*. London: Macmillan.
[xv] George Forbes. 1916. *David Gill Man and Astronomer*. London: John Murray.
[xvi] A.M. Clerke. 1888–1889. *Observatory* **11**, 428, and **12**, 134.
[xvii] Gill correspondence 1889. Cape/Royal Greenwich Observatory archives.
[xviii] A.M. Clerke. 1890. *The System of the Stars*. London: Longman.
[xix] Letter to Holden, 1 September 1889. Mary Lea Shane archives of Lick Observatory.
[xx] Hector Macpherson. 1907. *Popular Astronomy* **15**, 165.
[xxi] D.E. Osterbrock. 1984. *James Keeler, Pioneer American Astrophysicist*. Cambridge University Press.
[xxii] Obituary of Miss Agnes Mary Clerke. *Times*, 21 January 1907.
[xxiii] Obituary of Lady Huggins. 1916. *Observatory* **76**, 278.
[xxiv] R.A. Gregory. 1903. *Nature* **68**, 338; 1906. *Nature* **73**, 505.
[xxv] Agnes was not the only victim of *Nature's* surly critic. He described one of Sir Robert Ball's popular books as "different scraps of information joined together with an ingenuity that is only acquired after long practise" Quotation from W.H. Armytage 1957. *Sir Richard Gregory, his Life and Work*. London: Macmillan.
[xxvi] A.M. Clerke. 1905. *Modern Cosmogonies*. London: A. and C. Black.
[xxvii] Lady (Margaret) Huggins.1907. *Agnes Mary Clerke and Ellen Mary Clerke, an Appreciation*. London, privately printed.
[xxviii] Agnes M. Clerke. 2003. *A Popular History of Astronomy during the Nineteenth Century*. (reprint of edition of 1902). Decorah, IA: Sattre Press.

[xxix] Lady Margaret Huggins, op. cit. 1907.
[xxx] Mary Creese. *Ladies in the Laboratory*. 1998. Lanham Md. and London: The Scarecrow Press Inc. Chapter 13.
[xxxi] E.M. Clerke. 1899. *Fable and Song in Italy*. London: Grant Edwards.
[xxxii] E.M. Clerke. 1881. *The Flying Dutchman and other Poems*. London: W. Satchell & Co.
[xxxiii] Susan Elkin. *Oxford DNB*.
[xxxiv] E.M. Clerke. 1892. *Jupiter and his System*; 1893. *The Planet Venus*. London: Stanford.
[xxxv] E.M. Clerke. 1902. *Flowers of Fire*. London: Hutchison.
[xxxvi] Lady Huggins, op. cit.
[xxxvii] Williamina Fleming is included in the *Biographical Dictionary of Scottish Women* (eds. Elizabeth Ewan, Sue Innes, Siân Reynolds and Rose Pipes), Edinburgh University Press 2006.
[xxxviii] Scottish official Records; I thank Eric Percival for this information.
[xxxix] Anne McNaught. *The Scotsman*. 5 March 2004. Edinburgh.
[xl] Pamela E. Mack, 1990. Strategies and Compromises: Women in astronomy at Harvard College Observatory 1870–1920. *Journal for the History of Astronomy* **21**, 67–75.
[xli] Pamela E. Mack. ibid.
[xlii] J.B. Hearnshaw. 1986. *The Analysis of Starlight. One Hundred Years of Astronomical Spectroscopy*. p 126. Cambridge University Press. The catalogue was published in 1912, after her early death.
[xliii] Annie Cannon. 1911. *Publications of the Astronomical Society of the Pacific* **23**, 200–201.

Chapter 13
Slave-Wage Earners

The first serious effort in Britain to give women a scientific profession in astronomy was the "lady computer" scheme at the Royal Observatory Greenwich, introduced in 1890. It owed its existence to the Astronomer Royal of the day, W.M. (later Sir William) Christie, an outstanding director who was responsible for the greatest enlargement of the observatory during the whole of its time in Greenwich.[i] It constituted a praiseworthy innovation on his part, made perhaps in response to the campaign by forward-looking thinkers for the advancement of women.[ii] Though there had been a marked improvement in availability of university education for women during the previous decade, including training in the sciences, women remained disadvantaged in the matter of employment.

The status of the planned posts was not high, but the scheme opened a professional door by a small chink to university educated women of scientific bent who were prepared to work hard for low pay in order to gain a foothold in the male-dominated world.[iii] The Royal Observatory's "lady computers", as they were styled, were the first women to be employed there in any capacity. The Royal Observatory at Greenwich was the only institution in the British Isles to employ astronomers in any numbers. Its staff in the 1890s consisted of the Astronomer Royal, a chief assistant, a number of assistants (10 in 1892), and several computers. The computers were young men recruited by examination at the age of 13 or 14 who were apt to move on from Greenwich to better jobs as office clerks. They worked under the assistants, and their duties included the practical use of instruments as well as computations. Finally, at the bottom of the pecking order were the "supernumerary" or temporary computers who could be hired and fired as needs arose. It was to this lowly grade that the lady computers were appointed. The position of the lady computers was peculiar. In rank and salary they were at the level of boys straight from school. Yet in academic attainment they were superior to non-graduate assistants who had been recruited by examination in earlier days. Furthermore, they could not aspire to promotion: the Astronomer Royal, well-intentioned, was powerless to remedy this since the civil service under which the Royal Observatory operated debarred women from its permanent ranks. The computer posts were the only ones in the Astronomer Royal's own hands, and the only ones to which women could be hired quietly without violating the rules: their existence was not even mentioned in the annual reports of the observatory. The Astronomer Royal explained the position

M. Brück, *Women in Early British and Irish Astronomy*,
DOI: 10.1007/978-90-481-2473-2_13, © Springer Science+Business Media B.V. 2009

to the famous educationalist Ms. Dorothea Beale who was on the lookout for suitable candidates: "It is somewhat of an experiment. I do not think it would be wise to have more than four ladies to start with; in fact I have not funds available for more at present.... If the new departure turns out as successful as I hope, I may be able to get more funds for the lady computers."[iv]

There was keen interest at first in the Greenwich scheme from women graduates, products of the Cambridge, Oxford or London women's colleges, who heard of it through people in the world of education like Ms. Beale and Ms. C. Elder, secretary of the University Association of Women. The four women appointed in 1890–1891 came from the two women's colleges in Cambridge where they had studied mathematics.[v]

The young women worked under the same conditions as their male colleagues. Their hours were 9–1 every weekday and 2–4.30 on three afternoons a week, plus observing three nights a week for 3 or 4 h depending on the weather. On their appointment they were taught how to work with the Transit Circle (observing meridian transits of stars for timekeeping) and remained official transit observers throughout their years of service. Only the Astronomer Royal resided within the observatory grounds; others walked up through Greenwich Park to work. The fact that the women took part in night-observing in the 1890s, with no concession to their sex, is a remarkable aspect of the lady computer scheme. There were two shifts, and one wonders how the young women had the courage to be abroad at midnight, as they came on and off duty. One of Agnes Clerke's reasons for declining a post at Greenwich was the reputed peril for women of traversing Greenwich Park at night.

The computers were assigned to help the assistants in their various departments. The meridian and time departments were the principal ones; but there were also the meteorological, magnetic, photographic and solar, where a certain amount of astrophysical research went hand in hand with the routine recording of data. The women appear to have been given every encouragement and were allowed to make independent use of the Observatory's instruments according to their own ideas outside their formal duties. Nevertheless, the practicalities of everyday life were not easy for the less robust. One of the lady computers appointed at the beginning left after a year,[vi] her friend and colleague was obliged to resign a year later due to ill health. Though the latter, Edith Rix by name, disappears from the story at this point, her experience at Greenwich throws light on the hardship of life for a lone young woman away from home in an age when many were still chaperoned. It also illustrates the great effort made by the Astronomer Royal and his staff at the Royal Observatory to make the scheme work. Edith's troubles began when she got thoroughly wet in a November gale on her way to and from the observatory. The chimney pot of her lodgings was blown down; the house was full of smoke, and there was no food.[vii] She fell seriously ill and on the doctor's advice asked for a month's leave without pay. "I am very sorry indeed to hear that you have been knocked up, but I hope that a month's rest will set you up again", the Astronomer Royal wrote, offering her leave with half pay. This brought an effusive letter of thanks from her mother. She was living alone, the mother explained. The landlady was "motherly" and "if she could get good servants things might be more comfortable", but the one maid was "of the most inefficient and common kind" and meals were "bad and unpunctual". The Astronomer Royal promised to do something about easing her work load. In fact, Edith never fully recovered. Her sick leave was extended amid further expressions of gratitude from the parents who felt that she "had more indulgence than she could have expected". After more than three months she finally resigned. The patient Christie wrote in the warmest tone saying how sorry he was that she had to leave Greenwich. The work, he said, was too much for her.

It was not so for two robust young torch bearers, Alice Everett (Fig. 13.1) and Annie Russell, who persevered with great determination in their scientific aspirations.[viii] The level of activity of this pair was far above that expected from their nominal rank. They were enthusiastic members of the British Astronomical Association, and reported on and published in its Journal observations which they had made of phenomena such as comets, an eclipse, and the exciting "new star" Nova Aurigae 1892. They were both nominated for Fellowship of the Royal Astronomical Society by senior members of staff, though unfortunately and unexpectedly rejected (Chapter 10).

Alice Everett (1865–1949) was the daughter of Joseph David Everett FRS (1831–1904), Professor of Natural Philosophy (Physics) at Queen's University Belfast. She was born in Glasgow but was only 2 years old when her father took up his chair in Belfast where he remained until his retirement 30 years later. Her mother was Scottish, the daughter of a Church of Scotland minister; of the family of three sons and three daughters only Alice appears to have followed in the footsteps of her father, a strong supporter of women in science. She was educated at the Methodist

ALICE EVERETT.

Fig. 13.1 Alice Everett in her Royal University of Ireland academic robes 1887. (Methodist College, Belfast)

College in Belfast, a coeducational day school with a strong ethos of "plain living and high thinking" which after initial struggles had achieved a high level of academic distinction (one of its more recent alumni was Ernest Walton, Nobel Prizewinner in Physics in 1951). She was a prize pupil of the College. University level education for women in Ireland at that time was catered for by the Royal University of Ireland, a purely examining body which awarded degrees to any student who passed its examinations, conducted under the aegis of the provincial universities known as the Queen's Colleges. In 1882 Queen's College, Belfast, opened its doors to women students, enabling them to take lectures in preparation for the Royal University examinations. Alice Everett chose this option, and in 1884 took first place in the first-year scholarship examination in science, creating a dilemma for the authorities because of the fact that women students were not formally recognised. The university lawyers decided that women were ineligible for scholarships, and Alice's award was not allowed. It was not until 1895 that the statutes were changed to place women on an equality with men.

Alice Everett, aged 21, proceeded from Belfast to Girton College, Cambridge as a scholar in an intake of 29 students in 1886. One her classmates, Annie Russell, was to become her academic friend and colleague. Girton College, the first women's college in Britain of university rank, had been established in its Cambridge site in 1873. From 1882 women were permitted to sit the Cambridge Tripos (i.e. degree) examinations, though the university did not grant them degrees, a right not conceded until 1948. Alice took the mathematical Tripos examination in 1889 and while at Cambridge she also sat and passed with honours the external Royal University of Ireland's Bachelor of Arts degree in mathematics and mathematical physics in 1887. She was awarded its advanced Master of Arts degree two years later.

Alice Everett joined the Greenwich staff in January 1890. Her salary was £6 a month, the maximum for the rank of computer. She was assigned to the project of the Astrographic Catalogue (the *Carte du Ciel*), the international undertaking initiated in 1887 which aimed at surveying the entire sky photographically (as described later). The Royal Observatory at Greenwich installed its regulation astrograph in 1890.[ix] The work, under the elderly Mr. Criswick, one of the assistants, which had just commenced, involved taking the actual photographs, and subsequently measuring the positions of the stars from the photographic plates.[x] Following the preliminary setting up of comparison stars and taking experimental astronomical photographs, Alice spent two full years, 1893 and 1894, almost exclusively on the actual work of the Chart, making observations regularly twice a week (sometimes for up to seven hours a night at a stretch) with the astrograph, developing photographs, printing reseaux (transparent grids to be superimposed on the star field photographs for measurement purposes) and measuring the catalogue plates with the astrographic micrometer. In short, she was involved in every step in the development of this important programme. Her contribution is recorded in the published catalogue where she is shown as having made the first series of measurements of the photographs, from 1894 October to 1895 March.[xi] As part of her normal duties, Alice also did her stints observing by night with the transit circle as well as reducing the observations.[xii] As a sideline and for her own interest she learned to use the zenith telescope and worked with the observatory's equatorial telescope (for which she obtained a certificate of efficiency), observing occultations (eclipses of stars by the Moon, valuable for computation of the Moon's motion). She made the first recorded observation in 1892 of a dust storm on the planet Mars, a phenomenon that has become of interest again in the modern age.[xiii] She also took part in double-star work under the supervision of Thomas Lewis, the observatory's expert in this field, who communicated a paper by her to the Royal Astronomical Society which was published in its *Monthly Notices*.[xiv]

Annie Scott Dill Russell (1868–1947), the second of the two persevering lady computers, was, like her companion, educated in a Belfast school and at Girton College, Cambridge. She was born in Strabane, Co. Tyrone, the daughter of Reverend William Andrew Russell, minister of the Irish Presbyterian Church, and his second wife who was herself the daughter of a minister in the same Church. The six children in the Russell household had a devoutly Christian and serious minded upbringing; all were exceptionally talented and were high-level academic achievers. The elder of the two Russell daughters, Hester Dill Russell (later Smith) had an academic record that matched that of her sister. She studied medicine under the great pioneer Dr. Elizabeth Garrett Anderson at the London School of Medicine for Women, qualifying as first exhibitioner in the final MB examination in 1891. This was the same year that Annie took up her appointment at Greenwich so that the sisters lived together at 16 The Circus, Greenwich, before Annie's marriage. Hester became a medical missionary in India and married another medical missionary.

The sisters received their secondary education at the Ladies' Collegiate School, Belfast (renamed Victoria College in 1887), a pioneering institution founded in 1859 to provide a proper academic education for girls. Having won a prize in the public Irish Intermediate examination in 1886, Annie sat the Girton open entrance scholarship examination without any special preparation, and at age 18 was awarded a scholarship of £35 annually for three years.

At Girton, Annie found herself a contemporary of Alice Everett. Both women attained honours in the mathematical tripos examination in 1889. Annie was her college's top mathematician of her year, being ranked *Senior Optime* in the university class list, and occupying the highest place taken at Cambridge until that time by an Irishwoman. Her mathematics tutor, a Fellow of another (men's) college, praised her power in throwing herself into her work with such success, in spite of being "more than ordinarily handicapped – even for a woman – by an insufficiency of preliminary training". The Mistress of Girton College also testified to her "diligence, intelligence and conscientiousness." These characteristics were her hallmark throughout life.

Informed of a possible vacancy at Greenwich by her friend Alice who had joined on the first wave in January 1890, Annie wrote more than once to the Astronomer Royal, begging to be considered. Sir Robert Ball at Dunsink Observatory, Dublin, a family friend, wrote directly to Christie: "From all I know of the young lady and the talented family from which she comes, I feel sure that she would be highly qualified for the work". After a year in a job as mathematics mistress at the Ladies' High School on the island of Jersey which she did not enjoy,[xv] she was finally rewarded. The Chief Assistant, Herbert Hall Turner, on the Astronomer Royal's authority, offered her a post at £4 a month. She boldly protested that the salary was so small that she "could scarcely live on it" (as a teacher she was earning £80 a year with free residence). "Does the fact that I have taken the mathematical tripos at Cambridge make no difference?", she asked. The rather curt reply was that the salary offered "is simply all we can afford. And [the Astronomer Royal] would strongly urge very careful consideration before you leave tolerably remunerative employment for work here, which at present cannot be made very satisfactory in that respect". Neither, he said, could any increments be promised. It was a case of take it or leave it. She accepted, and began work on 1 September 1891.[xvi]

Annie Russell was assigned to the solar department, working on the famous Greenwich photoheliographic programme under Walter Maunder, who was also the recent founder of the British Astronomical Association (Chapter 10). The solar department at Greenwich had been set up in 1873 for regular photography of the Sun, and Maunder held the post of photographic assistant in charge of this work for his entire working life. The practice of daily photography of the Sun had begun with Warren de la Rue, one of the country's affluent Grand Amateurs and a pioneering astronomical photographer. From photography of the Moon, begun in 1857–1858 (Chapter 8), de la Rue went on to photographing the Sun. For this he designed a special camera called a photoheliograph, that had a 3-in. lens and a plateholder in place of an eyepiece. It was set up at the Meteorological Observatory at Kew, in London, in 1863, and maintained there at de la Rue's own expense until 1872. When the Royal Observatory founded its own solar and photographic Department his instrument was moved to Greenwich. It was later replaced by one of the photoheliographs that had been used to observe the Transit of Venus in 1874, now brought back from abroad. The photoheliograph had a lens of 4 in. aperture and produced an image of the Sun 4 in. in diameter.[xvii] This was the trusty instrument that was to serve solar astronomy well for half a century.

The early photographic work was done with the wet process, but within a few years the dry photographic plate came into use, one of the greatest boons of the century to practical astronomical work. In 1881 the new Astronomer Royal, W.M. Christie, took office. His first action was to reorganise and improve the solar Department. He upgraded the photoheliograph to give larger, 8-in. images of the Sun, and acquired a new measuring machine for analysing them. With the aim of establishing a continuous record of sunspots, he also arranged to have photographs of the Sun brought in from Mauritius and from Dehra Dun in India to fill inevitable gaps in the Greenwich observations due to bad weather. These stations had similar standard photoheliographs, left over from the Transit of Venus. Another photoheliograph was installed at the Royal Observatory at the Cape, South Africa, in 1910.

The routine which Annie followed as Maunder's amanuensis would have entailed photographing the Sun daily, weather permitting, developing the photographs and examining the negatives with a measuring micrometer. The same would be done with the imported photographs. The sunspots were numbered, and position of each one relative to the centre of the image, and its apparent area in units of tiny squares of a superposed glass grid, were measured; subsequent calculations converted these observations to heliocentric coordinates (i.e. relative to the centre of the Sun's disk) and true areas as fractions of the Sun's surface.[xviii] Annie was fortunate in taking up her duties at the approach of the sunspot maximum of 1894; she witnessed the famous giant spot of July 1892, source of a huge magnetic storm – the largest spot recorded at Greenwich up till that time. This event no doubt made a deep impression upon her: she was to become an expert in the field of solar-terrestrial phenomena.

The Astronomer Royal encouraged extra efforts in the solar department, and was able to report that in the year 1891–1891 the volume of recorded observations was greater by a factor of seven than the average over the 35 years since observations began.[xix] That was Annie's first year in her post, and one is surely justified in concluding – though her name was not mentioned by the Astronomer Royal – that she was responsible for much of this dramatic improvement. It is not surprising that her superior Walter Maunder should nominate her for Fellowship of the Royal Astronomical Society in 1892. The unexpected rejection – she called it "blackballing" – rankled, though she kept silent about her feelings for many years.

While solar work was her principal occupation, Annie did not escape the routine duties of her humble rank at the Transit instrument, with its night observing rota. Like her friend Alice Everett, she expanded her horizons by joining the British Astronomical Association, and was to be closely involved with it for the rest of her life. In November 1894, during Maunder's presidency, she was made editor of the Association's Journal, a duty which she discharged with notable success for 35 years.

These young women were greatly overqualified for the rank of supernumerary computer, and Alice for one was justified in hoping it would lead to better. In 1892 she heard "on good authority" that Dunsink Observatory of Trinity College Dublin was planning a programme of stellar photography with its recently acquired 15-in. reflector. With her experience in that very field she believed herself suitably qualified for the post of sole assistant at Dunsink which had become vacant because of the appointment of the director Sir Robert Ball to the Lowndean Chair at Cambridge and the promotion of his assistant Arthur Rambaut to replace him. She asked for a testimonial from the Astronomer Royal pointing out in a memorandum that some of the country's leading astronomers supported her application. They were Sir William Huggins, Sir Robert Ball, Professor Edward Stone of Oxford and Dr. Downing of the Nautical Almanac Office; these, she said, had "counselled my application and assure me that my qualifications abundantly entitle my application for the post". Her memorandum also dealt with the interesting question of her suitability, as a woman, for the post. She declared that she was very strong and was used to working with men in Greenwich where she and Ms. Russell "seem to work along quite naturally in the midst of them.... I do not think that my sex should be any real obstacle, though to some at first, the presenting of an unusual idea may prejudice against it". The Astronomer Royal wrote her a glowing testimonial,[xx] referring to her experience with the various instruments and her familiarity with computation. He described her as "a skilful photographer who has made herself expert in the manipulation of the new photographic equatorial" and who carried out her duties "with much intelligence and determination". Her experience would, he believed, "well qualify her for such a post as that of assistant at Dunsink Observatory".

Ms. Everett's expectations were in vain. The post was left temporarily unfilled (to avoid giving it to a woman, perhaps?). The telescope, intended for the *Carte du Ciel*, was never used for that purpose. Two years later, the assistant post was awarded to one of Alice's fellow computers at Greenwich, Charles Martin, a man 10 years her junior who had joined the Royal Observatory staff as a supernumerary computer at the age of 14. Now, aged only 19, he became assistant at Dunsink where he remained for the rest of his life. His work was confined to transit observations,[xxi] a field in which Ms. Everett was also amply qualified.

Alice also applied to the Royal Observatory Edinburgh and had a sympathetic reply from Professor Ralph Copeland who, however, had no vacancy on his small staff. "It is very unfortunate", he wrote, "that after so many years' work at Greenwich ways and means could not be found of retaining you on their permanent staff". It was a "Catch 22" situation. The temporary supernumerary computerships were the only appointments available to women; and being so, they were destined to remain temporary. Clearly, there were no prospects of advancement in Britain, notwithstanding the support of the highest echelons of the astronomical establishment.

As vacancies occurred, the lady computerships, originally so sought after, became difficult to fill. Everett and Russell were the last survivors. In October 1895 Alice Everett moved to the Astrophysical Observatory in Potsdam, Germany. Annie Russell was the last of the lady computers. On 31 October 1895 she resigned in accordance with the rules of the civil service, in order to marry Walter Maunder, and to continue her career in a different way. It would appear that the "lady computer" scheme, begun as an experiment, was destined to have a time limit and was not repeated. Thereafter enquirers were informed that "ladies are no longer employed at the Royal Observatory".

It was to be more than 40 years before before a woman astronomer would occupy a post of equal status with her male colleagues at the Royal Observatory.[xxii]

After five years at Greenwich, Alice Everett began work the Astrophysical Observatory in Potsdam in Germany on 1 October 1895. The Astrophysical Observatory at Potsdam was Europe's leading institution for astrophysical research set up in 1882 under the directorship of the pioneering stellar spectroscopist Hermann. C. Vogel, one of Agnes Clerke's many correspondents. She was the first woman to be employed at an observatory in Germany.[xxiii] Her post was that of scientific assistant working on the *Carte du Ciel*, the same work that she had been doing at Greenwich, and was a three year replacement for an astronomer who was away on military service. The rest of the staff of ten astronomers were men. Under the supervision of Julius Scheiner, another renowned spectroscopist, Alice Everett became involved in every stage of the work. In the year 1897, for example, she measured and reduced the positions of 22,000 stars on 72 photographic plates.[xxiv] She also made herself familiar with the advanced spectroscopic research being done at Potsdam, and indeed with every aspect of astronomy that she saw around her. She also became proficient in the German language. During this period she also published two papers, independently researched, in *Monthly Notices of the Royal Astronomical Society* on the orientations of the orbits of binary stars, and made some short contributions to the *Journal of the British Astronomical Association*, including a description of the Potsdam Observatory.[xxv]

On finishing at Potsdam, Alice Everett found employment as assistant for one year, 1898–1899, at the observatory of Vassar College, USA, the women's college where Mary Whitney, successor to Maria Mitchell, was professor of Astronomy. This small institution had only one member of staff besides Ms. Whitney, and Alice Everett no doubt was pleased to have the opportunity of working there, even temporarily. Her year was fruitful and resulted in two papers jointly written with Mary Whitney on observations of minor planets and a comet in the *Astrophysical Journal*.[xxvi]

On leaving Vassar in July 1899 Alice applied to James Keeler, Director of the famous Lick Observatory in California for a position on his staff. Keeler was most anxious to recruit her, and wrote to Mrs. Phebe Hearst, a benefactress of the observatory, asking if she would consider establishing a position for women at the Lick Observatory which he could offer to Ms. Everett, "a lady of distinction in astronomical science admirably qualified to aid us in a most important part of our work – the measurement of our photographs of star spectra in which were falling sadly behind". He suggested that the post should be at the grade of assistant astronomer with a salary of $1,200 per year, as "in the case of as capable a lady as Ms. Everett I should not like to offer less". Mrs. Hearst was not in a position to help at that time, but expressed regret at not being able "to secure the valuable service of so brilliant a scholar".[xxvii] It is not known if Alice tried other American observatories at this stage; if so, she had no greater luck, as in 1900, at the age of 35, she was back in England again, without employment. It was in fact the end of her career as an astronomer, and a sad loss to astronomy.

Alice did not, however, remain idle. She and her father, recently retired from his chair in Belfast, set about translating into English a monograph on Jena optical glass and its scientific applications, first published in German in 1900. The work, a very technical one which included references to the testing of the great refractor for Potsdam Observatory and other matters familiar to Alice from her German experience, appeared in 1902.[xxviii] Alice also made herself useful preparing the mathematics and physics entries for the Royal Society Catalogue of Scientific Papers.

From this time onwards Alice's principal scientific interest was optics. In December 1902 a short paper of hers was communicated to the Physical Society of London by her father, an active Fellow of that Society. It described experiments on zonal aberrations in lenses carried out privately in the Davy-Faraday Laboratory of the Royal Institution.[xxix] Brief though the paper was, it had the distinction of being the first by a woman to appear in the *Proceedings of the Physical Society*.

J.D. Everett died in 1904. Alice went with the British Astronomical Association to Cyprus to see the total eclipse of 1905, but without making serious observations. The next decade of her life is obscure. There was little opportunity for paid employment in astronomy in Britain: she could not, even if she had wished to, return to her old position at Greenwich, as the "lady computer" scheme had come to an end. At some stage she took an advanced course in practical Optics at the Technical College at South Kensington, mentioned in her Curriculum Vitae.[xxx] A paper on the spectre of the Brocken in 1913[xxxi] is the only glimpse we have of her from then until she emerges again in her second career as a physicist. Though not directly related to astronomy, her subsequent activities are worth recording as an example of a determined woman's perseverance in a competitive male-dominated scientific world.

The new opportunity came during World War 1 (1914–1918) when it became necessary to recruit women into formerly male occupations to replace men who were absent on active service. After a year employed in the optical laboratory of the well-known firm of Hilgers in London,[xxxii] Alice, now aged 52, joined the National Physical Laboratory as a junior assistant in the Physics Division in October 1917. The National Physical Laboratory was, and is, the Government institution responsible for maintaining scientific standards of all kinds. The staff numbers at the Laboratory rose in that year from 200 to over 500 on account of the war. The majority of scientists were engaged in testing of instruments and materials – of optical instruments alone almost 30,000 telescopes, sextants etc. were tested in the year 1917–1918 – but there were also research scientists, officially termed assistants, who were university graduates, with a Chief Assistant in charge of each division. Alice Everett had to start on the bottom rung of the scientific ladder on a salary scale of £175 to £235 per annum.[xxxiii] She was duly promoted after two years to the next higher rank, which she retained until her retirement.

Alice Everett was one of a team of 13 scientists in the Optical section of the Laboratory, of whom four out of five junior assistants were women during the war years. Their research was concerned with the design of optical instruments, photometry and spectrophotometry. Alice Everett worked mainly on theoretical problems, her special field being the calculation of aberrations in lens and mirror systems. Her success in these original researches may be judged from her publications in the scientific literature listed in the *Annual Reports of the National Physical Laboratory*.[xxxiv] She left the National Physical Laboratory on 15 May 1925 on reaching the statutory retiring age of 60.

Though now retired, her career was not over for this irrepressible woman. During the next two winters (1926–1928) she attended evening classes on practical wireless at Regent Street Polytechnic, London, and in Spring 1928 took and passed the College's examinations in "wireless, HF and AC measurements". In the session 1928–1929 she did research in the electrical engineering department of the City and Guilds College (a constituent of Imperial College).[xxxv]

There now followed a most fascinating period in Alice Everett's life, namely her association with the Baird Television Company and the Television Society.[xxxvi] On 26 January 1926 the Scottish engineer and pioneer inventor John Logie Baird (1888–1946), gave the first demonstration in Britain, and reputedly the first in the world, of a television image. The demonstration took place in a garret in Soho, London, to an invited audience of 40 guests, among them Sir William Bragg and members of the Royal Institution. At least two women scientists – unnamed – were present; Alice Everett may well have been one of them. In September 1927, the Television

Society (now the Royal Television Society) for the promotion of television research came into being. Alice Everett was one of its 325 foundation members, known as Fellows,[xxxvii] of whom only four or five were women. She remained a Fellow for the rest of her life.

The Baird Television Company was associated with the British Broadcasting Corporation (BBC) from 1929 until 1935. Among the receiving and transmitting apparatus developed by the Company during that period was a "mirror drum" for producing the necessary scanning light beam. Alice Everett suggested certain improvements to Baird's version of this device, and her ideas were evidently received with favour, as in 1933 a patent for the invention was applied for jointly by herself and the Company. For reasons to do with the business side, the new drum was never constructed.

However, Alice Everett continued to give service to the Television Society by translating foreign language publications for its library's index of current literature. In 1938 she was awarded a civil list (i.e., Government) pension of £100 a year in recognition of her own and her father's contributions to physical science (she had no pension from the National Physical Laboratory, having served for less than the required 10 years). She died in London on 29 July 1949, aged 84, leaving her library of scientific books to the Television Society and her canoe on the Thames to a friend.

Alice Everett's scientific life was a combination of high achievements and unmerited disappointments. Being a woman deprived her of an undergraduate scholarship at university in Belfast, of a full Cambridge degree, of permanent employment or promotion at Greenwich, of a career in her beloved astronomy in Dublin, even of a pension, since she was over 50 years of age before she gained an official professional post. Yet she never gave up or became disillusioned. She remained ever eager to acquire new skills and to take part in the new technology of the day – an enthusiast for science to the last.

The career of Alice's equally dedicated colleague, Annie Russell, now Mrs. Maunder, took a different path – in solar physics – as will be recounted in the next chapter.

The year 1890 – which saw the start of the Greenwich Lady Computer scheme and the foundation of the British Astronomical Association – marked an event in the history of astronomy which was particularly relevant to the role of women. It was the launch of the international project of the *Carte du Ciel* (*Chart of the Heavens)* and *Astrographic Catalogue* in which Alice Everett had a part. It was a hugely ambitious (indeed, as it transpired, over-ambitious) undertaking generating a vast amount of routine computations. Few jobs were immediately created thereby in Britain, but the scheme set a long lasting trend in the employment of women as inexpensive scientific assistants.

It was not the first instance in British astronomy of women employed in paid routine work. Mary Edwards, a century before was a computer for the Nautical Almanac (Chapter 4) and the unhappy Isis Pogson was a lowly assistant at Madras observatory in India for 25 years from 1873 (Chapter 10).

At Cambridge University Observatory a succession of women computers was employed between 1876 and 1904, most of them for short periods.[xxxviii] An exception, however, was Anne Walker who was appointed in 1879 at the age of 15 and served for 24 years under two successive famous professors, John Couch Adams and Robert Stawell Ball.[xxxix] She was the amanuensis of Andrew Graham, the Observatory's senior assistant, an almost legendary observer who devoted his 39 year career at Cambridge to the Observatory's star catalogue. Anne retired at the same time as Graham did, in 1903, apparently not happy to work under anyone else. The published history of the Cambridge Observatories gives no details of Ms. Walker's duties, but Roger Hutchins, who has researched the careers of the apparently faceless assistants in British observatories in the nineteenth century, has found that she did much more than routine computing: she also took part on occasion in Graham's intensive observational programme, making her the first recorded example in British astronomy (after Caroline Herschel) of a woman formally engaged in night-observing, earlier than the Greenwich lady computers.[xl] However, the first systematic entry of women to salaried employment in astronomy in Britain was the Astrographic Chart and Catalogue project, which may be said to mark the beginning of a new phase in their history.

The possibility of mapping the entire sky on photographs arose from the success of the dry gelatine process, already mentioned in connection with the spectroscopic work of the Hugginses. Previously, sensitive photographic emulsion took the form of a wet film spread on a glass plate before being exposed in the camera. It was awkward to use, and – its major disadvantage for astronomy – it dried out too quickly for the long exposures required for dim starlight to make an impression. Dry-plate photography and the improvement in the sensitivity of photographic emulsions "changed astronomical photography from a curious toy into a most important adjunct to an observatory".[xli]

The first to appreciate this was David Gill, who was struck by the amazingly large numbers of stars that showed up on a photograph taken at the Cape Observatory of the great bright comet of 1882. He made further experiments, and sent copies of the historic comet photograph to various colleagues pointing out its more far-reaching potential for astronomy. He urged that an international conference be called to see how best to implement the idea. The Director of the Paris Observatory, Admiral Amadée Mouchez, himself an experienced photographer, was deeply impressed, as were two members of his staff, the brothers Henry, who had experimented with photography when they found themselves frustrated by the overwhelming numbers of stars in the Milky Way. Mouchez called an international gathering of more than 50 distinguished astronomers from 16 nations to discuss the matter. The Paris Congress which met in Paris in 1887[xlii] was the first ever large-scale international astronomical venture. Eighteen of the world's best resourced observatories, capable of covering the entire sky, felt able to cooperate – from Europe, South America, Mexico, Australia, and South Africa. The North American observatories (the United States and Canada) did not take part; the former already had several major projects in hand. It was agreed that the sky should be shared between the various observatories, and that each should be responsible from start to finish for its allotted zone. The plan was to survey the entire sky photographically and to produce a catalogue of all stars down to magnitude 11. A standard instrument – a refractor of 13 in.

diameter and 11.25 ft focal length – was to be used by all observatories. In order to cover the planned range of magnitudes, each area of sky was photographed more than once, so that the number of individual images to be eventually measured was greatly multiplied.

Once the plates were obtained, the measurements and calculations were reduced to a routine operation. The plates, each 16 cm square, had a network of 5 mm squares imprinted on them, like the national grid on a geographical map. Positions of stars were measured under a microscope as x and y coordinates within these boxes. The plate was then turned though 180 degrees and remeasured to eliminate bias. These coordinates were then converted to celestial (angular) coordinates by formulae worked out by the astronomers from the positions of standard stars: the measurers had only to apply these without further query. The magnitudes of stars were estimated visually by comparing the sizes of their images with a standard set of spots, also supplied by the astronomers.

The *Carte du Ciel* project, which combined photography and mechanisation of measurement, led to vastly increased speed and precision. Teams of women worked on the *Carte du Ciel* in various observatories. The employment of women was a deliberate policy of Mouchez', but, as Charlotte Bigg who has made a special study of the origins of the project points out, it also brought with it "unsuspected social and material changes" in the role of women in astronomy. "[It] appears to have prefigured less an emancipation for them than the introduction of industrial methods of management into the observatory".[xliii] In Paris, the home of the project, a group of five women was set up under the direction of a *Directrice* Dorothea Klumpke (1861–1942), a pioneering and talented American woman astronomer educated at the Sorbonne who already held a responsible post on the staff of the Paris Observatory.[xliv] Britain has a tenuous claim on this distinguished woman. She took part with the British Astronomical Association in the eclipse expedition of 1896 in Norway where she met her future husband, Isaac Roberts, a wealthy amateur astronomer and renowned English astronomical photographer 30 years her senior whom she married in 1901. He died three years later, and she, now known as Klumpke-Roberts, retired to France where she devoted herself to publishing an atlas of Roberts' astronomical photographs. She spent her last years in her native California.

From Dorothea Klumpke's perspective as head of her Department, the scheme was a golden opportunity for women to enter the world of science. Similar teams were set up in the other participating French Observatories.[xlv] Nine out of the 16 "calculators" employed at the Cape Observatory were women; in Melbourne all seven were women, while in Perth, Australia, there were nine women.[xlvi] In Britain the two participating observatories, Greenwich and Oxford, did not employ women specifically for the work, though Alice Everett at Greenwich had considerable responsibility in the initiation of the project there, including the observational part.

At Oxford, one remarkable woman, Ethel Bellamy (1882–1960), was employed in measuring and reducing the plates obtained at that observatory. She served on the staff of the University observatory for almost 50 years, having begun in 1899 at the age of 17 as an assistant to her uncle F.A. Bellamy, senior of the observatory's two assistants. The *Carte du Ciel* work was carried out under the direction of the Professor of Astronomy Herbert H. Turner, formerly Chief Assistant at Greenwich, who played a key role in organising the British efforts. In his account of the Oxford contribution to the great project, Turner mentions "three or four measurers trained for the purpose", but they are not named, and there is no indication as to their sex.[xlvii] Ethel Bellamy was the chief participator and was a master of computational work of which she made a career.[xlviii] Her work for the *Carte du Ciel* was financed by government grants, as was a later computational task connected with the path of the minor planet Eros. She was appointed to a permanent post on the Oxford staff in 1912.

The participating British observatories, Greenwich and Oxford, completed their shares in the *Carte du Ciel* in record time, but other observatories found the enterprise more daunting than expected. Having completed every stage of its own celestial zone, the Oxford astronomers went to the aid of one of the original participants, the Vatican Observatory in Rome, which was held up at the computing stage of the task. It had followed "the example of the Paris Observatory where the use of women as measurers had proven valuable" by recruiting three nuns from a local convent to perform the measurements.[xlix] Turner "without seeking any recognition" supervised and aided in the subsequent reductions, the bulk of which was completed by Ms. Bellamy.[l] As a result, the Vatican Observatory could boast of being the fifth observatory after Greenwich, Oxford, Algiers and the Cape of Good Hope to fulfil its obligation, and Ms. Bellamy was honoured with a medal by the Pope in 1920.

The rest of Ethel Bellamy's working life was devoted to seismology; it became in fact a second specialist career that occupied her for a further quarter of a century. Turner had assumed responsibility for this relatively new branch of research at Oxford, and Edith Bellamy was to serve it well as editor of international Reports on earthquake records from stations all over the world, and other publications. Ms. Bellamy who did not have the advantage of a university education in her youth well deserved the award of an Honorary MA by the University of Oxford in 1936. She retired in 1947 and died aged 78 in 1960. She was an outstanding example of what can be achieved by devotion to science and a natural intellectual gift. "The contribution to science of workers characterised more by their love of the subject than by their preliminary training can scarcely be overestimated", wrote her obituarist of her.

Another observatory that found its resources inadequate to the burden of its 10 degree zone was Perth Observatory in Australia which was rescued by F. W. (later Sir Frank) Dyson, Astronomer Royal for Scotland, who undertook to have one third of the Perth plates brought to Edinburgh to be measured and reduced. Dyson was well experienced in this field: he had been formerly Chief Assistant at Greenwich and supervisor of the *Carte du Ciel* work there. The Perth zone was of course a southern hemisphere one; Edinburgh's share covered declinations −38 to −40 degrees inclusive. Dyson's staff at the Royal Observatory Edinburgh consisted in just three astronomers besides himself. For the new undertaking he decided to follow the French lead and to recruit and train young women who were referred to in the annual reports simply as 'measurers'.[li] No special qualifications were required of the women measurers. Their employment was intended "to economise on the time of the scientific staff" and "after some practice the ladies become very capable members of the staff".

In 1909, two sisters, the Misses Falconer, were appointed; one of them had a degree of MA from the University of Edinburgh. A year later they were joined by two more women, one of whom also had an MA degree. Two of the team were supported by a grant of £60 from a Government Grant for Scientific Investigations. Their salary of £30 a year (raised to £50 in 1919) was of the same order as that of the Greenwich lady computers in the 1890s. They worked in the forenoons only: Edinburgh's Blackford Hill was probably not deemed safe for ladies in the long northern hours of winter darkness.

The Falconer sisters were the longest serving of the Edinburgh measurers (Fig. 13.2). One left in 1915 to go to France to help the war effort, the other remained until 1918 and – not an uncommon occurrence in similar circumstances – married an astronomer at the observatory, Dr. Edwin Baker. Though the actual star measurements were completed by this time, reductions went on for several years. The final Perth catalogue was published – after many delays in funding – only in 1947. About 194,000 star images had been recorded, each measured twice over, on some hundreds of plates, a task that today would be accomplished in a matter of days with the observatory's modern automatic measuring machine.[lii] When their work for the catalogue came to an end, one or two of the measurers were kept on "for computing and general assistance". The Royal Observatory at Greenwich, following the demise of its own short-lived "lady computer" scheme, eventually resumed the employment of women in the same capacity.[liii]

Fig. 13.2 Edinburgh Measurers c. 1906. (Royal Observatory, Edinburgh)

Following the discovery of the asteroid Eros in 1898, the nearest by far to the Earth yet known, another cooperative programme among several observatories was organised by the international *Carte du Ciel* Committee to observe it during its opposition in the year 1900–1901 in which it was predicted to pass through perihelion (closest point the Sun), when it was only 0.27 Astronomical Unit (i.e., the unit of distance between the Earth and Sun), from Earth. The purpose was to determine its parallax, as had been done in earlier years with Mars and various asteroids. Cambridge was one of the participating observatories, and was the one where the final coordination of the large amount of data was done.[liv] Julia Bell (1879–1979) of Girton College worked as a computer on the reductions of the Eros data, for 5 years, from 1902–1907, under the direction of the astronomer Arthur Hicks.[lv] She had taken the mathematical Tripos examination in mathematics and was listed as a post-graduate student at Girton College. The Eros work, for which Julia Bell was paid from a Government grant, involved considerably more mathematical competence than the mechanical procedures of the *Carte du Ciel* routine. It was completed in 1907 and provided the best value of the Astronomical Unit until Eros' next close encounter in 1930–1931. (These results were eventually superseded by observations from space).

Julia Bell's skill as a mathematician brought her to the attention of Karl Pearson, professor of applied mathematics and director of the famous Galton Laboratory for National Eugenics at University College, London. Pearson was interested in the application of statistical methods to the study of heredity, and when the Eros results were finished, he invited Julia to join his laboratory and appointed her one of his assistants. After some years of statistical medical research she decided to acquire professional expertise in medicine, studying and qualifying as a doctor at the London School of Medicine for Women at the Royal Free Hospital. She had a highly distinguished career in medical genetics, gaining numerous distinctions and awards and continuing in research almost to the end of her life.[lvi] She lived until the age of 100.

Hers was an example of the potential of women in science if given the opportunity. As far as astronomy was concerned, there were no such opportunities. Female human computers became a feature of observatories throughout the world until they were replaced in the second half of the twentieth century by the electronic kind.

Notes

[i] W.H. McCrea. 1975. *The Royal Greenwich Observatory*, HM Stationary Office, London.

[ii] Daphne Bennett. 1990. *Emily Davies and the Liberation of Women*, chapter 24. London: Andre Deutch.

[iii] Mary R.S. Creese, 1991. *British Journal for the History of Science* **24**, 275; Mary R.S. Creese and Thomas M. Creese, 1994. *BJHS* **27**, 23. These surveys of the careers of women scientists pre-1906 illustrate the scarcity of salaried posts for women chemists and geologists many of whom conducted their research either unpaid or in low status posts. The first also lists women's colleges and universities admitting women before that date.

[iv] Christie to Miss Beale. 6 Feb 6 1890. RGO 7/140.

[v] M.T. Brück. 1995. Lady Computers in Greenwich in the 1890s. *Quarterly Journal of the Royal Astronomical Society* **36**, 83–95. The present account is based on this paper.

[vi] Her name was Harriet Maud Furniss. She proceeded to a successful career in teacher training and remained a member of the BAA.

[vii] File on Miss Rix's illness. RGO 7/140; Girton College Register.

[viii] Mary T. Brück. 1994. Alice Everett and Annie Maunder, torch-bearing women astronomers. *Irish Astronomical Journal* **21**, 280–291. The personal stories of these two women are taken from this paper.

[ix] W.H. McCrea. 1975. op. cit.

[x] Alice Everett 1893. *Journal of the British Astronomical Association* **3**, 24.

[xi] W.H.M. Christie. 1904. *Astrographic Catalogue 1900.0. Greenwich Section, Volume 1.* Edinburgh: HM Stationary Office.

[xii] H.H. Turner, May 8, 1893. Report to the Astronomer Royal on Miss Alice Everett. RGO archives, Cambridge University Library.

[xiii] Richard McKim. 1996. The dust storms of Mars. *Journal of the British Astronomial Association* **106**, 185–200.

[xiv] A. Everett, 1895. *Mon. Not. R. Ast. Soc.*, **55**, 440 (communicated by T. Lewis).

[xv] Girton College archives, communicated by Dr Kate Perry.

[xvi] Correspondence. RGO archives, Cambridge University Library.

[xvii] A.J. Meadows. 1975. *Greenwich Observatory*, volume 2, p 86. London: Taylor and Francis.

[xviii] H.W. Newton. 1958. *The Face of the Sun*. London: Penguin.

[xix] Meadows, op. cit, p 87.

[xx] Copy of the Astronomer Royal's testimonial, November 15, 1892. RGO 7/138.

[xxi] P.A. Wayman. 1987. *Dusink Observatory 1787–1985*. Dublin: Institute for Advanced Studies.

[xxii] She was Flora McBain (later Mrs Sadler) (1912–2000), a mathematician appointed to the staff of the Nautical Almanac (then a component of the Royal Observatory) in 1937. (George Wilkins. 2001. Obituary. *Astronomy and Geophysics* **42.4**).

[xxiii] Gudrun Wolfschmidt. 1995. Frauen in der Astronomie. *Astronomie und Raumfahrt* **32**, 11–13.

[xxiv] H.C. Vogel, 1898. Report of the Astrophysical Observatory, Potsdam, *Observatory* **21**, 279.

[xxv] A. Everett, 1895. *Mon. Not. R. Astr. Soc.*, **55**, 440; 1896. do., **56**, 462; 1903. *JBAA*, **13**, 74, 167, 275.

[xxvi] M.W. Whitney and A. Everett, 1900. *Astrophysical. J.* **20**, 47 and 76. A list of Alice Everett's papers is given by Mary Creese in *Ladies in the Laboratory?*

[xxvii] D.E. Osterbrock, *James Keeler, Pioneer American Astrophysicist*, Cambridge University Press, 1984. and Mary Lea Shane archives of the Lick Observatory.

[xxviii] H. Hoverstadt. 1902. *Jena Glass and its scientific and industrial Applications*, translated and edited by J.D. Everett M.A. FRS and Alice Everett M.A. London: Macmillan and Co Ltd.

[xxix] Katherine D. Watson, in Frank A. J. L. James. 2002 *The Common Purposes of Life. Science and Society at the Royal Institution of Great Britain*. Aldershot: Ashgate, Chapter 9. The paper was: A. Everett. 1903. *Proc. Physical Soc.* **18**, 376,1903. An account was also published in *JBAA* **13**, 74, 1903.

[xxx] Alice Everett's Curriculum Vitae, dated September 1928, Archives of Imperial College London. The date of the optics course is not stated. I am indebted to Mrs Barbara Strachan who generously gave me the benefit of her research into Alice Everett's career including her connection with Imperial College London. Mrs Strachan, a graduate of Imperial College, is the author of a very interesting account of the history of women in Imperial College (unpublished).

[xxxi] A. Everett, 1913. *Nature* **38**, 480–483.

[xxxii] Records of the Methodist College Belfast. Ronald Marshall. 1968. *Methodist College Belfast*. Belfast: Methodist College.

[xxxiii] National Physical Laboratory. *Supplementary Report of the NPL for the Year 1918*, 1920, p 14. London: HMSO.

[xxxiv] Seven papers in *Phil Mag*., also papers in *Trans. Optical Soc.* and *Proc. Physical. Soc.* between 1919 and 1929. The papers are listed in the annual reports of the National Physical Laboratory.

[xxxv] Register of Students at City and Guilds, Imperial College London 1934.

[xxxvi] M.T. Brück. 1994, *Irish Astronomical Journal* **21**, 281–90.

[xxxvii] First list of Fellows, 1928. Archives of the Royal Television Society, London.

[xxxviii] F.J.M. Stratton, 1949. The History of the Cambridge Observatories, *Annals of the Solar Physics Observatory Cambridge* Volume 1. There were no further female appointments at the Cambridge Observatories up to the date of Stratton's *History*.

[xxxix] Allan Chapman. 1998. *The Victorian Amateur Astronomer*. op. cit. p 157.

[xl] Roger Hutchins. Oxford DNB, 2004; entry on Andrew Graham.

[xli] J.L.E. Dreyer in Dreyer at al. 1923 (reprinted 1987). *History of the Royal Astronomical Society*, volume 1, p 213.

[xlii] Derek Jones. 2000. The Scientific Value of the Carte du Ciel. *Astronomy & Geophysics* **41**, 5.16–20; S. Debarat, J.A. Eddy, H.K. Richhorn and A.R. Upgren (eds.). 1988. *Mapping the Sky, past heritage and future direction*. Dortrecht: Kluwer.

[xliii] Charlotte Bigg. 2000. Photography and the labour history of astronomy. *Acta Historica Astronomiae* Volume 9. Thun and Frankfort am Main: Harri Deutsch.

[xliv] Dorothea Klumpke graduated in mathematics in 1886 and was awarded a doctorate in Astronomy in 1893. After her widowhood in France she returned to California where spent her last years. Kenneth Weitzenhoffer. 1996. The Triumph of Dorothea Klumpke. *Sky and Telescope* **72**, 109–110. J.H. Reynolds. 1994. Obituary Notice, *Monthly Notices of the Royal Astronomical Society* **104**. 92.

[xlv] Charlotte Bigg. 2000. op. cit.

[xlvi] P. Stroobant et al. 1907. *Les Observatoires Astronomiques et les Astronomes*. Brussels: Observatoire de Belgique.

[xlvii] H.H. Turner. 1904. *Astronomical Discovery*, p 133. London: Edward Arnold.

[xlviii] Peggy Aldrich Kidwell. 1984. Women Astronomers in Britain 1780–1930. *Isis* **75**, 534–46. Dr Kidwell points out that computing was so tedious that few women stayed more than five years.

[xlix] Sabino Maffeo S.J. 1991. *In the Service of Nine Popes: 100 years of the Vatican Observatory*. p 52. Vatican Observatory and Pontifical Academy of Sciences.

[l] H.H. Plaskett. 1961. Ethel F. Bellamy. *Quarterly Journal of the Royal Astronomical Society* **2**, 121–23.

[li] Mary T. Brück. 1998. Lady Computers. *Astronomy Now* **12**, 48–51.

[lii] Worldwide, the *Carte du Ciel* survey was finally brought to a close under the auspices of the International Astronomical Union only in 1970. Derek Jones, op. cit.

[liii] Peggy Aldrich Kidwell, op. cit. credits Dyson with this development in Greenwich as well as in Edinburgh. Dyson became Astronomer Royal in Greenwich in 1905. She also points out that Dyson favoured women in higher education: his wife and his daughters were all educated at Girton College Cambridge to which they were warmly attached.

[liv] Luisa Pigatto and Valeria Zanini. 2002. The 1900–1 opposition of 433 Eros. *Journal of Astronomical History and Heritage* **5**, 141–153.

[lv] Arthur R. Hicks. 1904. *Monthly Notices of the Royal Astronomical Society* **64**, 1904, p 701–727. This is the third and final of three papers.

[lvi] Greta Jones. *Oxford DNB*; *Girton College Register 1869–1946*. 1948. Cambridge: Girton College.

Chapter 14
Sunspots and Corona

Walter Maunder was a widower approaching 45 years of age when he married his young assistant Annie Russell, aged 27, whom we have met in the previous chapter (Fig. 14.1). The romance of close colleagues who worked so well together was perhaps inevitable. Walter, to judge from his photographs, was a handsome man, and, according to his friends, one of amiable personality, "a generous-minded gentleman, one who never willfully said an unkind word but often said kind ones."[i] They also shared a deep Christian faith. Both were children of evangelical ministers, Walter's father being a Wesleyan pastor. The couple were married in the Presbyterian Church, Greenwich on 28 December 1895.

Walter Maunder's first wife had died in 1888 leaving him with a family of five children whose ages at the time of his marriage to Annie ranged from 21 down to only seven. Walter and Annie had no children of their own. No doubt the rearing of the youngest stepchildren took up a great deal of Annie's time and energy; yet she was by no means cut off from astronomy. On the contrary: she carried on her editing of the *Journal of the British Astronomical Association* and soon found herself preparing to accompany her husband on an expedition to Norway to observe the total solar eclipse of 9 August 1896.

Maunder was an experienced eclipse observer, having taken part in an official British expedition in August 1886 to the tiny island of Carriacou in the West Indies where he obtained a series of photographs of the corona. The scientific authorities in Britain gave high priority to eclipse observations: a national Eclipse Committee, on which the Royal Observatory Greenwich was strongly involved (Maunder himself was a member), was responsible for financing and organising expeditions, and observers such as Maunder would be directed to travel and carry out observations as recommended by that Committee. For the 1896 eclipse, an independent expedition was organised by the British Astronomical Association for its amateur members and their friends, the first such venture by that society (Chapter 10). Annie's doctor brother accompanied them as a helper. The eclipse was unfortunately clouded out at their station at Nova Zembla, but the expedition was long remembered as an outstanding social success[ii] (and resulted in two romances – the Roberts', already mentioned, and the Eversheds', to be described later).

Another opportunity to observe a total solar eclipse arose on 22 January 1898, this time in India (Fig. 14.2). Annie offered some hints for observers in the pages of

M. Brück, *Women in Early British and Irish Astronomy*,
DOI: 10.1007/978-90-481-2473-2_14,

Fig. 14.1 Annie Maunder. (Armagh Observatory)

Fig. 14.2 Annie Maunder and eclipse observers in India, January 1898. (Institute of Astronomy, Cambridge)

the *Journal of the British Astronomical Association*,[iii] recommending suitable tents and equipment, doubtless advised by her sister who was a doctor in Poona. Maunder was not a designated member of the official British expedition on this occasion. He and Annie made their own private arrangements," hampered by no restrictions whatsoever, having received absolutely no financial help from any public body."[iv] With other observers from the British Astronomical Association, they had their site at Talni, a village in the Hyderabad District. They were favoured with excellent weather and reported their results in a charming illustrated book brought out by Walter Maunder on their return.[v]

The Maunders each had their separate apparatus of which Annie's produced the more original and far-reaching result. She had two cameras. One was of 2.5 in. (6 cm) aperture and focal length of 8 ft (2.4 m) mounted equatorially and driven by a clock. This compact and portable outfit, which Annie had seen in Norway on the 1896 clouded-out expedition, had been bequeathed to the Royal Astronomical Society by its owner, an amateur Fellow, and was lent to her by the Society. Annie mounted the second camera on the opposite side of the axis in place of the counterweight. (A telescope attached to one side of the axis of a mounting is normally balanced by a metal weight on the other side). This was a short-focus camera with a lens of only 1.5 in. (3.8 cm) diameter and 9 in. (23 cm) focal length which she had bought with a research grant from her Cambridge College. Its exquisite lens was made by the world-renowned London optician, T.R. Dallmeyer, well-known for his beautiful camera lenses including the telephoto lens which he invented in 1891.[vi]

Annie originally planned this special small camera for wide angle photography of the Milky Way. It had a field of view of over 40 degrees and would span the entire width of the belt. Photography of Milky Way panoramas was popular ever since the American astronomer Edward Emerson Barnard at Lick Observatory had tried a portrait lens on the sky in 1889, and expert astronomical photographers like Isaac Roberts were now producing stunning photographs of star clouds and nebulae. A serious challenge for all such astronomical optics was how to achieve sharp images at the outer parts of the field of view (the problem was eventually solved with the invention of the Schmidt telescope in 1930); the special high quality Dallmeyer lens which Annie acquired lost very little definition at the edges of the photograph. Long exposures were required, however, to show up images of faint stars, and a dark clear moonless sky was needed to prevent fogging of the photographic plate from background "light pollution". The eclipse expedition was an opportunity to try her Milky Way photography under clear Indian skies.

In the moonfree nights before the eclipse, Annie worked while other observers slept. The Zodiacal Light was visible for the first hour or so after sunset (the Maunders tried to photograph it without success) so Annie waited until the dead of night before beginning her solitary vigil. During the four-and-half hour exposures that were needed, the driving clock had to be wound every half-hour, and a constant eye kept on the guiding star by Annie, slow motion rods in hand to correct for the clock's imperfect movement.[vii] (These Indian photographs were not published but another excellent photograph of a Milky Way field taken by Annie in August 1900, perhaps in Algiers after the eclipse there, was reproduced in the Maunders' book *The Heavens and their Story*[viii]).

For the actual eclipse, Annie's plan was to take a series of photographs of the corona with each of the cameras, with exposures ranging from one to twenty seconds, which would cover a large range of brightness. She also varied the type of photographic emulsion. She herself operated her own new camera while a colleague from the British Astronomical Association operated the other. All was accomplished within the few minutes of totality. Some of the photographs she deemed "very successful", in particular those taken with her own small instrument which covered a large span of sky. One of these, centred well to one side of the eclipsed Sun, showed a streamer emanating from the corona which she described as "rod-like" extending to 10 million kilometres or 14 solar radii, by far the longest extension of the corona yet recorded, which was published in popular articles by Walter (under his own name).[ix] When the results of the various British expeditions were displayed at a Royal Society soirée Agnes Clerke, whose reports were highly respected, gave her verdict:" As regards the corona, Mrs. Maunder with her tiny lens has beaten all the big instruments". She reproduced this striking photograph (Fig. 14.3) in her well-known book *Problems in Astrophysics* (1903)[x] (as did Robert Ball in a new edition of his popular *Story of the Heavens* in 1901), but later writers appear to have overlooked this remarkable observation. The original drawing made from this historic photograph is preserved in the High Altitude Observatory in Boulder, Colorado. Only in recent years has its significance come to be appreciated by solar and geo-physicists (as is discussed later).

The Maunders took part in two further eclipse expeditions, in quick succession, both favoured with clear skies. In Algiers in May 1900, as members of a large

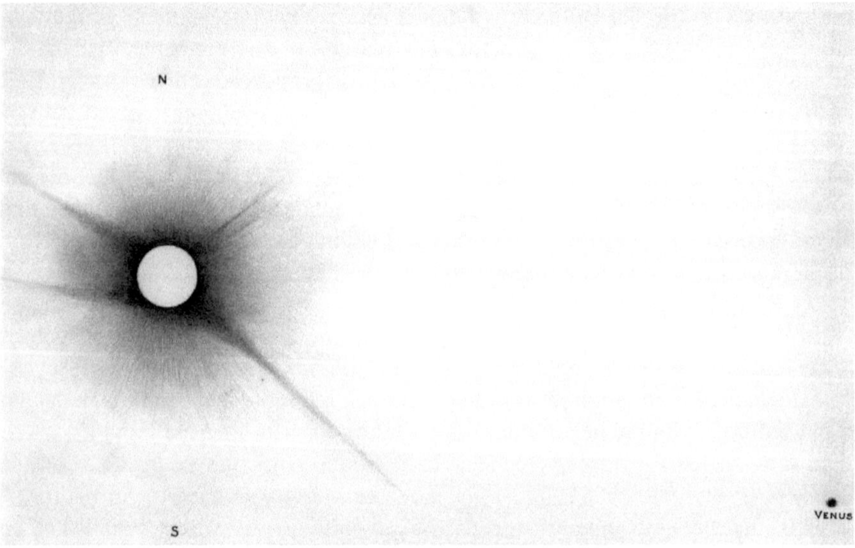

Fig. 14.3 Annie Maunder's "Longest Ray" 1898. (High Altitude Observatory, Boulder)

contingent from the British Astronomical Association, they were accompanied by Maunder's two daughters, then in their early twenties. Annie collaborated with her husband who used a small telescope of his own, a 4-in. refractor, to take photographs of the corona. A year later, in Mauritius in May 1901, Walter was sent as an official observer, equipped with Greenwich instruments to be used as instructed. He had hoped that Annie might be included as an official member of the Greenwich team; he raised the matter with the secretary of the Committee who seemed to be sympathetic, but, after much correspondence, the Maunders appear to have let the matter drop to avoid the embarrassment of a refusal.[xi] Annie nevertheless accompanied her husband at her own expense, and made her own entirely independent plans. They even observed from different locations. Annie's site was at the meteorological observatory at Pamplemousses, the same which supplied Greenwich with daily photographs of the Sun, where she had the help of the assistant astronomer there. She had brought her cherished small camera, intending to repeat the Indian programme – photographs of the southern Milky Way and of the outer parts of the corona – but encountered a serious drawback in the humid climate. Long exposures at night were impossible: the lens kept dewing up with moisture and had to be wiped every few minutes. She therefore got no Milky Way photographs, and only one short exposure of the corona. However, she had another instrument to fall back on – a camera lent by an amateur friend, G.J. Newbegin, with which she obtained a number of excellent photographs of the inner regions of the corona which had the characteristic round shape found in times of sunspot minimum, and showed a most delicate pattern of plumes and prominences.[xii] Two of the Mauritius corona photographs, an official one by Walter and an unofficial one by Annie, were included much later in the published record of the eclipse[xiii,xiv]; Annie's is distinctly superior in its remarkable detail.

The next suitable eclipse, in August 1905, passed over Canada, Spain and North Africa. The Maunders were invited by the Canadian Government to join its expedition to Labrador where it was planned that Annie, specifically, should employ a corona camera identical with that being used by the official British expedition in Egypt. The purpose was to verify that the corona is truly attached to the main body of the Sun by comparing its appearance at the two extremities of the eclipse path which were separated by a significant interval of time. A difference would be expected on account of the rotation of the Sun. Unfortunately the Canadian expedition was completely frustrated by clouds. It is worth noting that this was the only expedition on which Annie's expenses were paid – by the Canadian Government. Her excellent contributions to British astronomy in India and Mauritius were made at her own expense.

The principal contributions to solar physics associated with the name of Maunder are the analysis of the cyclical variation in sunspot latitudes, the discovery of a 27-day periodicity in terrestrial magnetic activity, both published in 1904,[xv] and the nature of the connection between large sunspots and strong magnetic storms. Annie was Walter's partner in these researches some of which were published under their joint names.

The 11-year cycle of sunspot numbers was established in the middle of the nineteenth century, and gave rise to a great interest in observing the Sun, including the establishment of the solar department in Greenwich under Maunder in 1873. It was found that sunspots never appeared on the Sun's equator or at its poles but were confined to a belt of about 30 degrees in latitude, between about 5 and 35 degrees north and south of the equator. The apparent movement of individual spots across the face of the Sun showed that the Sun rotated faster near the poles than at the equator, demonstrating that it was not a solid body. The spots themselves also tended to drift, a fact first noticed during the sunspot cycle of the 1860s by more than one solar observer. Accumulated records over many years also brought to light the fact that the favoured latitudes of spots changed steadily as the cycle advanced. Agnes Clerke summed up the general conclusions in 1902: "It may now be looked upon as established that the spot zones close in towards the equator with the advance of each cycle, their activity culminating as a rule at a mean latitude of about 16 degrees and expiring when it is reduced to 6 degrees. Before this happens, however, a completely new disturbance will have manifested itself some 35 degrees north and south of the equator, and will have begun to travel over the same course as its predecessor. Each series of spot is thus, to some extent, overlapped by the succeeding one".[xvi]

The analysis of the unrivalled Greenwich data, extending from 1877 to 1902 – more than two whole cycles – put these ideas on a firm basis. The Maunders plotted on graph paper the latitude of each individual spot against the date of observation – one calling out the numbers, the other plotting. They did this together at home, as Annie recalled many years later:" We made this diagram in a week of evenings, one dictating and the other ruling these little lines. We had to do it in a hurry because we wanted to get it before the [Royal Astronomical] Society at the same meeting as the other sunspot observers, whose views we knew to be heretical. As it turned out... the diagram wiped the other papers clean off the slate."[xvii] The striking diagram was more illuminating than a thousand words. The distinctive progress of the cycle was plain to see – the spots' upward drift in latitude as the cycle progresses, and the beginning of a fresh cycle before the old one vanishes. The same butterfly-shaped pattern repeats itself as records accumulate: today, a century later, a dozen "butterflies" have duly been born and passed on. It is one of the best known and delightful diagrams in astronomy.

The Butterfly Diagram (Fig. 14.4) was published in a paper by Walter Maunder in 1904, and is generally attributed to him alone.[xviii] It has taken 100 years for Annie's part in constructing it to become known through the efforts of Dr. Tom Bogdan of the High Altitude Observatory, Boulder, who had the original drawing restored and its history revealed in 2000, as is told in more detail later. Walter (it is not known if Annie was involved in this work also) followed up another remarkable fact noted by an earlier observer, the persistent and patient German astronomer Gustav Spörer. Spörer, an early observer of the sunspot latitude drift, noted that for a period of about 70 years previous to 1716, sunspots and aurorae had been extremely rare or, as he preferred to describe it, the sunspot cycle was submerged. This calm interlude is now known as the Maunder Minimum.

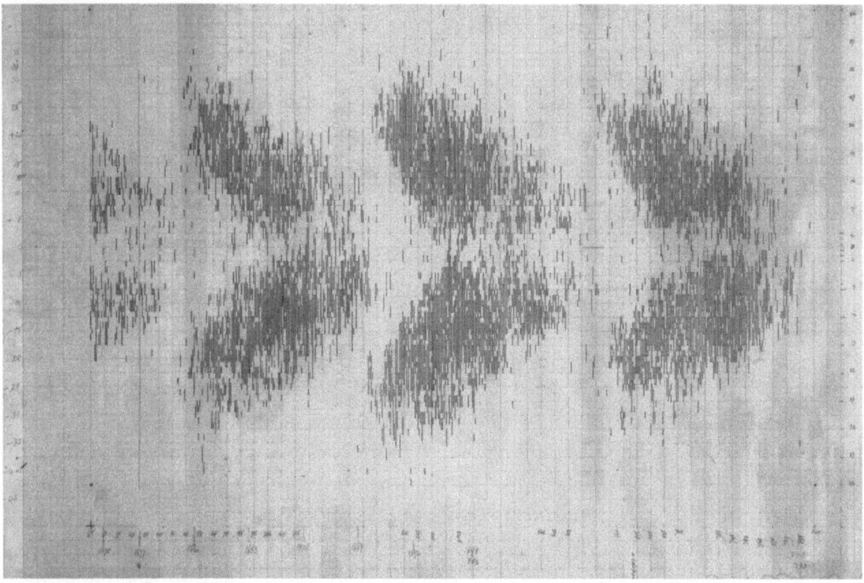

Fig. 14.4 The original Maunder Butterfly Diagram. (High Altitude Observatory, Boulder)

That there was a general correlation between sunspots, aurorae, and disturbances in the Earth's magnetism had long been known. In addition, extra strong magnetic effects (magnetic storms) caused by individual large sunspots were observed on a number of occasions in the nineteenth century. Annie had witnessed an especially dramatic example in February 1892 soon after she took up her post at Greenwich. A huge sunspot was seen close to the centre of the Sun's disk. The magnetic instrument at Greenwich and at magnetic stations worldwide vibrated wildly, and that same night there was a brilliant aurora seen over many parts of the globe. The cause of such an event – though not universally accepted until the end of the century – is that the Sun spews out from the sunspot or its neighbourhood a stream of electrically charged particles which, on passing close to the Earth several hours later, spiral down towards the Earth's magnetic poles. The moving charged particles themselves have their own magnetic influence, and thus disturb the magnetic compass needle which, in quiet circumstances, would point towards the north magnetic pole. Simultaneously, on ploughing through the atmosphere, they cause the air in the upper regions to glow, giving rise to the beautiful spectacle of the aurora in high northern and southern polar latitudes. (The detailed physics behind these phenomena is extremely complex, and is still the subject of much research). The Maunders could point out that large spots did not always produce magnetic storms, and that, conversely, magnetic storms sometimes occurred without a conspicuous spot. Here is Annie's explanation:

> "On a particular region of the sun some commotion occurs: sunspots, faculae, prominences are formed; above the disturbed area a great petal-like streamer of corona arises, its apex drawn out into a rod-like ray, which extends from the sun to distances which may be

> expressed in scores or hundreds or millions of miles. In these rays the particles, whatever their nature, are now not connected with the sun, though they once were; each still keeps the direction and motion which it had when it left the sun. The sun may go on spouting a coronal stream from the same region for months at a time, and in this region spots may break out and die, and again break out; for sunspots are but one symptom of the sun's activity, and perhaps not even the most important one. As the earth moves round the sun, which is himself turning on its axis, the same long stream may strike and pass it, may strike and pass again month after month, for many months at a time; or perhaps it may sometimes strike and sometimes miss."[xix]

This picture of the Sun's activity also explains another important discovery made by the Maunders – recurring magnetic storms in 27 day cycles. As seen from the Earth the Sun spins on its axis once in 27 days, bringing a disturbed area on the Sun round to face the Earth with that frequency, even though nothing special may be visible. The Maunders concluded that the culprit source was not on the Sun's obvious face but in the corona. They were entirely right. Their insights were years ahead of their time. It is only in recent decades that the Maunders' contribution to the understanding of the relations between solar and terrestrial phenomena are being reassessed and appreciated.[xx]

The book just referred to, *The Heavens and their Story*, was published nominally by the two Maunders in 1910, Annie's name appearing as the first author on the title page.[xxi] This delightful book, a model of what a science book for the general reader ought to be, displays Mrs. Maunder's own style and interests. In the space of a few hundred pages the essentials of astronomy and astrophysics are lucidly expounded. The highly readable text is spiced with literary quotations and imaginatively illustrated with photographs and drawings, including Annie's own photographs of the solar corona taken in India and at Mauritius. On the subject of the planets, the much discussed canals of Mars were given short shrift: both Maunders took their stand against the proposition that there were man-made water-carrying structures on our neighbouring planet. Annie Maunder elaborated on these deceptive Martian features in a separate article.[xxii]

In a Preface, Maunder states that the book "which stands in the joint names of my wife and myself, is almost wholly the work of my wife, as circumstances prevented me taking any further part in it soon after it was commenced". One may well ask why, that being the case, the husband could not have allowed the wife to publish entirely in her own name. It may have been for commercial reasons: his name was better known. But one suspects that it was part of the ethos of the day, when women knew their place as their husband's subordinates, though in the Maunders' case, the wife was academically the more qualified. Indeed, Annie's scientific career was in a way hampered by her husband, who had become somewhat disenchanted in his post at Greenwich. He had been in the same rank, with no prospects of further promotion, since his appointment 30 years earlier, while smart younger men, with university degrees, overtook him. He built an empire of his own among the amateur astronomers in the British Astronomical Association, in which Annie joined him. One result was that Annie's eclipse observations were published under Walter's name in popular journals rather than broadcast throughout the professional community in *Monthly Notices of the Royal Astronomical Society*. Not being a herself a Fellow of the Society, she was unable to submit her work there on her own account. Some professional scientists, such as Turner, the Chief Assistant at Greenwich, ignored the amateurs altogether.[xxiii] Yet, as Marilyn Ogilvie, a modern historian of astronomy and an admirer of Annie, points out, a career as an "obligatory amateur" was the only one open to her, and she made the most of it.[xxiv] She had, of course, the advantage over other amateurs that the Greenwich solar records were available to her. She published in an Appendix to the Greenwich Observations of 1907 an analysis of over 600 sunspot groups compiled from those records since their inception, which was not perhaps adequately valued at the time.

Walter Maunder retired in 1913 after 40 years of service. In 1914 the first World War broke out, and the Royal Observatory, in common with many institutions, found itself short of staff as its male workers left on active service. In 1916 Walter Maunder was recalled to his old post in charge of the Greenwich sunspot records. His wife joined him as a volunteer (i.e., without salary).[xxv] The Maunders kept up the solar work until 1920, well after the end of the war. These were surely happy years for them, as they resumed their old routines of 20 years earlier. The observatory was now under a different Astronomer Royal, Frank (later Sir Frank) Dyson, a man well known for his charm and friendly personality, with whom the Maunders felt at ease. They already knew him, as he had been Chief Assistant at Greenwich before taking up his previous appointment in Edinburgh. He was the same age as Annie, and his wife, a classicist, had been her student contemporary at Girton College, Cambridge.

In December 1914, when the ban on women Fellows of the Royal Astronomical Society was about to be lifted, Dyson invited her to put her name forward. Annie had not forgotten the earlier snub in 1892, when she and her two friends were proposed and rejected. She wrote thanking him, and continued [xxvi] :

> "But 22 years ago I was brought forward as a candidate for Fellowship in company with two other women, – one of whom, Elizabeth Brown, is now dead, – in just this way. The Society then declined to express a corporate opinion on the eligibility of women for Fellowship until some women had actually presented themselves for election. The election went against us. Of course, I recognise, and have done so all along, that the Society is perfectly within its right and competence to restrict its Fellowship as it deems fit; in other words, I feel that women cannot claim the election of women as a right, and I am sure that our rejection was not intended to convey the slightest discourtesy to us, either as women or as individuals. But just for these very reasons, I am bound to accept the decision then given, as far as I am concerned. If the Society should see fit, – seeing that I have already once applied, – to reverse its former decision, and grant me admission without further application, I should value it exceedingly.
>
> I have much appreciated the courtesy and honour which the Society has shown me, in inviting me to attend its meetings as a guest. This was a gift on the part of the Society, and I have gratefully accepted it for 22 years. If now, of its own initiative, it were to give me what it refused me before, that would be a gracious act. But if I make a second application for it, it would imply that I was dissatisfied with what has already been done for me; and further, if I were again blackballed – as might well be the case – I should feel that I had placed myself in quite a false position".

Behind the polite words lurks Annie's bitterness at her second-grade status among the professionals. Her suggestion was followed, and she was elected in 1916 on the reactivation of her husband's original nomination. The episode brought an unexpected revelation. Dyson's proposal to nominate her was supported by H.H. Turner, the same powerful figure who had been so dismissive, almost rude, when she applied for her "lady computer" post all those years ago, who made no secret of the fact that he disapproved of amateur science, and whom her husband mistrusted. She admitted to Dyson: "I appreciate the more highly the offer of his present support, because I had formed the impression, whether correct or not, that he had blackballed me on the former occasion".

A subject which Annie Maunder was to make very specially her own – the history of ancient cosmologies – first makes its appearance in her husband's book, *The Astronomy of the Bible*, published in 1908.[xxvii] Both the Maunders were deeply versed in the writings of the sacred scriptures, and in 1923 Annie was awarded a prize by Girton College Cambridge for an essay on a biblical topic.[xxviii] The book discusses and explains numerous astronomical allusions in the Old Testament, and is reported to have earned the commendation of the Rabbi, the Pope and the Archbishops of Canterbury and York.[xxix]

Among astronomical matters examined by the Maunders was the origin of the constellations. From a study of the uncharted regions of the southern sky on early celestial charts the position of the south celestial pole could be deduced as could the geographical latitude of the mapmakers' location. Taking precession into account, the date of the first mapping of the constellations was calculated to have been made about 2700 BC from a place of latitude about 37 degrees north. This was the first serious attempt to work out this interesting date.[xxx]

In 1910 the Maunders wrote a joint paper on the origin of the planetary symbols, an interest triggered by their study of old astronomical texts in the library of the Royal Observatory at Greenwich[xxxi] Virtually all of Annie Maunder's later work was dedicated to researches of a similar kind. She made a detailed study, using translations, of astronomical allusions in Iranian and Indian sacred texts in a lecture delivered to the Victoria Institute in London in 1916.[xxxii] The Victoria Institute, which still flourishes, was founded in 1867 to examine the relation between science and religion from the Christian point of view; with a membership of 600 men and women, the Institute had many distinguished scientists in its ranks including the astronomers Sir David Gill and Sir Frank Dyson. E.W. Maunder was the secretary. Being a reluctant public speaker Mrs. Maunder's lecture was delivered by her husband while Sir Frank Dyson, the Astronomer Royal, took the chair. Her paper displays a vast amount of research in what was then a little-studied field. She also looked into cometary records in the writings of the seventeenth century English traveller and explorer Peter Mundy[xxxiii] and wrote an astronomical appendix to a new printing of his travels. As late as 1936 she returned to the subject of the origin of the constellations in a paper in the *Observatory* where she revised her estimated date of their establishment to 2900 ± 100 BC,[xxxiv] a date which agrees well with modern opinion. The large volume of her historical work put Mrs. Maunder ahead of her time in the now popular field of ancient and archaic astronomy. In her lifetime she was looked upon as an expert on this subject; like her eclipse observations, her efforts may not have been sufficiently appreciated by later generations.

Maunder died in 1928 after a long illness. Annie, though bereft of her partner of 33 years, continued to devote herself to the work of the British Astronomical Association and to her historical researches. She also wrote numerous popular articles on astronomy.[xxxv] The years of the Second World War were stressful, but she and Mrs. Evershed (next chapter) responded to a request to write a history of the British Astronomical Association which was now half a century in existence. It was "a long and difficult matter",[xxxvi] but was accomplished. Annie, aged 74, read her paper on 30 September 1942.[xxxvii] She recollected their first meeting and the spirit of the Association which she herself did so much to sustain: "Men and women astronomers came in on equal terms; so also with the rich and the poor; those who worked with their hands and those with their heads; and all pooled their varying knowledge for the public good."

An event that gave Annie great pleasure in later years was her association with the solar High Altitude Observatory in Colorado to which she donated the original Butterfly diagram and a reproduction of her "Longest Coronal Ray". The chain of events that led to this began in 1937 when she received a request from an amateur astronomer in USA for permission to copy a certain illustration from Maunder's *Astronomy of the Bible*. She gladly agreed – incidentally discovering that the book in her correspondent's possession was a pirated edition of which she knew nothing.[xxxviii] An exchange of letters followed, and in 1941, when war conditions in London were grim, Annie decided to send the original Butterfly Diagram for safe keeping to her American friend. He in turn, with Annie's permission, gave it on long-term loan to the High Altitude Observatory in Climax, Colorado, in the care of Walter Orr Roberts, then a young solar physicist.[xxxix] "I am especially glad", Annie wrote to Roberts in 1943, "to get the "Butterfly" into a safe and kind home . . . . I am tempted to ask you to take in another "evacuee". It is the enlarged drawing made by the late Mr. Wesley of my "Longest Coronal Ray", which I photographed during the eclipse of 1898 January 22 in India".[xl] Both these historic drawings are now lovingly preserved in the High Altitude Observatory.

In the same year, Annie had the satisfaction of listening to Sydney Chapman, President of the Royal Astronomical Society, who chose the Butterfly Diagram as the subject of his 1943 Presidential address (the actual title was "Magnetism in the Earth's atmosphere"). Chapman was a key thinker in the interpretation of magnetic storms in terms of particles emanating from the Sun, a problem for which the Maunder diagrams were highly relevant. His work led in 1955 – though Annie did not live to hear it – to Eugene Parker's solution of a hot corona and the solar wind.

The Maunders' conclusions regarding the influence of the Sun's activity on earthly disturbances were revisited and confirmed by the American solar physicist John Eddy who in 1975 renamed the prolonged dearth of sunspots in the eighteenth century the "Maunder Minimum" in his honour (or in honour of the couple). The expression has entered the language. No doubt this naming was done with the best intentions; but one wonders by what right a scientist may deprive someone of an earlier age of his priority, in this case Gustav Spörer, acclaimed by Maunder himself as the discoverer of the historic phenomenon now bearing the latter's name.[xli] (Spörer has been given the consolation prize of an earlier, shorter, sunspot lull.)

Annie survived her husband by almost 20 years, and at the end of her long and active life had a home close to her brother's in Surrey. She died on 15 September 1947 in her 18th year. The Maunders are individually commemorated by craters on the Moon – a unique case for a couple. Annie herself would probably choose to be remembered by her most cherished pieces of work – the Butterfly Diagram and the Longest Coronal Ray.

Notes

[i] H.P. Hollis. Obituary, *Journal of the British Astronomical Association*, May 1928, 229–233.
[ii] R.A. Marriott. 1991. *JBAA* **101**, 162.
[iii] A.S.D. Russell, 1895. *JBAA* **3**, 29.
[iv] E.W. Maunder 1898. *Observatory* **21**, 279.
[v] E.W. Maunder 1898. *The Indian Eclipse 1898*, Hazell, Watson and Viney, London.
[vi] Henry C. King. 1979. *The history of the telescope*. New York: Dover Publications (reprint), p 273.
[vii] A.S.D. Maunder. 1898. Some photographs of the Milky Way and of the solar corona of January 22, 1898. *The Girton Review* **50**, 1–6. August 1898.
[viii] Annie S.D. Maunder and E. Walter Maunder. 1910. *The Heavens and their Story*. London: Charles H. Kelly.
[ix] E.W. Maunder, 1899. *Observatory* **22**, 315. The photograph was also published the magazine *Knowledge*, and in the Maunders' book.
[x] Agnes M. Clerke. 1903. *Problems in Astrophysics*, Adam and Charles Black, London.
[xi] M.B. Ogilvie, 2000. Obligatory Amateurs. *British Journal for the History of Science* **33**, 67–84.
[xii] Reproduced in the Maunders' book, op. cit.
[xiii] F.W. Dyson and W. Christie. *Memoirs RAS* **64**,1925–29; reprinted in W.W.-H. Soon and S.H. Yaskell, 2003. *The Maunder Minimum and the Variable Sun-Earth Connection*. p 157. Singapore: World Scientific.
[xiv] M.T. Brück. 1994. Alice Everett and Annie Russell Maunder, torch bearing women astronomers. *Irish Astronomical Journal* **21**, 281–90.
[xv] E.W. and A.S.D. Maunder. 1905. *Monthly Notices of the Royal Astronomical Society* **65**, 2–34 and 813.
[xvi] Agnes Clerke. 2003. *History of Astronomy* (reprint edition), p 148.
[xvii] Annie Maunder. Letter to a friend to whom she gave the diagram in 1940. Archives of the High Altitude Observatory, Boulder, Colorado. Courtesy Dr T. Bogdon.
[xviii] For example in H.W. Newton. 1958. *The Face of the Sun*, 51–52. Penguin Books. Newton was a successor of Maunder as head of the Solar Department at Greenwich.
[xix] Annie S.D. Maunder and E. Walter Maunder. 1910. *The Heavens and their Story*. op. cit.
[xx] Willie Wei-Hock Soon and Steven Yaskell. 2004. *The Maunder Minimum and the Variable Sun-Earth* Connection. Singapore: World Scientific.
[xxi] Annie S.D. Maunder and E. Walter Maunder. 1910. op. cit.
[xxii] A.S.D. Maunder, 1907. *Knowledge* **4**, 169.
[xxiii] M.T. Brück. 2002. *Agnes Mary Clerke and the rise of Astrophysics*, Chapter 13. Cambridge University Press.
[xxiv] Marilyn Bailey Ogilvie. 2000 Obligatory amateurs: Annie Maunder (1868–1947) and British women astronomers at the dawn of professional astronomy. *British Journal of the History of Science* **33**, 67–84.
[xxv] Royal Observatory Greenwich, 1916 et seq. Annual Reports to the Board of Visitors.
[xxvi] Mrs A.S.D. Maunder to Dyson. 1914 December 4. RGO 8/150.

[xxvii] E. Walter Maunder FRAS, 1908 (Fourth edition 1922). *The Astronomy of the Bible*. London: The Epworth Press.

[xxviii] Girton College Report 1923. The prize, founded in 1889 by Mrs Y.J. Gibson, was for the best essay on a subject connected with the Greek testament.

[xxix] H.P. Hollis, 1928. Obituary of E.W. Maunder. *Observatory* **38**, 229.

[xxx] Alex A. Gurshtein. 1993. *Vistas in Astronomy* **36**, 171.

[xxxi] E.W. Maunder and A.S.D. Maunder. 1920. *JBAA* **30**, 219.

[xxxii] A.S.D. Maunder. 1915. On Astronomical Allusions in Sacred Books of the East, *Journal of the Transactions of the Victoria Institute* **47**, 181–232. A popular account appeared in *Observatory* **39**, 48.

[xxxiii] A.S.D. Maunder. 1934. *Observatory* **57**, 279. The work was done for the Hakluyt Society, a society dedicated to publishing rare or unpublished voyages and travels, named after a famous 16th century historian of discoveries, Richard Hakluyt.

[xxxiv] A.S.D. Maunder. 1936. *Observatory* **59**, 367.

[xxxv] Anthony Kinder who is writing a biography of Walter Maunder has a collection of magazine articles.

[xxxvi] A.S.D. Maunder to Roberts. 1943 July 17. Archives of the University of Colorado at Boulder Libraries. Annie's wartime worry, as she mentioned in a letter to Roberts, was on behalf of her cousin who was on active service. The cousin was Field Marshall Sir John Dill, sent by Churchill in 1941 to Washington as his representative on the American Joint Chiefs of Staff. He died before the end of the war and was accorded burial by President Roosevelt in Arlington Cemetery, "the only foreigner to be so honoured." (David Stafford. *Roosevelt and Churchill, men of secrets. 1999*. p 128. London: Little, Brown and Co.).

[xxxvii] A.S.D. Maunder. 1943. Reminiscences of the British Astronomical Association, *Journal of the British Astronomical Association* **42**, 268.

[xxxviii] Maunder Family papers. I thank Anthony Kinder for kindly showing me these letters.

[xxxix] W.O. Roberts correspondence, Archives of the University of Colorado at Boulder Libraries. Courtesy T.G. Bogdon.

[xl] A.S.D. Maunder to Mr and Mrs Roberts. 1942 November 19. Archives of the University of Colorado at Boulder Libraries. Courtesy T.G. Bogden.

[xli] Agnes Clerke, following Maunder, states that Spörer's researches showed that the sunspot cycle "was in abeyance during some seventy years previously to 1716, during which period sun-spots remained persistently scarce, and auroral displays were feeble and infrequent even in high northern latitudes. An unaccountable suspension in solar activity is, in fact, indicated". (*History* op. cit., p 148).

Chapter 15
Mountain Paradise

Mary Evershed (née Acworth Orr) (1867–1949), Annie Maunder's collaborator as historian of the British Astronomical Association in the last phases of their lives, had a career which in many ways ran parallel to her friend's. They were close contemporaries in age; both married solar physicists, and themselves contributed to the same field as their husbands. For a major part of their active lives, however, they were separated by two continents. Mary Evershed spent 17 years in India, where her husband John Evershed was the director of the Observatory at Kodaikanal. She also established an independent reputation under her maiden name M.A. Orr, as an expert on the astronomy of the poet Dante (Fig. 15.1).[i]

Mary Acworth Orr was born at Plymouth Hoe on 1 January 1867, the fifth child and third daughter of Andrew Orr, an officer in the Royal Artillery, and his wife Lucy née Acworth. The father died when Mary was only 3 years old and the family was raised in the home of the maternal grandfather, a clergyman, in a large country vicarage near Bath. Mary and her youngest sister Lucy, the close companion of her childhood, were educated entirely at home by a governess, an exceptional woman who was in charge of them for 10 years. When Mary was 20 the sisters travelled abroad to study languages and the arts in Germany and in Italy. They spent the years 1888–1890 in Florence where they studied the writings of Dante and where Mary, who from an early age had an interest in astronomy, became fascinated by the astronomical references in Dante's poetry, a subject which she was to make her own.

Following those years of education in Europe the Orr family – mother and daughters – lived from 1890 to 1895 in Australia, in the vicinity of Sydney in New South Wales where the eldest daughter was already living. There the sisters continued their studies of Dante, and Mary resumed her active interest in astronomy. She had the encouragement of John Tebbutt (1834–1916), Australia's leading astronomer, well-known as the discoverer of the great comet of 1861 which bears his name, who had a well-equipped private observatory at Windsor.[ii]

Endeavouring to familiarise herself with the southern constellations, Mary found that there existed no simple maps of the southern sky. She prepared *An Easy Guide to the Southern Stars*, a booklet similar in format to Gall's guide of the northern constellations,[iii] designed to fit handily in the pocket, containing maps of recognisable naked-eye star groups observable from latitude 34°S. Her motive, she wrote to Tebbutt,[iv] was "to get people (children and adults) on the track of observing

DOI: 10.1007/978-90-481-2473-2_15,

Fig. 15.1 John and Mary Evershed in India. (Royal Astronomical Society)

for themselves the movements of the heavenly bodies, to help them recognise and admire the stars and constellations on the Australian skies, and then to interest them generally in astronomy". The book, with a Foreword by John Tebbutt recommending the work of "the enterprising authoress", was published in Britain by Gall and Inglis (publishers of James Gall's book and of the famous Norton's *Star Atlas*, beloved of generations of amateur watchers of the sky) after her return to England.[v] A second edition of this useful book was printed in 1911.

In 1891 Mary Orr joined the California-based Astronomical Society of the Pacific[vi] recently founded by Edward Holden, Director of Lick Observatory, of which Tebbutt was also a member. The publications of this fast growing society kept its members – who included women – informed of progress in astronomy worldwide but especially in the United States. The British Astronomical Association, set up in London in 1890 and also open to women, extended its activities overseas with the founding of the New South Wales branch in 1895, with Tebbutt as President. In that same year the Orr family returned to Britain, to a home in Kent, when Mary was able to join the parent Association in London.

The British Astronomical Association offered Mary Orr a very congenial milieu. Its membership included intellectual women who were debarred from the all-male Royal Astronomical Society, among them Agnes Clerke (Chapter 12) and Annie Maunder (Chapter 14). These women were highly competent in astronomy and deeply interested in its historical as well as in its scientific aspects. Agnes Clerke, 25 years older than Mary Orr, was a recognised authority on the history of astronomy and the foremost commentator on astronomy in the English-speaking world. The second edition of her famous *History of Astronomy during the Nineteenth Century* and her treatise *The System of the Stars* had not long been published (1893). Agnes Clerke welcomed and encouraged the newcomer. Many years later, on the centenary of Clerke's birth in 1942, Mrs. Evershed paid tribute to their 20 years' friendship: "As one who had the privilege of knowing her personally since 1895 I can testify that her influence was inspiring to beginners of the science she so much loved."[vii] Annie Maunder, formerly Annie Russell, at 28 the same age as Mary Orr, was the newly married wife of the astronomer Edward Walter Maunder.

The British Astronomical Association became a centre of Mary Orr's life and activity. An early exciting event was the Association's expedition to observe the total solar eclipse in Norway of August 1896, a major venture for that young society, as already mentioned (Chapter 10). The eclipse was visible in the extreme northern counties of Finnmark and Lapland. The central line ran through Vadsö, at latitude 70° on the east coast of Finnmark, chosen as the site of the BAA expedition[viii] and of groups from other countries. Navigation companies organised cruises from Great Britain: the "splendid pleasure yacht" Norse King, carried 165 passengers, including the official Government-funded parties and the BAA group of 58 amateurs astronomers and their friends, led by Maunder. Many of the participants had brought their own instruments; some joined forces, or (like Mary Orr) were ready to act as assistants, timekeepers or notetakers to others. Unfortunately the crucial morning of August 9 was cloudy. Though scientifically disappointing, this, the BAA's first eclipse expedition, was a practice run for later successful expeditions. It was also a memorable social event, as it was on this expedition that Mary Orr first met her future husband John Evershed whom she married 10 years later.

The next expedition organised by the BAA was to Algiers, to observe the total eclipse of 1900 May 28.[ix] On this occasion Mary had her own instrument, a 7.6 cm (3-in.) refractor, and was in a contingent of four women who had their observation post on the roof of the British Consul's villa. "Ms. Orr (Mary's sister), high up among the chimneys, watch in hand, gave the time"; Mary reported on the appearance of the landscape during totality, and observed Baily's Beads, the striking row of bright spots in the last cusp of the Sun before it is completely obscured by the Moon,[x] but her observations were not published. Evershed was also in Algeria observing the eclipse more seriously with his spectroscopic apparatus, at a site outside the city.

In 1900, Mary Orr settled in Frimley, Surrey, where she set up a little observatory equipped with her 7.6 cm (3 in.) refractor with which to observe the Moon and variable stars, as recorded in the list of the world's observatories and astronomers compiled in 1902.[xi] She also, as she wrote to Tebbutt, had taken up the study of mathematics and the history of astronomy. Variable stars were the subject of her first scientific paper published at the end of 1904 which shows her as already a practised observer.[xii] In it she gave a survey of observations and characteristics of long-period variable stars especially those showing irregular variations. She clearly had ambitious to make a career in this field. In early 1906 Edmund T. Whittaker FRS, mathematician at Cambridge and Secretary of the Royal Astronomical Society, whose wife was a cousin of hers, was appointed Professor of Astronomy at Trinity College Dublin and Director of Dunsink Observatory near Dublin. Whittaker took up his post in June of that year, and arrangements were being made for Ms. Orr to work with him at Dunsink. Being about the same age as her cousin she would presumably have lived with the family in their large observatory residence and worked as a volunteer. On arriving at Dunsink, Whittaker commenced a programme of observation of red stars, many of which were variable, and began a systematic search for variable stars using photographs taken with the 15-in. Roberts reflector, a programme which would have well suited her. One wonders how her career would have developed had the Dunsink scheme come to fruition; but by June she was already engaged to be married.

Mary Orr and John Evershed were married on 4 September 1906 at St Mary's Parish Church, Cloughton near Scarborough in Yorkshire. She was 39; her husband was 42. John Evershed, who until then had been employed as an industrial chemist, had been interested in astronomy from childhood. In his recollections, written when he was 90 years old, he described his excitement as a small boy at being shown a projection of a partial eclipse of the spotted Sun.[xiii] Later, having read Sir Norman Lockyer's articles on solar spectroscopy in *Nature* he constructed a spectroscope attached to small telescope with which to view prominences at the Sun's limb. In 1891 he read about George Ellery Hale's invention of the spectroheliograph, and set out to construct one for himself. Hale, then only 22 years of age, paid his first visit to Britain that same year when he addressed the British Association and was received with acclaim by leading astronomers including Sir William Huggins, the Association's President. This was the occasion when Hale met and became a lifelong friend of Agnes Clerke. On Hale's next visit to London two years later Evershed was introduced to him. It was the beginning of another lasting and fruitful friendship. Hale, who was shown Evershed's equipment, declared that Evershed was the only one besides himself to have built a true spectroheliograph by 1893.

The spectroheliograph was an ingenious instrument (Fig. 15.2) that was to give service in abundance to solar observation for many decades. An image of the Sun was scanned by the narrow slit of a spectroscope which produced a spectrum, from which any desired wavelength could be isolated. The isolated wavelength chosen in practice was one which a prominence emitted naturally and copiously – either the deep violet line of ionised calcium or the red line due to hydrogen. A second slit allowed this light to pass and to fall on a photographic plate behind. As the image of the Sun was scanned, the photographic plate recorded it in that one particular wavelength. The instrument thus acted as a narrow band colour filter. The satisfying result was an image of the Sun that showed the prominences surrounding it, and also the mottled Sun's face which was in fact its lower atmosphere or chromosphere. Later,

Fig. 15.2 Mary Evershed observing the Sun. (Royal Astronomical Society)

the instrument was adapted for visual observation, so that an observer could see the Sun in the light of the red hydrogen alpha spectrum line, effectively monitoring the behaviour of the ever changing hot hydrogen gas in the Sun's chromosphere. This version of Hale's instrument was called a spectrohelioscope.

Evershed, whose employers allowed him leave of absence for the purpose, took part in the British Astronomical Association's total solar eclipse expeditions to India in 1898 and to Algeria in 1900. Using his own improved solar spectrograph, he obtained ultra-violet spectra of prominences which went deeper into the ultra-violet than had ever been achieved before. The results aroused the interest of Sir William Huggins, and it was through Huggins' influence that he was offered the post of assistant astronomer at the Kodaikanal Observatory, India, in 1906, when it became vacant. He took up duties there in 1907, accompanied by his bride. Mary was thus by her husband's side throughout his entire professional career.

The Eversheds travelled to India via the United States and Japan, visiting Harvard, Princeton, Washington and Yerkes Observatories on the way. Their most important call was at Mount Wilson where Hale had for 2 years been busy erecting the famous observatory, and where the solar installation with his new spectroheliograph had not long been in operation. The Eversheds spent some weeks at Mount Wilson, "studying the instruments and methods and the work being carried out under [Hale's] inspiring direction". The Mount Wilson work was imitated successfully by the Eversheds in Kodaikanal, and became, in fact, the subject of Mrs. Evershed's own research.

Evershed's arrival at Kodaikanal heralded the observatory's "golden age"[xiv] which was shared by his wife. For her, the 16 years spent in India were the happiest of her life. During the first few years she was able to bring to fruition her study of astronomy as found in Dante's writings.[xv] At the same time she had the opportunity of learning and practising a new branch of astronomy – solar physics.

Astronomy in India had a long history. During the early nineteenth century the principal activities there were classical meridian astronomy at the Madras Observatory, and the Trigonometrical Survey of India. In the late 1870s the desirability of making solar observations from India came up for discussion in various quarters. The upshot was a decision in 1893 to establish a solar physics observatory in Kodaikanal in the Palani Hills of South India. Astronomical activity at the Madras Observatory was to be transferred to Kodaikanal. The new observatory, under the control of the Central Government, was opened in 1899. It had a variety of instruments including a photoheliograph for daily photography of the Sun and a grating spectroscope, to which were added in 1903 a spectroheliograph, a solar telescope and a siderostat. The Director of the Observatory, who masterminded the actual move, was the successor of the unhappy Norman Pogson, Charles Michie Smith, a physics graduate of the University of Edinburgh. Evershed was Michie Smith's assistant until 1911 when he succeeded him as Director of the Madras and Kodaikanal Observatories. Their relations seem to have been cordial. Michie Smith, who was not married, was very popular and very happy in India. He loved Kodaikanal so much that when it came to his retirement he stayed on with his sister Lucy who, having abandoned any hope of marriage, joined him there. They both lived until 1922 and are buried in Kodaikanal.[xvi]

The Kodaikanal instrumentation was enhanced by Evershed's own spectroscopic equipment which he brought with him. Adding to the existing spectroheliograph, Evershed built a number of spectroscopes, and continued his work, begun in England, of sunspot spectroscopy. It was in this latter field that he discovered the outward radial motions of material in the spots known as the "Evershed Effect" (1910). A year later he became Director of the Observatory, and in 1915 was elected a Fellow of the Royal Society. Some years later he built a second spectroheliograph for recording the Sun in the light of the red hydrogen alpha line thereafter taking daily spectroheliograms of the Sun in both calcium and hydrogen light. He was awarded the Gold Medal of the Royal Astronomical Society for this work in 1918.

Evershed's staff at Kodaikanal consisted of a Chief Assistant, Dr. T. Royds, and four Indian astronomers – as well as his wife, who assisted in an unofficial capacity, though her name rarely appears in the formal observatory Reports. Indeed, Mrs. Evershed seems to have taken part in all her husband's activities, such as observing and photographing Halley's Comet with its tail extending 100 degrees upwards "in the beautiful early dawns" of May 1910.[xvii] She took a special interest in solar prominences; Evershed refers to this in his recollections. "My late wife was greatly interested in the prominences I recorded at Kenley [his home in England] and in those we studied together with the fine equipment at Kodaikanal. She was much occupied at this time in writing her important work on *Dante and the Early Astronomers*. She nevertheless was able to find time to study in detail the prominences recorded in the years 1890–1914."[xviii] In 1913 she published a substantial paper in *Monthly Notices of the Royal Astronomical Society*[xix] in which she analysed observations of prominences associated with sunspots made between 1908–1910. She was able to classify the prominences into various active and eruptive types, and to draw conclusions regarding their motions in the fields of spots. The work, illustrated by excellent spectroheliograms, was the first ever systematic classification of prominences. The paper, illustrated by 40 slides, was read by Mary in person at the meeting of the Royal Astronomical Society on 11 April 1913. The Eversheds were at the time presumably in London on leave, when Mary also arranged for the publication of her book on Dante's astronomy.

Mary pursued the same research topic in a joint paper with her husband in 1917.[xx] The practical work was, in fact, entirely hers; Evershed noted in the Introduction that "the Memoir was drawn up by Mrs. Evershed under the supervision of the Director" with the help of two other members of staff. It was the usual formula of husbands when they published their wives' work: one recalls Mrs. Sabine's translations, nominally ascribed to her husband, and Annie Maunder's book which ostensibly had the husband as primary author. This substantial piece of work filling 126 pages involved almost 60,000 individual prominences observed over an entire 11 year sunspot cycle. The huge and important sample of data was analysed in great detail using statistics, diagrams and photographs. With this, and her earlier paper, Mary Evershed was effectively the pioneer of prominence classification. Sir Harold Spencer-Jones in his popular textbook *General Astronomy*, attributes the original classification of prominences into quiescent and active to John Evershed alone rather than to Mary or to the Eversheds jointly.[xxi]

The years in India brought opportunities for travel. In response to discussions in Britain about setting up a permanent British solar observatory in a favourable location abroad, Evershed was requested to test the quality of a site in Kashmir for this purpose. Thus it was that in 1915 the Eversheds spent three months at a temporary observatory in the Valley of Kashmir. Conditions for solar research were found to be excellent, but the scheme to place an observatory there permanently did not materialise. The total solar eclipse of 1922, their last year in the East, was the occasion of another absorbing journey for the Eversheds who made their observations at Wallal, Western Australia.

Mary Evershed's enthusiasm for Dante began, as already mentioned, in her days as a student in Italy. She started to outline her book as early as 1896[xxii] but the work came to fruition in Kodaikanal. "An observatory on a mountain top is an ideal place in which to write on astronomy and poetry", she wrote in the introduction to her book, published in 1913.

Dante Alighieri (1265–1361) is universally known as the author of the *Divine Comedy*. That great and unique work of poetic imagination narrates the poet's fictitious journey through Hell, Purgatory and Heaven, finally reaching the abode of God in the Empyrium beyond space and time. The poem incorporates Dante's vision of the physical Universe, derived from the cosmology of his day. It also includes numerous references to the positions and movements of the heavenly bodies on the celestial sphere, and the use of astronomical descriptions to indicate date, hour, or passage of time. These references can be baffling to readers not conversant with the elements of spherical astronomy. "It is a matter of regret that even students of ability and culture often refuse so much as to attempt to understand Dante's astronomical references", wrote a distinguished Dante scholar in 1866. Mrs. Evershed's book was the first, at least in English, expressly written to help readers to understand these references and to appreciate the poetry and symbolism of Dante's allusions to the heavenly bodies.

In her book Mary Evershed examined Dante's considerable knowledge of astronomy by surveying the textbooks which he is known to have used in his studies and in his own various writings, both prose and poetry, in Latin and in Italian. The cosmology of Dante's day was Ptolemy's earth-centred model with the Sun, Moon and planets each occupying its own sphere. Two further spheres took account of the stars and finally, embracing them all, was the Empyrean of the thirteenth century Christian theologians. Mary Evershed saw Dante as one of the world's greatest poets who, a 1,000 years after Ptolemy's death, immortalised his work. She devoted the first part of her book to the history and development of astronomy from its beginnings up to Dante's day.

She then examined specific astronomical allusions in Dante's poetry. The entire sky appears to turn around the poles of the sky, and therefore the apparent position of the Sun, Moon, or constellations depends on the time of day or night. Dante is always exact in these particulars. The action of the Divine Comedy is set at Easter-time, close to the equinox and at the time of full moon. The poet's 8 day-long journey takes him through the imaginary underground levels of Hell, out again to the island of Purgatory at the antipodes, and finally through the ever higher spheres of the planets to Heaven. In Dante's imaginative scheme, Purgatory is an island in the middle of the ocean in the antipodes of Jerusalem, i.e., latitude 32 °S, effectively the same as that of New South Wales. The poet's itinerary may be reconstructed step by step from his astronomical allusions, of which Mary Evershed lists 58 chronologically in a Table.

In the famous verses in the first Canto of the Purgatory, for instance, the poet emerging from the gloom of Hell sees the sky once again:

> The fair planet which inspires love
> Was making all the orient smile,
> Veiling the Fishes which were in her train.[xxiii]

The planet is Venus; the time of year is Spring when the Sun is in Aries and when the Fishes, the next zodiacal sign east of Aries, is in the dawn. It is a beautiful sight which Mrs. Evershed recalled enjoying while observing Halley's comet in 1910.[xxiv]

In another example, the period before dawn is indicated by the clue: *The Fishes are quivering on the horizon and all the Wain lies over Caurus*. The sign of the Fishes is the zodiacal sign ahead of the Ram (where the Sun is at the time of the equinox) and rises two hours before the Sun. Caurus is a name for the north west-wind. The Wain is the constellation of the Plough, and at this time of year is in the north west).

There are many such riddles, some extremely elaborate, to indicate time of night or day. Mary explained them all, giving a new dimension to the reading of Dante's perfectly constructed masterpiece.

Dante's imaginary journey had always been supposed, from historical and other clues, to have taken place in the year 1300. The question of whether the astronomical particulars described in the course of that journey conformed to the actual situation in that year was much debated by certain literal-minded nineteenth century Dantists. Mary Evershed had a close look at the alleged problem but took the view that such scientific treatment was really not appropriate: the astronomy is there for the poetry, not the other way round. "No-one will dispute a poet's right to arrange the skies as he thinks fit. It is the poet's artistic use of the astronomy of his day which merits our admiration quite as much as the scholar's proficiency".[xxv] The traditionally accepted date of 1300 is the one which prevails among modern scholars.[xxvi] Dante chose his astronomical configuration to represent symbolically "an ideal Easter week in an ideal universe".[xxvii]

Dante and the Early Astronomers was a fascinating and unusual work. The explanations of astronomical phenomena were brilliantly clear and free from intimidating jargon, and – apparently for the first time in Dante studies – made use of diagrams. Also remarkable was that the work was accomplished almost in isolation, though she was in touch with the famous astronomer G.V. Schiaparelli (1835–1910) of Milan, an authority on medieval astronomy[xxviii] (also known as the first observer of the markings on Mars which he called channels, unfortunately translated as "canals" in English, thus giving rise to the speculation of intelligent beings on our neighbour planet).

The manuscript of *Dante and the Early Astronomers* was completed in 1913. Perhaps to distinguish it from her strictly astronomical research, Mary published this book under her maiden name but added her married surname, styling herself "M.A. Orr (Mrs. John Evershed)". The manuscript was prepared for the printers by her sister and early collaborator Lucy, to whom to book is dedicated. The book came out in 1914.

A full page review of *Dante and the Early Astronomers* by J.L.E. Dreyer, astronomer and historian of astronomy, appeared in *Nature* in December 1914.[xxix] and a short note in *The Observatory* magazine.[xxx] These seem to have been the book's only notices on its first appearance in Britain, though it was not so quickly forgotten in Italy.[xxxi]

The Eversheds returned to England in 1923 on the husband's official retirement. They settled in Ewhurst, Surrey, where John Evershed set up a private observatory and continued his researches with undiminished vigour for a further 30 years. Together they observed two further solar eclipses, in England in 1927 and in Greece in 1936, and attended meetings of the International Astronomical Union. Mary resumed her participation in the work of the British Astronomical Association and became a Fellow of the Royal Astronomical Society in 1924 serving for many years on the Society's library committee. Though she continued to enjoy observing prominences with the spectrohelioscope, her chief interest in this last phase of her life was in matters historical. Within months of her return to England, in June 1924, she contributed a long essay on *The Astronomy of Dante* to the *Journal of the BAA*[xxxii] in which, ultra-modestly, she made no mention of her own book.

Mary Evershed's most far-reaching piece of work for the British Astronomical Association was the foundation of its Historical Section in 1930. The Association was traditionally organised in Sections, each with its own Director. The historical section was set up in the same manner, with Mrs. Evershed as Director and with the support of knowledgeable enthusiasts such as Mrs. Maunder. She outlined its aims: "to study the history of astronomy and to co-operate in research, helping to bring new facts to light and unearthing facts now buried in old books and papers".[xxxiii] This placing of research into the history of astronomy on an organised basis was a milestone in the annals of the Association and indeed of astronomy generally. It has to be recalled that "history of science" was not a formal branch of knowledge in those days. Journals devoted to the history of science did not exist (The *Journal for the History of Astronomy* was founded only in 1970). Mrs. Evershed suggested that members should start with a systematic course of study, beginning with the Babylonians and progressing, period by period, to medieval and modern science. Those with knowledge would recommend reading lists, and members would meet, correspond and exchange notes. The educational and cultural value of historical research was emphasised. On the practical level, it was suggested that members of the Section should make an index of articles of historical interest in old numbers of the journals "which often seem to be sunk without trace". Mrs. Evershed's principal interest for the rest of her life was in history. The Historical Section, which she directed for 14 years, thrived, and remains active today.

Over the years Mrs. Evershed made her own erudite and charmingly written contributions to the Section's work with articles in its Journal covering a range of periods of history: *Dante's Virgil* (i.e., astronomy in the Roman Empire), *Galileo and Meyer*, *Kepler, Flamsteed* and *Arab Astronomy*.

As head of the Historical Section, Mrs. Evershed's most ambitious project was a directory named *Who's Who on the Moon*[xxxiv] which identified every person commemorated by name in lunar formations. The Introduction, written by her with her usual care for accuracy, traced the history of lunar nomenclature from the map of Langrenus of Brussels published in 1645 to date. A number of catalogues and maps had been published in the nineteenth century, between which there had been a certain amount of confusion. A collated list had been drawn up by Mary Blagg (next chapter) for the International Astronomical Union in 1913, which was followed by a catalogue and map by Mary Blagg and K Müller, published by the International Astronomical Union in 1935. Using this catalogue as its basis, Mrs. Evershed and her team undertook to identify all 672 names listed there of which 609 were personal names, and to write mini-biographies of every one of these persons. Besides members of the BAA the team of 30 had outside contributors and advisers, including Pio Emanuelli of Rome, already known to Mrs. Evershed through studies of Dante. The dictionary was published as a *Memoir of the BAA* and distributed as a separate booklet. It was extremely useful not only in satisfying the curiosity of lunar observers but as a biographical dictionary of astronomers, many of them extremely obscure, who had been honoured in the distant past by places on the Moon.

A most happy occurrence at the end of Mrs. Evershed's life was the belated discovery of her *Dante and the Early Astronomers* by those for whom it had been principally intended – readers of Dante's poetry. Dr. Barbara Reynolds, a leading Italianist in Cambridge, came across the book quite by chance while browsing among the stacks in Cambridge University Library. She had not heard of it before, nor had any of her colleagues: having been published by a scientific firm it had failed to reach any of the Dante bibliographies. "From then on" she wrote, "I was able to explain the astronomical references to my students, instead of saying, as T.S. Eliot did, that they don't matter and you can skip them."[xxxv] She also by good fortune met the astronomer David Thackeray (later Director of the Radcliffe Observatory Pretoria) at the Cambridge Observatory and learned that he was Ms. Orr's nephew, her sister's son. In the summer of 1949 Dr. Reynolds visited the Eversheds, conversed with Mrs. Evershed who was already seriously ill, and was shown round the observatory by her husband. Mrs. Evershed died a few months later.

After Mrs. Evershed's death Dr. Reynolds wrote her Obituary Notice together with a retrospective review of *Dante and the Early astronomers* in the Journal *Italian Studies*.[xxxvi] For most Italianists it was their first intimation of the existence of the book. To Dr. Reynolds the author was "a Dantist no less than an astronomer" who had "come to the rescue" of non-scientific readers while never losing sight of the fact that "the *Divina Commedia* is a poem and not a scientific treatise". The admirable clarity of the author's style and her use of excellent diagrams and illustrations had made the history of astronomical ideas intelligible and attractive to the layman. The writer Dorothy L. Sayers, translator of the *Divine Comedy* described it as "quite the best guide available to Ptolemaic Astronomy and to Dante's handling of celestial phenomena".[xxxvii]

The book was by now out of print, and Barbara Reynolds prepared a welcome second edition which was published in 1956. Dorothy Sayers died in 1957 before completing her translation of the third book of the *Divine Comedy* which was brought to conclusion in 1962 by Barbara Reynolds who also wrote the commentaries. Readers of the Sayers-Reynolds translation of the *Divine Comedy*[xxxviii] find the allusions to astronomy and cosmology elucidated entirely on the basis of M.A. Orr's work. *Dante and the Early Astronomers* remains the only book in English devoted specifically to the subject. It is cited as an authority on Dante's astronomy and recommended to students.

Mary Evershed died at her home in her 83rd year, on 25 October 1949. Her legacy to astronomy was considerable: her collaboration with John Evershed in establishing Kodaikanal Observatory as a major scientific institution, where their old home is now Evershed House, a residence for visiting astronomers; her scholarly work on Dante that has introduced many readers and students to that great poet's view of the universe; and her enthusiasm for the history of astronomy which bears fruit in profusion in the British Astronomical Association. She was remembered by all who knew her for her willingness to enlighten and her unfailing courtesy.[xxxix]

Notes

[i] M.T. Brück. 1998. *Journal of Astronomical History and Heritage* **1**, 45–59.
[ii] Wayne Orchiston. 1998. *Journal of the British Astronomical Association* **98**, 75–84 and other publications.
[iii] James Gall. 1866.*Guide to the Constellation.*, Edinburgh: Gall and Inglis.
[iv] M.A. Orr to John Tebbutt, 5/7/1895, Tebbutt letters, Mitchell Library, Sydney.
[v] M.A. Orr. 1897. *An Easy Guide to the Southern Stars*. London and Edinburgh: Gall and Inglis; second edition 1911.
[vi] List of Members. 1892. Publications of the Astronomical Society of the Pacific **4**.
[vii] M.A. Evershed. 1943. *J. Br. Astron. Assoc.* **53**, 43.
[viii] R.A. Marriott. 1991. Norway 1896: the BAA's first organised eclipse expedition, *J. Br. Astron. Assoc.* **101**, 3.
[ix] E.W. Maunder (editor). 1900. *The Total Solar Eclipse 1900*. Knowledge Office for British Astronomical Association, London.
[x] Ibid. p 184.
[xi] P. Stroobant et al. 1907. *Les Observatoires Astronomiques et les Astronomes*. Brussels.
[xii] M.A. Orr. 1904. Variable Stars of Long Period, *J. Br. Astron. Assoc.* **15**, 129.
[xiii] J. Evershed. 1955. Recollections of Seventy Years of Scientific Work. *Vistas in Astronomy* (ed. A. Beer). Volume 1, 33–4.
[xiv] R.H. Kochhar. 1991. The Growth of Modern Astronomy in India 1651–1960. *Vistas in Astronomy* **34**, 69–105.
[xv] M.A. Orr (Mrs John Evershed). 1913. *Dante and the Early Astronomers*, Gall and Inglis, London and Edinburgh.
[xvi] Alice Thiele Smith.2004. *Children of the Manse: Growing up in Victorian Aberdeenshire*. Edinburgh: The Bellfield Press.
[xvii] Quoted from her book *Dante and the Early Astronomers*.
[xviii] J. Evershed, op. cit. p. 36.
[xix] M.A. Evershed. 1913. Some types of Prominences associated with sunspots. *Mon. Not. Roy. Ast. Soc.* **73**, 6.
[xx] J.E. and M.A. Evershed. 1917. Results of Prominence Observations. *Memoirs of Kodaikanal Observatory, Volume 1, Pt 2*. Madras.
[xxi] H. Spencer-Jones. 1951. *General Astronomy*, p. 170. (third edition). London: Edward Arnold & Co.
[xxii] B. Reynolds. 1956. Preface to the second edition of the book.
[xxiii] Dante. *Purgatory* 1, verses 13–21. Most examples are considerably more complex.
[xxiv] M.A. Evershed. 1914. Dante and Halley's Comet. *Journal of the BAA* **24**, p 362.
[xxv] M.A. Evershed. Part 2, chapter 4 of her book.
[xxvi] C. Gizzi. 1974. *L'Astronomia nel Poema Sacra*. 2 vols. Naples: Loffredo.
[xxvii] Patrick Boyde. 1981. *Dante Philomythes and Philosopher. Man in the Cosmos*, 163–65. Cambridge University Press.
[xxviii] A letter in the possession of Dr. Barbara Reynolds.
[xxix] H.L.E. Dreyer. *Nature* **44**, 359–60.
[xxx] *The Observatory* 1914. **23**, 332–33.
[xxxi] Pio Emanuelli, *Corriera della Sera*, 3 February 1931, quoted by B. Reynolds in the introduction to the second edition of M.A. Orr's book.
[xxxii] M.A. Evershed FRAS. 1924. The Astronomy of Dante. *Journal of the BAA* **34**, 314–321.
[xxxiii] M.A. Evershed, 1930. *Journal of the BAA* **40**, 203–205.
[xxxiv] Historical Section, Director Mrs John Evershed 1938. *Who's Who on the Moon. Memoirs of the British Astronomical Association*. Printed for the BAA, Edinburgh.
[xxxv] Barbara Reynolds. 1950. Notice and Retrospective Review of M.A. Orr. *Italian Studies* V, 72.
[xxxvi] ibid.

[xxxvii] Dorothy L Sayers. 1955. *The Comedy of Dante Alighieri, Cantica II, Purgatory*, (*Bibliography*). London: Penguin Books.
[xxxviii] D.L. Sayers and B. Reynolds. 1962. *The Divine Comedy 3: Paradise*. London: Penguin.
[xxxix] Obituary. 1950. *Nature* **165**, 98, January 21.

Chapter 16
Mapping the Moon

As the stories told so far show, astronomy until well into the twentieth century has had its share of willing helpers who were happy to contribute to the cause of science with no thought of reward. Mary Adela Blagg (1858–1944) is an example of such a helper. Scores of astronomers, unfamiliar with her name, are unknowingly in her debt, for she was responsible for the lunar maps and identifications officially adopted for universal usage by the International Astronomical Union at its foundation in 1920. They comprised all recognised named features on the Moon, and remained the standard reference until space age photographs added a crop of additional details on its near face and a whole new territory on the far side. She was also responsible for a large volume of published work on variable stars.

Mary Blagg preferred to work at home, away from the limelight: her obituarist described her as something of a recluse. She belonged to the British Astronomical Association and was elected a Fellow of the Royal Astronomical Society in 1916 as soon as it was opened to women, but was not a socially conspicuous member of the community of votaries of the heavens. Little is recorded of her apparently uneventful private life, spent entirely in Cheadle in Staffordshire, a small town not far from Stoke-on-Trent, where she was born. She was the daughter of a solicitor, and was educated at home and at a private boarding school. She nurtured a natural aptitude for mathematics by borrowing her brother's schoolbooks – as Mary Somerville did in her day – since mathematics did not form part of her female school curriculum. She was already in her forties when she came across astronomy at University extension lectures, and was captivated.

The University extension movement aimed to bring modern knowledge, including acquaintance with the sciences, to adults, of whom most were denied higher education or who lacked the opportunity of acquiring it. This very successful scheme dated back to the middle of the nineteenth century: and by the time Mary Blagg studied in Cheadle, it had spread to most towns of any size in Britain. The lecturers taking part were academic experts of standing: an important aspect of the movement was its help in promoting higher education for women.[i] Mary Blagg's instructor in astronomy was J.A. Hardcastle (1868–1917), who was a neighbour of Samuel A. Saunder (1852–1912), a mathematics lecturer, Fellow of the Royal Astronomical Society, and an amateur astronomer and selenographer with a private observatory at Crowthorne, Berkshire.

M. Brück, *Women in Early British and Irish Astronomy*,
DOI: 10.1007/978-90-481-2473-2_16,

Mary Blagg was led to her serious work through these associations. In 1905, Saunder drew the Society's attention to the confused state of lunar nomenclature which he discovered in the course of his own work. He realised that though excellent maps of the Moon existed, they did not all agree on the names they assigned to its various features.[ii]

The naming of features on the Moon goes back a long way. Galileo, pioneer of the astronomical telescope, was the first to draw a map of the Moon which showed bright and dark areas and a rough surface "like the face of the Earth itself, which is marked here and there with chains of mountains and depths of valleys".[iii] Galileo was followed by others who, as instruments improved, could distinguish more and more features on its surface. The great observer, Hevelius of Gdansk (1611–1687), whose wife Elizabeth was the first modern woman astronomer, was responsible for the label "mare" (sea) on the large smooth lunar plains, which he saw as counterparts of oceans on our Earth. His beautiful names for these "maria" or seas – the Sea of Serenity, the Sea of Tranquility – were permanently retained, even after astronomers realised that there is no water on the Moon. It was Hevelius, too, who named the great lunar mountain ranges after the Alps, the Apennines and other terrestrial ones; in fact, his system was to name all lunar features after earthly counterparts. His acclaimed set of beautiful maps, the *Selenographia*, showing the Moon full and at different phases, remained the principle source of information on the lunar surface for 150 years, though few of his original names were eventually retained.[iv]

The end of the eighteenth century saw the publication of many more exquisite charts of the Moon, some using a different naming system.[v] The most common feature on the Moon are the circular craters, so-named for their resemblance to volcanic craters on the Earth; and these were now named after distinguished philosophers, mathematicians and astronomers of the past. As telescopes improved, each new student of the Moon added more features to those already known, and attributed names to them.[vi] Further complexities first noted in the 18th century were "rills" – apparent clefts or cracks in the lunar surface, which added another category to the list. The number of lunar features thus multiplied.

What became, and long remained, the most reliable atlas of the Moon was that of two German astronomers, Wilhelm Beer (1797–1850) and Johann Mädler (1894–1874), published in 1837. It was the result of seven years' observation and recording[vii] and consisted of a large-scale map, the *Mappa Selenographica*, showing more than 400 formations, encompassing all those recorded on previous maps from Hevelius onwards, and 133 newly observed and named ones, together with a catalogue.

The next major event was a new large-scale map of the Moon published in 1868 by the Director of the Athens Observatory, Julius Schmidt (1825–1884), the fruit of three decades of observation. It has been claimed that Schmidt's energy and talent for observation "has perhaps never been equalled".[viii] His final topographical map, 6 ft in diameter, showing tens of thousands of craters and rills, "greatly surpassed in completeness and accuracy anything previously attempted". Dozens of new names were added, mainly of nineteenth century scientists, but the minute craters and rills were not all given individual names; his purpose was not naming but accurate mapping.

Meantime, in 1864 the British Association embarked on an ambitious project to study the surface of the Moon for possible changes with time. Areas of the Moon were to be shared out among observers, who would make accurate drawings on a specified large scale. The plan was never completed – the independent amateur astronomy tradition in Britain was not particularly suited to such a ponderous collaborative task[ix] – though it inspired a renewed interest in selenography among amateur astronomers in Britain. (Charles Piazzi Smyth, whom we have already met, was one of those who conscientiously carried out his part in the project – a detailed drawing in colour of the Mare Crisium area). The task of coordinating the work was undertaken by an amateur astronomer William Birt (1804–1881). Some new features were recorded, which Birt and the prominent amateur John Lee (Chapter 8), took upon themselves to name. One of these was a crater Somerville, after Mary Somerville who was then active and living in Italy. Unfortunately it was one of the features that did not survive later tidying, though the name was subsequently restored elsewhere.

Another British study of the Moon was by Edmund Neilson (1851–1938), a Fellow of the Royal Astronomical Society, who regarded Beer and Mädler's great work as "the foundation on which all lunar maps must be constructed". He revised that map for increased detail, adding some new features of his own to bring the total to over five hundred. His impressive book[x] of 1876 incorporated a list, numerous maps, and explanatory notes on the larger formations.

The task of assembling and identifying all known features on the Moon might seem in principle straightforward, but it was not so in practice. The aspect of craters and mountains varies with conditions of illumination; and accurate drawings demanded repeated viewing under various conditions. If the true dimensions of the features are needed, the observations have to be corrected for foreshortening. The Moon, though presenting approximately the same face to the Earth at all times, undergoes librations whereby one can sometimes see around the edges, left and right, top and bottom: altogether, 59 percent of the Moon's surface is visible at one time or another. This situation makes it difficult to decide on a uniform system of co-coordinates on the Moon's face, and to distinguish small neighbouring features near the Moon's edge. Not surprisingly, inconsistencies existed between even the most careful observers, and there were many cases of features being differently named by different observers, or of the same name being used more than once.

By the end of the nineteenth century, photography had become a well-established technique in astronomy, superior to drawings for fixing exact positions. A major programme of lunar photography, using a 24-in. refractor, was undertaken at the Paris Observatory (by two astronomers, P. Puiseax and M. Loewy) which resulted in a large published photographic Lunar Atlas. Independently in Germany, a set of photographs taken at Lick Observatory in California was used by Julius Franz (1847–1913) at the observatory of Breslau, who made many measurements of positions, and even named a few more new features.[xi] It was at this stage that Samuel Saunder drew attention to the urgent need for sorting out the confusion and to establish an agreed identification and nomenclature. He, too, had obtained some high-quality lunar photographs from Paris and also from Yerkes Observatory in the

United States, and had measured the positions of over 3,000 points in the Moon's face; but realised the need for an agreed system of names.

Saunder brought the matter up with the Royal Astronomical Society in 1905, as a result of which an international committee of lunar observers was set up to examine the whole question. All agreed with Saunder's ideas as to what needed to be done. A standard coordinate system on the Moon was adopted. But who would do the work? Mary Blagg's name was put forward by her friend Saunder. "Mary Blagg does the donkey work"[xii] is how Ewan Whitaker, modern historian of the subject, referred to the tedious task that she took upon herself of sorting the names, once for all. The outcome, published three years later, in 1913, was her *Collated List of Lunar Formations*, based on the three chief authorities – Beer and Mädler, Schmidt and Neison.[xiii] The title page described it as "compiled and annotated for the Commission by Mary A. Blagg under the direction of the late S.A. Saunder" (who had died the previous year); but Professor H.H. Turner, the Lunar Commission's Chairman, stated in the Introduction that "the list is throughout the work of Ms. Blagg". There are 4,789 formations listed, each with a symbol to denote its type – craters and rings; mountains and plateaus; rills; valleys and gaps. The work entailed over 40,000 comparisons between the three catalogues, not to mention consulting innumerable references. It has truly been described as "the fruits of an indefatigable enthusiasm".[xiv]

The Great War of 1914–1918 now intervened, during which Mary Blagg did voluntary social and church work which included the care of refugee children.

After the war, the International Astronomical Union was founded for the purpose of coordinating the efforts of working astronomers worldwide, especially in collaborative programmes. It was to meet – as it continues to do – every three years, and its activity was divided among Commissions, devoted to various specialities. Mary Blagg was made a member of the Commission on the Moon, one of the four first women to belong to the International Astronomical Union at its initial meeting in 1922 in Rome, though she was not present.[xv] She attended meetings in Cambridge in 1925 and in Leiden in 1928, submitting at the latter a list of ambiguously named lunar features not included in her first catalogue. It was decided that she and another lunar astronomer, Karl Müller (1866–1942) of Vienna, should examine photographs of the Moon, exclude all doubtful objects, and prepare a definitive list. Müller was a retired Czechoslovakian government official and an amateur astronomer with a particular interest in the Moon.

The two astronomers kept in constant communication while they worked, checking and comparing. Their final catalogue, on which they spent seven years, consisted of more than 6,000 formations, each with a number, a designation, rectangular coordinates and diameters of circular features. Former designations and their sources were also given. A second volume was made up of maps in Ms. Blagg's hand. The opus (which, it may be remarked, was performed entirely by two unpaid amateurs) was presented at the International Astronomical Union meeting in Cambridge, USA, in 1932, and published in 1935 under the title *Named Lunar Formations*.[xvi] The Commission recommended that the two authors should have formations on the Moon named after them. Both declined, but their objections were overruled, and their names are to be found in the published list. "Blagg" is a small crater of 5 km diameter near the Moon's centre (1.3 N, 1.5 E).

Named Lunar Formations remained the authoritative reference source for all astronomers until manmade spacecraft reached the Moon and revealed the unseen far side. A new edition of Blagg and Müller's list was published in 1967 to include the principal far-side craters. The photographic survey of the Moon by Lunar Orbiter satellites recorded the entire surface of the Moon in much greater detail, and the list that had served so well for over 30 years finally required up-dating. The revised catalogue added the newly revealed features, named in the traditional manner after past astronomers by the International Astronomical Union. The system of designating some of the minor craters by Greek symbols or Roman numerals, was also tidied up. The *NASA Catalogue of Lunar Nomenclature*,[xvii] containing 681 named formations and more than 5,400 others, with their accurate coordinates and dimensions, replacing Blagg and Müller's, was published in 1982.

Mary Blagg was well over 70 years of age when she finished her work on the Moon. However, it was not her only interest. While still working on the lunar catalogue she volunteered to help H.H. Turner in a task requiring similar skill and patience. It was to reduce a large set of raw observations of variable stars which had come into Turner's hands and which he deemed worth preserving. They had been made many years earlier by an astronomer, Joseph Baxendell (1815–1887), who had died leaving much unpublished material behind. Baxendell, a largely self-educated astronomer and meteorologist, held an official post in charge of the City Observatory in Manchester, and also had his own private observatory where he made copious observations of variable stars.[xviii] Mary Blagg's analysis of these observations resulted in a set of ten papers in *Monthly Notices of the Royal Astronomical Society*[xix] published jointly by her and Turner between 1912 and 1918. Turner recorded in the final paper that "practically the whole of the work of editing has been undertaken by Ms. Blagg".[xx] (It is worth noticing the part played by Turner in the lives of women astronomers. At Oxford where he was Professor of Astronomy, he encouraged Edith Bellamy, while earlier he played a part in the appointment of the "lady computers" at the Royal Observatory at Greenwich (Chapter 13). He was also a friend and admirer of Agnes Clerke).

Working on her own account, Mary Blagg continued with a series of papers on variable stars, using observations collected by members of the British Astronomical Association. These papers, which gave scope to her skill and originality, were also published by the Royal Astronomical Society.[xxi] She was more than 70 years of age when she published the last of these papers. She suffered from heart trouble in the last 8 years of her life, but kept up her hobby of chess until a few weeks before her death in 1944 at the age of 86.

Mary Blagg's career in astronomy is remarkable for beginning so late in life, yet lasting so long. Her catalogue of Lunar Formations which laid the foundations of the modern authoritative version, is Mary Blagg's permanent memorial. She is also one of the first women to be commemorated on the Moon (Fig. 16.1) – "thus", as her obituarist wrote, "ensuring her a well-deserved place among the lesser immortals".

Fig. 16.1 The Blagg crater on the Moon. *Blagg* is the small crater (5 km diameter) at the centre of the photograph. The large crater to its right is *Rheticus*. (Orbiter Photographic Atlas of the Moon, Lunar and Planetary Institute)

Notes

[i] *Encyclopaedia Britannica*. 1961. 16th edition. volume **22**, 882.

[ii] A.S. Saunders. 1905. *Monthly Notices of the Royal Astronomical Society* **56**, p 41.

[iii] Galileo Galilei. 1989. *Siderius Nuntius* or *the Sidereal Messenger*. p 40. Translated by Albert van Helden, Chicago and London: University of Chicago Press.

[iv] Ewan A. Whitaker. 1999. *Mapping and Naming the Moon, a history of lunar cartography and nomenclature*, Chapter 4. Cambridge University Press.

[v] Ewan Whitaker, op. cit,. Chapters 5 and 6 show a confusing variety of beautiful maps from the late 18th century.
[vi] See e.g. Agnes Mary Clerke. *A History of Astronomy during the Nineteenth Century*, p 263; reprinted fourth (1902) edition, Decorah, USA: Sattre Press.
[vii] *Der Mond*. Mädler, a scholar of great distinction was the principal author. Beer was a wealthy amateur.
[viii] Joseph Ashbrook. 1984. *The Astronomical Scrapbook. Skywatchers, Pioneers and Seekers in Astronomy*. Cambridge University Press.
[ix] Allan Chapman. 1988. *The Victorian Amateur Astronomer. Independent Astronomical Research in Britain 1820–1920*, p 296. Chichester: Praxis and Wiley.
[x] Edmund Neilson. 1896. *The Moon, and the conditions and configurations of its surface*. London: Longmans Green.
[xi] These were published some years later. Franz concentrated on the border regions of the Moon, whereas Saunders concentrated more on the middle. The observations of the two selenographers thus complemented each other.
[xii] Ewan Whitaker, op. cit.
[xiii] Mary Adela Blagg. 1913. *Collated List of Lunar Formations named or lettered in the maps of Neison, Schmidt and Madler*. Neill & Co. Cambridge.
[xiv] Ewan A. Whitaker. 1999. *Mapping and Naming the Moon. A history of lunar cartography and nomenclature*. Cambridge University Press.
[xv] Barbara Welther. 1995. The first women appointed to IAU Commissions. *Highlights of Astronomy* **10**, p 124. IAU General Assembly. The Hague: Kluwer. Two of the women were American and two were British. The second British woman was Mrs Fiametta Wilson (1864–1920) who served on the Commission on Meteors. (Kristine Larsen. 2006. Shooting Stars: the women directors of the meteor section of the BAA. in press.)
[xvi] Mary A. Blagg and K. Müller. 1935. *Named Lunar Formations*. London: Percy Lunds, Humphries and Co.
[xvii] L.E. Andersson and E.A. Whitaker. 1982. *NASA Catalogue of Lunar Nomenclature*, NASA. The corresponding photographic chart of the Moon is available in modern style on the Internet.
[xviii] Oxford DNB.
[xix] Volumes **73–78**.
[xx] *Monthly* Notices*of the Royal Astronomical Society* **78**, 528. 1912.
[xxi] *Monthly* Notices*of the Royal Astronomical Society*, volumes **84–85, 88–91**.

Chapter 17
The End of an Era; the Beginning of a New

Eudosia's dream that women should have access to learning like their brothers, took more than 100 years to come true. By the close of the nineteenth century all universities in Britain and Ireland admitted women to their courses. In principle, therefore, there was no reason why women should not now take part as professionals in scientific research on equal terms with men. Yet, though women were very active in the "leisured amateur" world of astronomy, there were none (apart from a few computers) to be found in salaried employment in Britain or Ireland as the new century rolled on. It might appear, alas, that women in these islands were destined to remain forever excluded from cosmic discoveries. Amateur astronomical societies were their only outlet, where many enjoyed observing transient phenomena like meteors and aurorae and could give, as well as listen to, popular lectures.[i]

The situation seemed very different in the United States, where a tradition of women astronomers was well established in the independent women's colleges. As already mentioned, the leading members of the Harvard group of women astronomers were graduates of such colleges, some of the second or third generation. The women's colleges also provided careers on their teaching staffs, while women graduates worked in assistant posts in major observatories like Mount Wilson. The first decades of the twentieth century saw women astronomers in USA with PhD degrees and honorary doctorates from male and co-educational universities.

The truth was that science itself had changed during the lifetimes of Annie Maunder, Mary Evershed and their generation. It had become formalised and professionalised. The golden days of the family based "gifted amateur" were over. Opportunities for female helpers to make themselves useful disappeared. Academic scientists were expected to have university degrees and to have learned to do research under experienced masters. Modern telescopes and spectroscopes were built by advanced professional instrument makers, not by enthusiasts in their back gardens. Research in astrophysics required properly (i.e., expensively) equipped observatories, such as had sprung up under Government patronage on the continent of Europe or through the largesse of millionaire patrons in the United States. Furthermore, physics – and with it astrophysics – was being transformed. Einstein's famous formula $E = mc^2$ was annunciated in 1905; Bohr's theory of the atom was put forward in 1913 to explain the mysteries of spectra; and in 1915 Einstein announced the General Theory of Relativity. This was a different, a more complex

M. Brück, *Women in Early British and Irish Astronomy*,
DOI: 10.1007/978-90-481-2473-2_17,

and a more theoretically demanding universe from that chronicled by Agnes Clerke and explored by Margaret Huggins.

In Great Britain, only a few dozen suitably qualified astronomers – mostly mathematicians – were needed to fill the official posts available at that time. Unlike medicine or the law or accountancy, astronomy was not a skilled occupation employing hundreds or thousands of practitioners. It offered only limited prospects even to men graduates, and none at all to women, even to those with good university degrees. Women graduates in mathematics or physics mostly settled for careers in school teaching. Some high-minded women were drawn to causes of social justice, of universal education, of factory women's working conditions. Others re-trained as medical doctors or devoted their lives to charitable work. Many were actively involved in the suffrage movement: it puts matters into perspective to contemplate that women were without a vote until 1920 when it was granted to those over the age of 30 and were to wait a further 8 years before it was extended universally. By the time the Royal Astronomical Society decided belatedly to allow women access to Fellowship (1915) which gave them some status in the professional world, there was a gory war raging in Europe that made such a matter seem trivial. Women from every walk of life had other things to think about.

Where do we end our story? The non-existent prospect for women in astronomy at this time was decisively illustrated by the experience of a young English woman and aspirant astronomer, Cecilia Payne (later Payne-Gaposhkin) (1900–1979) (Fig. 17.1). She was a student at Cambridge University, who set her heart firmly on becoming a real research astronomer. The exact occasion of her almost mystical vocation which she recalled vividly to the end of her life was a lecture given by Sir Arthur Eddington in which he described the results – revelations might be a better word – of observations made at the total solar eclipse of 29 May of that year. These observations were a crucial test of Einstein's General Theory of Relativity, enunciated in 1915, which postulated that light is bent in its path by the

Fig. 17.1 Cecilia Payne-Gaposhkin in 1936. (The Schlesinger Library, Radcliffe Institute, Harvard University)

gravitational field of a massive body such as the Sun. According to the theory, the positions of stars close to the Sun on the sky appear slightly shifted as a result of this bending. The effect, not observable in the normal daytime sky, ought to be seen in a darkened sky during a total eclipse of the Sun. It was indeed found exactly as calculated at the total solar eclipse of 29 May 1919. The eclipse was observed from the island of Principe off the west coast of Africa. Several photographs were secured during the approximately five minutes of totality and compared with photographs of the same region of sky taken by night at a different time of year. This confirmation of Einstein's revolutionary theory of space was announced publicly at a meeting of the Royal Society in London by the Astronomer Royal, Sir Frank Dyson, and caused world wide excitement. Sir Arthur Eddington, who had led the successful eclipse expedition from the Royal Observatory Greenwich came to Cambridge to tell the story to a packed hall. There was great demand for tickets for this lecture, and the young Cecilia was fortunate to get one that another student was unable to use. She was 19 years old, a student of Natural Sciences at Newnham College who had recently begun her chosen courses in physics, chemistry and botany.

Eddington's lecture was the turning point of her life. His exposition of the theory of Relativity and the nature of motion struck her like a "thunderclap". She had a phenomenal memory, and as soon as she returned to her room she wrote down Eddington's lecture word for word. So excited was she that she could not sleep for three nights, and there and then determined to be an astronomer. She attended the undergraduate astronomy lectures, including Eddington's, and was encouraged in her private studies by him and other astronomers at the university observatory. However, to learn from others was not enough for her: she wanted to "explore the frontiers" for herself. Even before completing her degree in Physics she realised that there was no future for a woman in astronomy in Britain: the Royal Observatory in Greenwich employed only mathematicians in professional posts, who, furthermore, were all men. On good advice from a young astronomer friend,[ii] she was introduced to the Director of the Harvard College Observatory who was visiting London, and became aware of the possibility of joining the famous women's team at that great institution. This she succeeded in doing after graduation in 1923.

In Harvard, in that all-women's group, began her long and glowing career in research, teaching and writing. It was not a smooth path. Real though the opportunities for women were in the United States compared with Britain, they still had their limitations. There were career heights that convention decreed were unattainable by women, and Cecilia was aware that, with her intellectual gifts, she might have gone further in worldly terms had she been a man. Yet she never lost sight of her early vision. Her dedication to astronomy remained pure and unquenchable. She was a revered scientist, and an inspiration to women. She was the first woman PhD in astronomy at Harvard University, and the first woman to hold the rank of Professor at that most prestigious of academic seats of learning. She received numerous awards and honorary degrees. Most of all, she was a great astronomer, recognised as the first modern woman astronomer in the world, "an astronomer who happened to be a woman", and not a curiosity in a world of men.

Cecilia Payne Gaposhkin wrote a moving and uplifting autobiography, published posthumously by her daughter, which describes her youthful aspirations, her struggles and her achievements. Her brilliant scientific career spanning more than half a century belongs to America, but her influence and example were felt beyond its shores, not least in her native country where her personal story symbolises the transformation in the role of women in the academic world during her lifetime. When the time was ripe, women would hold their own in a new and different era.

That era finally blossomed in the mid-twentieth century with the rise of radio astronomy, the space age, the great international telescopes and the computer revolution. With it, thanks to the vastly greater number of posts that accompanied these changes and the improved accessibility of educational opportunities after World War II, has emerged a new generation of women astronomers who shine today without discrimination or prejudice. Their evolving success story will be for a future telling.

But as we admire and marvel at the wonders of the modern cosmos, we recall how far this mighty edifice is built on the labours of the past; and pause to give credit to those who contributed to it, not least the generations of women who for 200 years served Urania in the different ways that came to hand – helpers, encouragers, sustainers of learning., creators of intellectual taste, propagators of knowledge, educators of the young; transcribers, translators, financial backers, companions for discussion and debate. They deserve to be remembered with admiration and affection.

Let the new faces play what tricks they will
In the old rooms; night can outbalance day,
Our shadows rove the garden gravel still.
The living seem more shadowy than they.

W. B. Yeats

Notes

[i] Allan Chapman devotes an entire chapter in his book (1998. *The Victorian Amateur Astronomer*. op. cit.) to the activities of women in the amateur societies in the early part of the 20th century. The Meteor Section of the British Astronomical Association was directed by a series of energetic women in that period. Their exploits are recorded by Kristine Larsen. 2006. Shooting Stars: the women directors of the meteor section of the BAA. *The Antiquarian Astronomer* **3**, 75–82. Other dedicated women were the widow and daughter of Sir Norman Lockyer who successfully kept his memory and his work alive in their private observatory in Sidmouth, as described by George Wilkins. 2006. The Lockyer Ladies. *The Antiquarian Astronomer* **3**,101–106.

[ii] The friend was Leslie John Comrie (1893–1950), a graduate student from New Zealand who became famous in the science of computing and in designing and building machines for astronomical computation. Dr. Comrie was also responsible for introducing women computers into the Nautical Almanac Office for the first time in the 1930s. He it was, also, who encouraged Flora McBain (Mrs. Sadler) to apply for a scientific post at the Nautical Almanac Office in 1935, to which she was appointed, thus becoming the Royal Observatory's first woman scientist of equal rank with her male colleagues. The Director of Harvard College Observatory was the renowned Harlow Shapley.

Bibliography

Barbara Becker. Dispelling the myth of the able assistant. William and Margaret Huggins at the Tulse Hill Observatory. In H.M. Pycor et al (eds.) *Creative Couples in the Sciences*. New Brunswick: Rutgers University Press 1996.

Barbara Becker. Margaret Huggins, William Huggins *Oxford DNB* 2004.

John L Birks. *John Flamsteed, the First Astronomer Royal at Greenwich*. London: Avon Books 1999.

Mary Adela Blagg. *Collated List of Lunar Formations Named or Lettered in the Maps of Neison, Schmidt and Madler*. Neill & Co. Cambridge 1913.

Mary A. Blagg and K. Müller. *Named Lunar Formations*. London: Percy Lunds, Humphries and Co. 1935.

H.A. and M.T. Brück. *The Peripatetic Astronomer, the Life of Charles Piazzi Smyth*. Bristol: Adam Hilger 1988.

Mary Brück. *Agnes Mary Clerke and the Rise of Astrophysics*. Cambridge University Press 2002.

Margaret Bryan. *Lectures on Natural Philosophy, the result of many years' practical experience of the facts elucidated, with an Appendix containing a great number and variety of Astronomical and Geographical problems and some useful tables*. London: George Kearsley 1806.

Margaret Bryan. *A Compendious System of Astronomy*. (second edition). London: J. Wallis, and Wynne and Scholey 1799.

Frances Burney. *Journals and Letters*. London: Penguin Books 2001.

Günther Buttmann. *The Shadow of the Telescope, a biography of John Herschel* (tr B. Pagel). Guildford and London: Lutterworth Press, 1974.

Allan Chapman. *The Victorian Amateur Astronomer*. Chichester: Praxis Publishing Ltd. and John Wiley 1998.

Allan Chapman. *Mary Somerville*. Bristol: Canopus Publishing 2004.

Desmond Clarke. *The Ingenious Mr Edgeworth*, London: Oldbourne 1965.

A.M. Clerke. *Familiar Studies in Homer*. London: Macmillan 1892.

A.M. Clerke. *Modern Cosmogonies*. London: A and C Black 1905.

Agnes Clerke. *History of Astronomy during the Nineteenth Century*. Decorah: Sattre Press 2003 (reprint of edition of 1902).

Agnes M. Clerke. *The System of the Stars*. London: Longman 1890; 2nd edition London: Adam and Charles Black 1905.

Agnes M. Clerke. *Problems in Astrophysics*. London: Adam and Charles Black 1903.

Ellen M. Clerke. *Jupiter and his System*. London: Stanford 1892.

Christina Colvin (ed.). *Maria Edgeworth. Letters from England 813–1844*. Oxford: Clarendon Press 1871.

Mary R.S. Creese. *Ladies in the Laboratory?* Lanham Md and London: Scarecrow Press 1998.

J.L.E. Dreyer and H. H. Turner. *History of the Royal Astronomical Society 1820–1920*. Oxford: Blackwell Scientific Publications 1987.

Richard Lovell Edgeworth. *Memoirs of Richard Lovell Edgeworth Esq.*, begun by himself and concluded by his daughter Maria Edgeworth. Reprinted with an Introduction by Desmond Clarke. Shannon, Ireland: Irish University Press 1969.

Maria Edgeworth, (Claire Colvin ed.). *Letters for Literary Ladies,* London: Everyman 1993. (originally published 1795).

Patricia Fara. *Pandora's Breeches, Women Science and Power in the Enlightenment.* London: Pimlico 2004.

James Ferguson FRS. *Astronomy Explained upon Sir Isaac Newton's principles.* Sixth Edition. London: W. Strahan 1778.

James Ferguson FRS. *The Young Gentleman and Lady's Astronomy. An Easy Introduction to Astronomy for Young Gentlemen and Ladies.* (first edition 1768); Fourth Edition. London: T. Cadell, 1779.

Margaret Flamsteed and J. Hodgson. *Atlas Coelestis.* London 1729.

Eric Forbes. *Greenwich Observatory*: Volume 1. London: Taylor and Francis 1975.

George Forbes FRS. *David Gill, Man and Astronomer.* London: John Murray 1916.

Cecilia Payne-Gaposchkin (edited by Katherine Haramundanis). *An Autobiography and other Recollections.* Cambridge University Press 1996.

Mrs Isobel Gill. *Six Months in Ascension: An Unscientific Account of a Scientific Expedition.* London: John Murray 1878.

Gerard Gilligan et al. *The History of the Liverpool Astronomical Society.* Liverpool Astronomical Society 1996.

I.S. Glass. *Victorian Telescope Makers, the lives and letters of Thomas and Howard Grubb.* Bristol and Philadelphia: Institute of Physics Publishing 1997.

Ian Glass. *Revolutionaries of the Cosmos. The Astrophysicists.* Oxford University Press 2006.

Philip Gosse. *St. Helena 1502–1938.* Oswestry: Anthony Nelson Ltd. 1989.

Robert Percival Graves. *Life of Sir William Rowan Hamilton,* 3 volumes. Dublin: Hodges Figgis and Co; London: Longmans Green and Co.1882.

James Hamilton. *Faraday, a Life.* London: Harper Collins 2002.

James Hamilton. *Turner, a Life.* London: Hodder and Stoughton/Sceptre 1997.

Sir Harold Hartley. *Humphry Davy.* Wakefield: EP Publishing 1972.

J.B. Hearnshaw. *The Analysis of Starlight.* Cambridge University Press 1986.

John Herschel. *Outlines of Astronomy.* London: Longman Green 1887.

Mrs John Herschel. *Memoir and Correspondence of Caroline Herschel.* London: John Murray 1879 (Reprint edition Bath: The William Herschel Society 2000).

Derek Howse. *Nevil Maskelyne, the Seaman's Astronomer.* Cambridge University Press 1989.

Derek Howse. Greenwich Observatory: Volume 3. London: Taylor and Francis 1975.

Michael Hoskin (editor). *The Cambridge Illustrated History of Astronomy.* Cambridge University Press 1997; *The Concise Cambridge History of Astronomy.* Cambridge University Press 1999.

Michael Hoskin. *William Herschel and the Construction of the Heavens.* London: Oldbourne 1963.

Michael Hoskin. *The Herschel Partnership.* Cambridge: Science History Publications 2003.

Michael Hoskin (editor). *Caroline Herschel's Autobiographies.* Cambridge: Science History Publications 2003.

Alexander von Humboldt. *Cosmos: Sketch of a Physical Description of the universe.* Translated under the superintendence of Lieut.-Col. Edward Sabine, R.A. For. Sec. R.S. London: Longman, Brown, Green, and Longmans; and John Murray 1846.

Lynette Hunter and Sarah Sutton (eds.) *Women Science and Medicine 1500–1700.* Stroud: Sutton Publishing 1997.

Lady Margaret Huggins. *Agnes Mary Clerke and Ellen Mary Clerke, an Appreciation.* London, privately printed 1907.

Anthony Hyman. *Charles Babbage Pioneer of the Computer.* Princeton University Press 1982.

Phebe Mitchell Kendall. *Life, Letters and Journals of Maria Mitchell.* Houghton Mifflin, New. 1896.

Henry C. King. *The History of the Telescope.* New York (Dover edition) 1979.

Capell Lofft. *Eudosia: or a Poem on the Universe.* London: W. Richardson and C. Dilly 1781.

Constance A. Lubbock. *The Herschel Chronicle*. Cambridge University Press 1933. (Reprint edition Bath: The William Herschel; Society 1997.)

W.H. McCrea. *The Royal Greenwich Observatory*. London: Her Majesty's Stationery Office 1975.

Hector MacPherson. *Astronomers of Today and their Work*. London and Edinburgh: Gall and Inglis 1905.

A.J. Meadowes. *Greenwich Observatory*: Volume 2. London: Taylor and Francis 1975.

Lesley Murdin. *Under Newton's Shadow, Astronomical Practices in the Seventeenth Century*. Bristol: Hilger 1985.

Mrs Jane Marcet. *Conversations on Natural Philosophy: in which the Elements of that Science are familiarly explained and adapted to the comprehension of young persons*. London: Longman, Brown, Green and Longmans. (twelfth edition) 1851.

E.W. Maunder. *The Indian Eclipse 1898*, London: Hazell, Watson and Viney 1898.

Annie S. D. Maunder and E. Walter Maunder. *The Heavens and their Story*. London: Charles H. Kelly 1910.

John R. Millburn, in collaboration with Henry C. King. *Wheelwright of the Heavens, the Life and Work of James Ferguson FRS*. London: The Vade-Mecum Press 1988.

C.E. Mills and C. F. Brooke. *A Sketch of the Life of Sir William Huggins KCB, OM*. London (privately printed) 1936.

Patrick Moore. *The Astronomy of Birr Castle*. Birr: The Telescope Trust 1971.

Vanda Morton. *Oxford Rebels, the life and friends of Nevil Story Maskelyne*. Gloucester: Alan Sutton 1987.

Mary Mulvihill (ed.). *Stars Shells and Bluebells, Women Scientists and Pioneers*. Dublin: WITS 1997.

H.W. Newton. *The Face of the Sun*. London: Penguin 1958.

M.A. Orr (Mrs John Evershed). *Dante and the Early Astronomers*. London and Edinburgh: Gall and Inglis 1913; 2nd edition (with an Introduction by Barbara Reynolds). London: Allan Wingate 1956.

Elizabeth Chambers Patterson, *Mary Somerville and the Cultivation of Science, 1815–1840*. The Hague: Nijhoff 1983.

Colin A Ronan. *Edmund Halley, Genius in Eclipse*. London: Macdonald 1969.

H. Schellen.(tr. Jane and Caroline Lassell) *Spectrum Analysis*. 2nd enlarged edition, London: Longman Green and Co. 1872.

William H. Smyth. *The Bedford Catalogue* (Reprint edition) (Foreword by George Lovi). Richmond, Virginia: Willman-Bell, Inc. 1986.

W.H. Smyth. *Aedes Hartwellianae*. London (privately printed) 1851.

C. Piazzi Smyth. *Teneriffe, an Astronomer's Experiment*. London: Lovell Reeve 1858.

C. Piazzi Smyth F.R.S.S.L. and E. *Life and Work at the Great Pyramid* volume 1. Edinburgh: Edmonton and Douglas 1867.

C. Piazzi Smyth. *Madeira Meteorologic*. Edinburgh: David Douglas 1882.

Mary Somerville. *Personal Recollections from early life to old age (with selections from her correspondence by her daughter Martha Somerville)*. London: John Murray 1874.

Mary Somerville (Dorothy McMillan, ed). *Queen of Science. Personal Recollections of Mary Somerville*, edited and introduced by Dorothy McMillan. Edinburgh: Canongate Classics 2001.

Mary Somerville. *The Connexion of the Physical Sciences*. (first edition 1834). 2nd edition. London: John Murray. 1835.

Mary Somerville. *Physical Geography*. London: John Murray. 1848.

Willie Wei-Hock Soon and Steven H. Yaskell. *The Maunder Minimum and the Variable Sun-Earth Connection*. Singapore: World Scientific. 2004.

E.G.R. Taylor. *The Mathematical Practitioners of Hanoverian England 1714–1840*. Cambridge University Press 1966.

The Hon. Mary Ward. *The Telescope*, illustrated by the author's original drawings. Sixth Edition. London: Groombridge and Sons. n.d. [1858].

Brian Warner. R*oyal Observatory, Cape of Good Hope 1820–1831. The founding of a colonial observatory, incorporating a biography of Fearon Fallows*. Dordrecht/Boston/London: Kluwer Academic Publishers 1995.

Ewan A. Whitaker. *Mapping and Naming the Moon, a history of lunar cartography and nomenclature*. Cambridge University Press 1999.

Helen Wright. *Sweeper in the Sky.* Nantucket: The Maria Mitchell Association. 1969.

Biographical Dictionaries

Dictionary of National Biography 1910 (abbr. DNB).

Oxford Dictionary of National Biography 2004 (abbr. Oxford DNB).

Larousse Dictionary of Scientists (editor Hazel Muir). Edinburgh and New York: Larousse 1994.

Dictionary of Nineteenth Century British Scientists (editor Bernard Lightman). London: Thoemmes 2003.

The Encyclopaedia of Ireland (General editor Brian Lalor). Dublin: Gill and Macmillan 2003.

Biographical Encyclopedia of Astronomers (Editor-in-chief Thomas Hockey). New York: Springer 2007.

Afterword

by Peter Brand

Mary Brück, eldest of a large and talented family, was born 1925 in Ireland. She gained her BSc in Physics at University College Dublin and her Ph.D in astronomy at Edinburgh. She married Hermann Brück, then Director of Dunsink Observatory, before moving in 1957 with him to Edinburgh where he became Regius Professor and Astronomer Royal for Scotland, running the Royal Observatory Edinburgh.

After raising their children, she returned to science as a lecturer in astronomy at Edinburgh University. She formed an important part of a small core of teachers in a subject that was to expand rapidly, and set her particular stamp of intellectual rigour and warm humanity on the methods and style of the teaching. One of the major new teaching tasks that she undertook was to run the undergraduate laboratory. This work gave rise to two texts, both much used in developing astronomy teaching.

Her research at Edinburgh was into the nature of stars in clusters, and she focussed her efforts on those in the Small Magellanic Cloud, a nearby galaxy caught in our own gravitational field. She was able to use her data to infer the star formation history and dynamics in this intriguing astronomical neighbour. In her research career she supervised many research students, notably many female, some of whom now occupy important positions in astronomy.

Her full retirement in 1987 led to a new career: astronomical history, which she started in collaboration with her husband. The major work from this time is *The Peripatetic Astronomer: the life of Charles Piazzi Smyth* (1988). She quickly achieved a reputation in her own right for her work, concentrating on the lives of women in astronomy. Her outstanding work was her book *Agnes Clerke and the Rise of Astrophysics* (2002). She contributed to several journals on the history of science, and to the Dictionary of National Biography, and was on the editorial board of the journal of the Society for the History of Astronomy.

However, her publications were not all in learned journals. She was a charming public speaker, had broadcast children's programmes on RTE in Ireland, and wrote an influential Ladybird book *The Night Sky* (1965). In 2001 the Edinburgh Astronomical Society awarded her the Lorimer Medal for her public outreach. One of her last public acts was to deliver a very successful public lecture in May 2008 at the

Royal Astronomical Society in London on the subject of women in astronomy in the last two centuries, the basis for this book.

Without any feminist flag flying she did a tremendous amount, by her scholarship and by her example for the place of women in science. It is intriguing that the subject of this book finds an echo in the life she lived with her husband: whilst he was the major figure and she saw her role as one of support, she was a significant contributor to our human understanding in her own right.

She died in December 2008 at the age of 83, lively and intent - and organising! - until the very end. She was sustained through her life and at her death by a strong religious faith, which she was not frightened to lay alongside her scientific understanding.

Index

MIX
Papier aus verantwortungsvollen Quellen
Paper from responsible sources
FSC® C105338

If you have any concerns about our products,
you can contact us on
ProductSafety@springernature.com

In case Publisher is established outside the EU,
the EU authorized representative is:
Springer Nature Customer Service Center GmbH
Europaplatz 3, 69115 Heidelberg, Germany

Printed by Libri Plureos GmbH
in Hamburg, Germany